Introduction to Modern Traffic Flow Theory and Control

Boris S. Kerner

Introduction to Modern Traffic Flow Theory and Control

The Long Road to Three-Phase Traffic Theory

Springer

Boris S. Kerner
Daimler AG
GR/PTI
HPC: G021
71059 Sindelfingen
Germany
boris.kerner@daimler.com

ISBN 978-3-642-02604-1   e-ISBN 978-3-642-02605-8
DOI 10.1007/978-3-642-02605-8
Springer Heidelberg Dordrecht London New York

Library of Congress Control Number: 2009933980

© Springer-Verlag Berlin Heidelberg 2009
This work is subject to copyright. All rights are reserved, whether the whole or part of the material is concerned, specifically the rights of translation, reprinting, reuse of illustrations, recitation, broadcasting, reproduction on microfilm or in any other way, and storage in data banks. Duplication of this publication or parts thereof is permitted only under the provisions of the German Copyright Law of September 9, 1965, in its current version, and permission for use must always be obtained from Springer. Violations are liable to prosecution under the German Copyright Law.
The use of general descriptive names, registered names, trademarks, etc. in this publication does not imply, even in the absence of a specific statement, that such names are exempt from the relevant protective laws and regulations and therefore free for general use.

Cover image: © WONG SZE FEI - Fotolia.com

Cover design: deblik, Berlin

Printed on acid-free paper

Springer is part of Springer Science+Business Media (www.springer.com)

Preface

The understanding of empirical traffic congestion occurring on unsignalized multi-lane highways and freeways is a key for effective traffic management, control, organization, and other applications of transportation engineering. However, the traffic flow theories and models that dominate up to now in transportation research journals and teaching programs of most universities cannot explain either traffic breakdown or most features of the resulting congested patterns. These theories are also the basis of most dynamic traffic assignment models and freeway traffic control methods, which therefore are not consistent with features of real traffic.

For this reason, the author introduced an alternative traffic flow theory called three-phase traffic theory, which can predict and explain the empirical spatiotemporal features of traffic breakdown and the resulting traffic congestion. A previous book "The Physics of Traffic" (Springer, Berlin, 2004) presented a discussion of the empirical spatiotemporal features of congested traffic patterns and of three-phase traffic theory as well as their engineering applications.

Rather than a comprehensive analysis of empirical and theoretical results in the field, the present book includes no more empirical and theoretical results than are necessary for the understanding of vehicular traffic on unsignalized multi-lane roads. The main objectives of the book are to present an "elementary" traffic flow theory and control methods as well as to show links between three-phase traffic theory and earlier traffic flow theories. The need for such a book follows from many comments of colleagues made after publication of the book "The Physics of Traffic".

Another important objective of this book is to give an introduction to methods of spatiotemporal traffic congestion recognition and prediction, on-ramp metering, speed limit control, and some other freeway control and dynamic management methods whose theoretical basis is three-phase traffic theory. The importance of this subject can be explained as follows. Almost all other traffic flow theories and the associated freeway control and dynamic management methods assume the existence of a *particular* (fixed or stochastic) highway capacity of free flow at a highway bottleneck and, therefore, they use the highway capacity as a basic parameter of dynamic traffic management models. In this book we show and explain how and why the application of a particular highway capacity in methods for dynamic freeway traffic management like on-ramp metering, speed limit control, or dynamic traffic assignment, is not consistent with features of real traffic.

Through an application of the principle "no more results than are necessary", I hope to present traffic flow theory and control in a manner understandable to a broad audience of readers interested in traffic phenomena. With this aim, the book also includes an extended glossary with definitions and explanations of terms used.

I thank Ralf G Herrtwich and Matthias Schulze for their support as well as many other my colleagues at the Daimler Company, in particular, Hubert Rehborn, Gerhard Nöcker, Andreas Hiller, Achim Brakemaier, Ines Maiwald-Hiller, Winfried Kronjäger for fruitful discussions and advice. I thank also Dietrich Wolf for useful suggestions. Particular thanks are to Achim Brakemaier, Viktor Friesen, Sergey Klenov, Gerhard Nöcker, Andreas Hiller, Winfried Kronjäger, Jochem Palmer, and Hubert Rehborn who have read the book and made many useful comments. I thank also Hesham Rakha, Hani Mahmassani, and Jorge Laval for helpful discussions about approaches to traffic flow modeling in Woods Hole in July 2008. Many thanks to Rüdiger Hain, Oliver Baumann and all other friends who have encouraged me while writing this book. I am grateful to Sergey Klenov for his help with numerical simulations and the preparation of illustrations for the book. Finally, I thank my wife, Tatiana Kerner, for her help and understanding.

Stuttgart, May 2009 *Boris Kerner*

Contents

1 **Introduction** 1
  References 5

Part I Three-Phase Traffic Theory

2 **Definitions of The Three Traffic Phases** 9
  2.1 Traffic Variables, Parameters, and Patterns 9
  2.2 Free Flow (F) and Congested Traffic........................... 13
  2.3 Methodology of Three-Phase Traffic Theory 18
  2.4 Two Traffic Phases in Congested Traffic:
      Wide Moving Jam (J) and Synchronized Flow (S) 20
  2.5 Characteristic Parameters of Wide Moving Jam Propagation 26
  2.6 Microscopic Criterion for Traffic Phases in Congested Traffic 30
  2.7 Motivation for Traffic Phase Definitions 38
  References 39

3 **Nature of Traffic Breakdown at Bottleneck**....................... 41
  3.1 Induced Traffic Breakdown 41
  3.2 Explanation of Nature of Traffic Breakdown at Bottleneck through
      Fundamental Hypothesis of Three-Phase Traffic Theory 45
  3.3 Nucleation Features of Traffic Breakdown at Bottleneck 57
  3.4 Dual Role of Lane Changing in Free Flow:
      Maintenance of Free Flow or Traffic Breakdown 69
  References 72

4 **Infinite Number of Highway Capacities of Free Flow at Bottleneck** .. 73
  4.1 Definition of Highway Capacity of Free Flow at Bottleneck 74
  4.2 Characteristics of Highway Capacities 75
  References 79

5 Nature of Moving Jam Emergence 81
5.1 Pinch Effect in Synchronized Flow 81
5.2 Nucleation Features of Wide Moving Jam Emergence in Synchronized Flow 83
5.3 Dual Role of Lane Changing in Synchronized Flow: Maintenance of Synchronized Flow or Wide Moving Jam Emergence 90
5.4 Comparison of F→S and S→J Transitions 93
5.5 Empirical Double Z-Characteristic for Phase Transitions in Traffic Flow 95
References 98

6 Origin of Hypotheses and Terms of Three-Phase Traffic Theory 99
6.1 Hypotheses of Three-Phase Traffic Theory as The Result of Empirical Criteria for Traffic Phases 99
6.2 Are Terms of Natural Science used in Three-Phase Traffic Theory Needed for Transportation Engineering? 100
References 105

7 Spatiotemporal Traffic Congested Patterns 107
7.1 Simplified Diagram of Congested Patterns at Isolated Bottleneck ... 107
7.2 Variety of Congested Patterns at Isolated Bottleneck 109
7.3 Complex Congested Patterns at Adjacent Bottlenecks 126
References 135

Part II Impact of Three-Phase Traffic Theory on Transportation Engineering

8 Introduction to Part II: Compendium of Three-Phase Traffic Theory 139
References 142

9 Freeway Traffic Control based on Three-Phase Traffic Theory 143
9.1 Reconstruction and Tracking of Congested Patterns 143
9.2 Feedback On-Ramp Metering 150
9.3 Speed Limit Control 155
9.4 Cooperative Driving for Improving of Traffic Flow and Safety 159
9.5 Traffic Control based on Wireless Car Communication 162
9.6 Adaptive Cruise Control 168
References 170

10 Earlier Theoretical Basis of Transportation Engineering: Fundamental Diagram Approach ... 173
  10.1 Traffic Description and Control based on Fundamental Diagram of Traffic Flow ... 173
  10.2 Congested Traffic Description in the Framework of Lighthill-Whitham-Richards (LWR) Traffic Flow Theory ... 177
  10.3 Traffic Breakdown Description through Free Flow Instability in General Motors (GM) Model Class ... 182
  10.4 Common Features of earlier Traffic Flow Models ... 200
  10.5 Empirical Tests of earlier Traffic Flow Models ... 203
  10.6 Applications of Highway Capacity Definitions in Transportation Engineering ... 205
  10.7 Comparison of Feedback On-Ramp Metering Methods ... 214
  References ... 216

11 Three-Phase Traffic Flow Models ... 221
  11.1 Overview of Three-Phase Traffic Flow Models ... 221
  11.2 Deterministic Acceleration Time Delay Three-Phase Traffic Flow Model ... 222
  11.3 Stochastic Three-Phase Traffic Flow Model ... 229
  11.4 Cellular Automata Three-Phase Traffic Flow Model ... 237
  11.5 Methodology of Empirical Test ... 238
  11.6 What Three-Phase Traffic Flow Model is better to Use? ... 239
  References ... 242

12 Linking of Three-Phase Traffic Theory and Fundamental Diagram Approach to Traffic Flow Modeling ... 245
  12.1 Three-Phase Traffic Models in the Framework of Fundamental Diagram Approach ... 245
  12.2 What Features of Three-Phase Traffic Theory are Missing in Earlier Traffic Flow Theories and Models? ... 249
  References ... 252

13 Conclusions and Outlook ... 253

Glossary ... 255

Index ... 263

Acronyms and Symbols

F	Free traffic flow
C	Congested traffic
S	Synchronized flow phase of congested traffic
J	Wide moving jam phase of congested traffic
Line J	Characteristic line in the flow–density plane representing a steadily propagation of the downstream front of a wide moving jam. The slope of the line J is determined by the mean velocity of the downstream jam front
F→S transition	Traffic breakdown, i.e., phase transition from the free flow phase to the synchronized flow phase
S→J transition	Phase transition from the synchronized flow phase to the wide moving jam phase
F→S→J transitions	Sequence of an F→S transition with a following S→J transition
SP	Synchronized flow pattern
LSP	Localized SP
WSP	Widening SP
MSP	Moving SP
GP	General congested traffic pattern
EP	Expanded traffic congested pattern
ESP	Expanded synchronized flow pattern
EGP	Expanded general pattern
FOTO	**Fo**casting **O**f **T**raffic **O**bjects is a macroscopic model for automatic traffic phase reconstruction and tracking of synchronized flow
ASDA	**A**utomatische **S**tau**d**ynamik **A**nalyse (automatic tracking of moving traffic jams) is a macroscopic model for tracking of moving jams
ANCONA	**A**utomatic o**n**-ramp **co**ntrol of co**n**gested p**a**tterns is a control approach in which congestion is allowed to set in at a bottleneck. The basic idea is to maintain congestion conditions at the bottleneck to the minimum possible level; in particular, a congested pattern should not propagate upstream
ACC	Adaptive cruise control

v	vehicle speed [km/h] or [m/s]
v_ℓ	speed of the preceding vehicle [km/h] or [m/s]
$\Delta v = v_\ell - v$	speed difference between the speed of the preceding vehicle and the vehicle speed [km/h] or [m/s]
a	vehicle acceleration (deceleration) [m/s^2]
q	flow rate [vehicles/h] or [vehicles/min]
ρ	vehicle density [vehicles/km]
g	space gap between vehicles [m] that is also called as net distance or space headway
τ	time headway between vehicles [s] that is also called as time gap or net time distance
d	vehicle length [m]
o	occupancy [%]
v_g	The mean velocity of the downstream front of a wide moving jam
q_{out}	The flow rate in free flow formed in the outflow of a wide moving jam
ρ_{min}	The vehicle density in free flow formed in the outflow of a wide moving jam
$q_{\text{th}}^{(B)}$	Threshold flow rate for traffic breakdown at a bottleneck that is the minimum highway capacity of free flow
$q_{\text{max}}^{(\text{free B})}$	Maximum flow rate in free flow downstream of a bottleneck that is the maximum highway capacity of free flow
$q^{(\text{cong})}$	The average flow rate within a congested pattern upstream of a bottleneck averaged during a time interval that is considerably longer than time distances between any moving jams within the congested pattern
q_{in}	The flow rate in free flow on the main road upstream of a bottleneck
q_{on}	The on-ramp inflow rate at an on-ramp bottleneck
q_{sum}	The flow rate in free flow downstream of a bottleneck under condition that free flow is at the bottleneck
q_{off}	The flow rate of vehicles leaving the main road to off-ramp at an off-ramp bottleneck
$v_{\text{cr, FS}}^{(B)}$	A critical speed required for traffic breakdown at a bottleneck
$\rho_{\text{cr, FS}}^{(B)}$	A critical density required for traffic breakdown at a bottleneck
$v_{\text{cr}}^{(\text{SJ})}$	A critical speed in metastable synchronized flow required for wide moving jam emergence
$\rho_{\text{cr}}^{(\text{SJ})}$	A critical density in metastable synchronized flow required for wide moving jam emergence
$\tau_{\text{del}}^{(a)}$	The mean time delay in vehicle acceleration
$\tau_{\text{del, jam}}^{(a)}$	The mean time delay in vehicle acceleration at the downstream front of a wide moving jam
g_{safe}	A safe space gap between vehicles
τ_{safe}	A safe time headway (safe time gap) between vehicles
G	A synchronization space gap between vehicles

τ_G	A synchronization time headway (synchronization time gap) between vehicles
τ_{reac}	The mean driver reaction time
T_{av}	Averaging time interval for traffic variables
$P_{FS}^{(B)}$	Probability of traffic breakdown at a bottleneck
$P_C^{(B)}$	Probability that free flow remains at a bottleneck during time interval T_{av}
τ_{step}	A time step in traffic flow models with a discrete time
$r = \text{rand}(0,1)$	A random number uniformly distributed between 0 and 1

Chapter 1
Introduction

Vehicular traffic is an extremely complex dynamic process associated with the spatiotemporal behavior of many-particle systems. The complexity of vehicular traffic is due to nonlinear interactions between the following three main dynamic processes (Fig. 1.1):

(i) travel *decision* behavior, which determines traffic demand,
(ii) *routing* of vehicles in a traffic network, and
(iii) *traffic congestion* occurrence within the network.

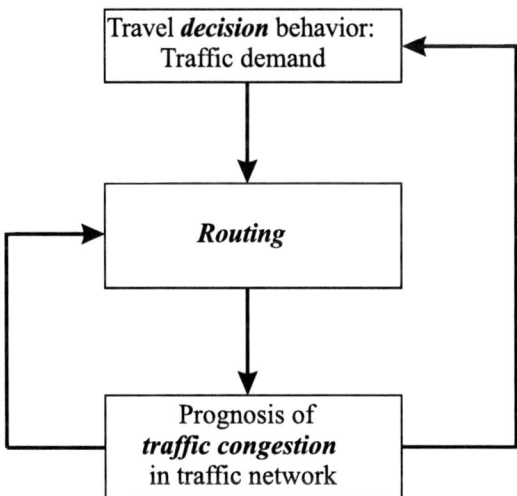

Fig. 1.1 Explanation of complexity of vehicular traffic

Travel decision behavior determines travel demand. Traffic routing in the network is associated with traffic supply. However, traffic congestion occurring within the traffic network restricts free flow travel. This influences both travel decision behavior and traffic routing in the network. Indeed, because of traffic congestion, a person decides to stay at home or travel by train rather than by car. A feedback between *traffic congestion* and *travel decision* is symbolically shown by arrow on the right hand side in Fig. 1.1. In turn, because of traffic congestion on a route from an origin to a destination usually used, a person changes the route of travel. A feedback between *traffic congestion* and *routing* is symbolically shown by arrow on the left hand side in Fig. 1.1.

Empirical traffic congestion, i.e., traffic congestion observed in real measured traffic data is a *spatiotemporal effect*: The traffic congestion occurs in space and time in the form of spatiotemporal congested traffic patterns propagating within a traffic network. These empirical congested traffic patterns exhibit a variety of complex spatiotemporal features. For this reason, the complexity of traffic management is associated with this variety of the empirical congested traffic patterns as well as with the necessity in the optimization of these patterns. This optimization should ensure either the dissolution of traffic congestion or, if this is not possible to achieve, the minimization of the influence of traffic congestion on travel costs.

- We see that the understanding of *empirical traffic congestion* is the key for effective traffic management, control, organization, and all other applications of transportation engineering.

Inputs to travel decision behavior models are the typical regional model data about social, economic, and demographic information of potential travelers and land use information to create schedules followed by people in their everyday life. The output are detailed lists of activities pursued, times spent in each activity, and travel information from activity to activity. A review of travel decision behavior models has recently been done by Goulias [1].

Routing based on a traffic optimization, which is associated with a minimization of chosen travel "costs", together with a prognosis of traffic congestion is called *dynamic traffic assignment* in the traffic network. A dynamic traffic assignment model should find the link inflows for the traffic network. The model includes usually a traffic flow model, which makes a prognosis of traffic in the network, and a routing model associated with the problem of traffic optimization. The router model computes the sequence of roadways that minimize travel costs of the traffic network. Examples of the travel costs are travel time, fuel consumption, or HC and CO_2 emissions. The traffic flow and router models are connected by a feedback loop (see the feedback loop between *traffic congestion* and *routing* in Fig. 1.1). As a result, traffic congestion in the network predicted by the traffic flow model changes results of dynamic traffic assignment considerably. For this reason, the traffic flow model should model traffic congestion as close as possible to real traffic congestion found in empirical observations. A review of models for dynamic traffic assignment in traffic networks has recently been done by Rakha and Tawfik [2]. Approaches to traffic prognosis have recently been reviewed by Rehborn and Klenov [3].

A complex spatiotemporal behavior of empirical traffic congested patterns was studied during the last 75 years by several generations of researchers (see references in reviews and books [4–26]). It was found that traffic congestion in the traffic network results from traffic breakdown in an initially free flow: vehicle speeds decrease abruptly to lower speeds in congested traffic [4, 5, 27–29]. Traffic breakdown is observed mostly at highway bottlenecks. A bottleneck can be a result of road works, on- and off-ramps, a decrease in the number of road lanes, road curves and road gradients, bad weather conditions, accidents, etc. [4, 5, 27–29]. Beginning from the classic work of Greenshields [30], the most of the traffic flow theories and models have been made in the framework of the so-called *fundamental diagram of traffic flow*. The fundamental diagram is a flow rate–density relationship, i.e., a correspondence between a given vehicle density and the flow rate in traffic flow. The fundamental diagram reflects the obvious result of empirical observations that the greater the density, the lower the averaged speed in vehicular traffic.

However, the puzzle of empirical spatiotemporal features of traffic congestion has been solved only recently [25]. As will be discussed in this book, it turns out that earlier traffic flow theories and models reviewed in [4–24] cannot explain either traffic breakdown or most features of the resulting spatiotemporal congested patterns. These traffic flow theories and models, which dominate up to now in transportation research journals and teaching programs of most universities, are also the theoretical basis for dynamic traffic assignment models and methods for freeway traffic control. Therefore, the associated methods for dynamic traffic management are not consistent with features of real traffic.

To explain empirical spatiotemporal features of vehicular traffic, the author introduced an alternative traffic flow theory called three-phase traffic theory. A consideration of empirical spatiotemporal features of congested traffic patterns and three-phase traffic theory that explains these pattern features as well as some resulting engineering applications have been presented in the previous book [25].

Rather than a comprehensive discussion congested traffic patterns, in the present book the author gives only an introduction to traffic flow theory and control on multi-lane roads[1] including no more empirical and theoretical results than are necessary for the understanding of vehicular traffic as well as to make a more detailed consideration of links between three-phase traffic theory and earlier traffic flow theories. The main objectives of this book are as follows:

(1) To explain why rather than the fundamental diagram of traffic flow, *spatiotemporal* analysis of empirical congested traffic patterns is the key for the understanding of traffic flow characteristics as well as for the development of dynamic traffic management methods (including methods for dynamic traffic control and assignment) that are consistent with real traffic.

---

[1] In the book, we limit attention to dynamic traffic phenomena determined by driver interactions in traffic. These traffic phenomena play the most important role on freeways and highways. In contrast, in city traffic, light signals and other traffic regulations at road intersections can often almost fully determine traffic dynamics. A review about urban traffic control has recently been done by Gartner and Stamatiadis [31]. See also the UTA model for the urban traffic analysis and prognosis in Sect. 22.4 of the book [25].

(2) To explain why classic traffic flow theories and models cannot explain either traffic breakdown or most features of the resulting spatiotemporal congested patterns.

(3) To give a new basis for the development of models for dynamic traffic operation methods, dynamic traffic assignment models, and highway traffic control methods, which are consistent with features of real traffic.

The importance of these objectives can be explained as follows. Most earlier traffic flow theories and the associated freeway control and dynamic management methods assume the existence of a *particular* fixed or stochastic highway capacity of free flow at a highway bottleneck. Therefore, they use the highway capacity as a basic parameter of dynamic traffic management models. In this book we show and explain how and why the application of a particular highway capacity as a control parameter in methods for dynamic freeway traffic management like on-ramp metering, speed limit control, or dynamic traffic assignment, is not consistent with features of real traffic.

The book consists of two parts. Part I is devoted to a consideration of empirical spatiotemporal features and characteristics of traffic and three-phase traffic theory that explains these traffic features and characteristics. In Part II, we discuss the impact of three-phase traffic theory on traffic control and management. Because simulations of the prognosis of traffic congestion with mathematical traffic flow models can be considered a part of traffic control and management models (see Fig. 1.1 and related explanations made above), a critical discussion of the impact of three-phase traffic theory on these models has also been included in Part II.

Part I begins with traffic phase definitions made in three-phase traffic theory (Chap. 2). Explanations of empirical spatiotemporal traffic flow characteristics with three-phase traffic theory are the subject of the subsequent Chaps. 3–7.

In particular, in Chaps. 3 and 4 we explain why the fundamental empirical features of traffic breakdown at a bottleneck leads to the conclusion of three-phase traffic theory that rather than a particular highway capacity, there are the infinite number of highway capacities at the bottleneck. A consideration of the spontaneous emergence of wide moving jams is the subject of Chap. 5. The origin of some of the hypotheses and terms of three-phase traffic theory used in previous chapters of the book is discussed in Chap. 6.

In Chap. 7, we discuss a variety of spatiotemporal congested patterns arising from traffic breakdown and wide moving jam emergence.

Part II begins from a compendium of three-phase traffic theory (Chap. 8). In Chap. 9, we briefly discuss methods for spatiotemporal congested pattern reconstruction, tracking, and control.

Earlier theoretical basis of transportation engineering is discussed in Chap. 10. Here we discuss both the achievements and drawbacks of earlier traffic flow theories in explanations of real measured spatiotemporal features of traffic congestion. In particular, the critical part of this consideration contains the following subjects:

(i) That and why many of the fundamental empirical spatiotemporal features of traffic patterns are *lost* in the fundamental diagram of traffic flow.

(ii) Why the earlier traffic flow theories and models in the framework of the fundamental diagram of traffic flow, which are up to now the basic approaches in transportation research [4–24], have failed to show empirical features of traffic breakdown.

(iii) That and why well-accepted definitions of highway capacity as a particular (either fixed or stochastic) value are also not consistent with the fundamental empirical features of traffic breakdown and, as a result, methods for traffic flow control, dynamic traffic assignment as well as other methods of dynamic traffic management, which are based on these capacity definitions, are not consistent with real measured spatiotemporal traffic flow characteristics.

In Chap. 11 we discuss some mathematical traffic flow models in the framework of three-phase traffic theory. These models can simulate traffic breakdown and all resulting spatiotemporal traffic flow characteristics as they are observed in real measured traffic data.

A discussion of links between three-phase traffic theory and the fundamental diagram approach to traffic flow modeling is the subject of Chap. 12. In this discussion we would like to answer the question what features of three-phase traffic theory are missing in the earlier traffic flow theories of Chap. 10 resulting in the failure of these theories in the explanation of traffic breakdown as observed in real measured traffic data.

References

1. K.G. Goulias, in *Encyclopedia of Complexity and System Science*, ed. by R.A. Meyers. (Springer, Berlin, 2009), pp. 9536–9565
2. H. Rakha, A. Tawfik, in *Encyclopedia of Complexity and System Science*, ed. by R.A. Meyers. (Springer, Berlin, 2009), pp. 9429–9470
3. H. Rehborn, S.L. Klenov, in *Encyclopedia of Complexity and System Science*, ed. by R.A. Meyers. (Springer, Berlin, 2009), pp. 9500–9536
4. A.D. May, *Traffic Flow Fundamentals*, (Prentice-Hall, Inc., New Jersey, 1990)
5. *Highway Capacity Manual 2000*, (National Research Council, Transportation Research Boad, Washington, D.C., 2000)
6. F.A. Haight, *Mathematical Theories of Traffic Flow*, (Academic Press, New York, 1963)
7. I. Prigogine, R. Herman, *Kinetic Theory of Vehicular Traffic*, (American Elsevier, New York, 1971)
8. W. Leutzbach, *Introduction to the Theory of Traffic Flow*, (Springer, Berlin, 1988)
9. N.H. Gartner, C.J. Messer, A.K. Rathi (eds), *Traffic Flow Theory: A State-of-the-Art Report*, (Transportation Research Board, Washington DC, 2001)
10. M. Cremer, *Der Verkehrsfluss auf Schnellstrassen*, (Springer, Berlin, 1979)
11. G.B. Whitham, *Linear and Nonlinear Waves*, (Wiley, New York, 1974)
12. R. Wiedemann, *Simulation des Verkehrsflusses*, (University of Karlsruhe, Karlsruhe, 1974)
13. G.F. Newell, *Applications of Queuing Theory*, (Chapman Hall, London, 1982)
14. C.F. Daganzo, *Fundamentals of Transportation and Traffic Operations*, (Elsevier Science Inc., New York, 1997)
15. M. Papageorgiou, *Application of Automatic Control Concepts in Traffic Flow Modeling and Control*, (Springer, Berlin, New York, 1983)
16. M. Brackstone, M. McDonald, Transportation Research F **2**, 181 (1998)

17. D.C. Gazis, *Traffic Theory*, (Springer, Berlin, 2002)
18. N. Bellomo, V. Coscia, M. Delitala, Math. Mod. Meth. App. Sc. **12**, 1801–1843 (2002)
19. D. Chowdhury, L. Santen, A. Schadschneider, Physics Reports **329**, 199 (2000)
20. D. Helbing, Rev. Mod. Phys. **73**, 1067–1141 (2001)
21. R. Mahnke, J. Kaupužs, I. Lubashevsky, Phys. Rep. **408**, 1–130 (2005)
22. T. Nagatani, Rep. Prog. Phys. **65**, 1331–1386 (2002)
23. K. Nagel, P. Wagner, R. Woesler, Oper. Res. **51**, 681–716 (2003)
24. B. Piccoli, A. Tosin, in *Encyclopedia of Complexity and System Science*, ed. by R.A. Meyers. (Springer, Berlin, 2009), pp. 9727–9749
25. B.S. Kerner, *The Physics of Traffic*, (Springer, Berlin, New York, 2004); in *Encyclopedia of Complexity and System Science*, ed. by R.A. Meyers. (Springer, Berlin, 2009), pp. 9302–9355; 9355–9411
26. S. Maerivoet, B. De Moor, Phys. Rep. **419**, 1-64 (2005)
27. F.L. Hall, K. Agyemang-Duah, Trans. Res. Rec. **1320**, 91–98 (1991)
28. L. Elefteriadou, R.P. Roess, W.R. McShane, Transp. Res. Rec. **1484**, 80–89 (1995)
29. B.N. Persaud, S. Yagar, R. Brownlee, Trans. Res. Rec. **1634**, 64–69 (1998)
30. B.D. Greenshields, in *Highway Research Board Proceedings*, **14**, pp. 448-477 (1935)
31. N.H. Gartner, Ch. Stamatiadis, in *Encyclopedia of Complexity and System Science*, ed. by R.A. Meyers. (Springer, Berlin, 2009), pp. 9470–9500

Part I
Three-Phase Traffic Theory

Chapter 2
Definitions of The Three Traffic Phases

2.1 Traffic Variables, Parameters, and Patterns

Traffic flow phenomena are associated with a complex dynamic behavior of spatiotemporal traffic patterns. The term *spatiotemporal* reflects the empirical evidence that traffic occurs in *space and time*. Therefore, only through a *spatiotemporal* analysis of real measured traffic data the understanding of features of real traffic is possible. In other words, spatiotemporal features of traffic can only be found, if traffic variables are measured in real traffic in space and time.

The term a *spatiotemporal traffic pattern* (traffic pattern for short) is defined as follows:

- A spatiotemporal traffic pattern is a distribution of traffic flow variables in space and time.

Examples of traffic variables are the flow rate q [vehicles/h], vehicle density ρ [vehicles/km], and vehicle speed v [km/h] or [m/s] (see, e.g., [1–3]).

The term *empirical* features of a spatiotemporal traffic pattern means that the features are found based on an analysis of real measured traffic data.

A spatiotemporal traffic pattern is limited spatially by pattern fronts. There are downstream and upstream fronts of a traffic pattern. The downstream pattern front separates the pattern from other traffic patterns downstream. The upstream pattern front separates the pattern from other traffic patterns upstream.

The term *front of traffic pattern* is defined as follows:

- A front of a traffic pattern is either a moving or motionless region within which one or several of the traffic variables change abruptly in space (and in time, when the front is a moving one).

Traffic variables and traffic patterns can depend considerably on *traffic parameters*.

- Traffic parameters are parameters, which can influence traffic variables and traffic patterns.

Examples of traffic parameters are a traffic network infrastructure (including, e.g., highway bottleneck types and their locations), weather (whether the day is sunny or rainy or else foggy, dry or wet road, or even ice and snow on road), percentage of long vehicles, day time, working day or week-end, other road conditions, and vehicle technology.

Considering traffic flow patterns, we distinguish between *macroscopic* and *microscopic* descriptions of the patterns.

In the macroscopic pattern description, the behavior of macroscopic measured traffic variables and macroscopic characteristics of traffic flow patterns in space and time should be studied and understood.

Examples of the macroscopic traffic variables are the flow rate, vehicle density, occupancy, and average vehicle speed (see, e.g., [1–3]).

An example of macroscopic characteristics of a traffic pattern is the mean velocities of the downstream and upstream fronts of the pattern. We see that the macroscopic traffic variables and pattern characteristics are associated with an averaging behavior of many vehicles in traffic, i.e., the variables and characteristics are averaged during an averaging time interval for traffic variables denoted by T_{av}.

As an example of the term an *averaging time interval for traffic variables*, we consider *1-min average data* that means the following: all macroscopic traffic variables associated with a traffic pattern under consideration are averaged with the use of the same averaging time interval $T_{av} = 1$ min.

In contrast with the macroscopic description of traffic patterns, the microscopic description of traffic flow patterns is associated with a study of microscopic traffic variables and microscopic pattern characteristics that reflect the behavior of individual (called also *single*) vehicles in traffic flow.

Examples of the microscopic traffic flow variables are single vehicle space coordinates and their time-dependence, a time headway (net time distance) τ [s] and a space gap (net distance) g [m] between two vehicles following each other (Fig. 2.1), a single vehicle speed v [km/h] or [m/s], a vehicle length d [m] [1–3]. In particular, vehicle space coordinates and their time-dependence can be used for the reconstruction of vehicle trajectories, i.e., the trajectories of vehicles in the space–time plane[1]. Note that measured traffic data in which microscopic traffic variables can be identified are also called *single vehicle data*[2].

There are many measurement techniques of traffic flow variables based on road detectors (see, e.g., [1–3]), video camera measurements (see, e.g., [4]), etc. We briefly discuss measurements of traffic variables with induction double loop detectors installed at some road locations.

Each detector consists of two induction loops spatially separated by a given small distance ℓ_d (Fig. 2.1). The induction loop registers a vehicle moving on the road by producing a pulse electric current that begins at some time t_b when the vehicle reaches the induction loop and it ends some time later t_f when the vehicle leaves the

---

[1] An example of empirical vehicle trajectories is shown in Fig. 2.3 of Sect. 2.2.1.

[2] Naturally, there is also an intermediate description of traffic called as a *mesoscopic* description of traffic phenomena in which both macroscopic and microscopic traffic flow variables and/or characteristics of traffic patterns are studied.

2.1 Traffic Variables, Parameters, and Patterns

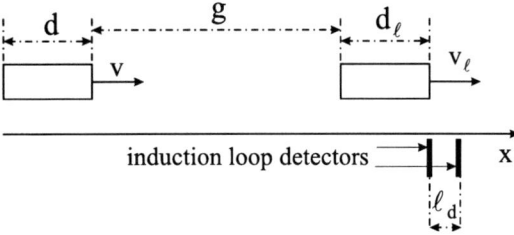

Fig. 2.1 Qualitative scheme of induction loop detector measurements

induction loop. The duration of this current pulse

$$\Delta t = t_f - t_b \tag{2.1}$$

is therefore related to the time taken by the vehicle to traverse the induction loop.

Every vehicle that passes the induction loop produces a related current pulse. This enables us to calculate the gross time gap between the vehicle with a speed v and the preceding vehicle with a speed v_ℓ that have passed the induction loop one after the other:

$$\tau^{(\text{gross})} = t_b - t_{\ell, b}, \tag{2.2}$$

where subscript ℓ is related to the preceding vehicle (Fig. 2.1). We can further calculate the flow rate q as the measured number of vehicles N passing the induction loop during a given averaging time interval for traffic variables T_{av}:

$$q = \frac{N}{T_{av}}. \tag{2.3}$$

Because there are two different induction loops in each detector, separated by a known distance ℓ_d from one another, the detector is able to measure the individual vehicle speed. Indeed, due to the distance ℓ_d between two loops of the detector, the first (upstream) loop registers the vehicle earlier than the second (downstream) one. Therefore, if the vehicle speed v is not zero, there will be a time lag δt between the current pulses produced by the two detector induction loops when the vehicle passes both. It is assumed that by virtue of the small value of ℓ_d, the vehicle speed does not change between the induction loops. This enables us to calculate the single (individual) vehicle speed v:

$$v = \frac{\ell_d}{\delta t} \tag{2.4}$$

and the vehicle length d

$$d = v\Delta t. \tag{2.5}$$

From Eqs. (2.2) and (2.5) it is possible to calculate the time headway:

$$\tau = \tau^{(\text{gross})} - \frac{d_\ell}{v_\ell}. \tag{2.6}$$

At a given time instant $t = t_1$, the time headway between vehicles $\tau(t_1)$ is *defined* as a time it takes for a vehicle to reach a road location at which the bumper of the preceding vehicle is at the time instant t_1. In single vehicle data measured at a road detector (Fig. 2.1), t_1 is the time at which the preceding vehicle leaves the detector whose location is therefore related to the location of the bumper of the preceding vehicle in the time headway definition; the time headway is equal to $\tau(t_1) = t_2 - t_1$, where t_2 is the time at which the vehicle front has been recorded at the detector. The time headway τ in (2.6) is related to the time instant t_1.

Single vehicle speeds also enable us to calculate the average (arithmetic) vehicle speed v of N vehicles passing the detector in time interval T_{av},

$$v = \frac{1}{N} \sum_{i=1}^{N} v_i, \qquad (2.7)$$

where index $i = 1, 2, ..., N$.

The vehicle density (the number of vehicles per unit length of a road, e.g., vehicles per km) can be roughly estimated from the relation

$$\rho = \frac{q}{v}, \qquad (2.8)$$

where v is the average speed. However, it should be noted that the vehicle density ρ is related to vehicles on a road section of a given length whereas the vehicle speed is measured at the location of the detector only and is averaged over the averaging time interval T_{av}. As a result, at low average vehicle speeds, the vehicle density estimated via (2.8) can lead and does usually lead to a considerable discrepancy in comparison with the real vehicle density. For a more detailed consideration of the criticism of measured data analyses associated with a considerably error in the density estimation with formula (2.8) see [5] and a recent review [6].

A road detector can also measure a macroscopic traffic variable called *occupancy*, which is defined through the formula (e.g., [1]):

$$o = \frac{T_{veh}}{T_{av}} 100\%, \qquad (2.9)$$

where T_{veh} is the sum of the time intervals when the detector has measured vehicles during the time interval T_{av}:

$$T_{veh} = \sum_{i=1}^{N} \Delta t_i, \qquad (2.10)$$

Δt_i is defined via (2.1).

2.2 Free Flow (F) and Congested Traffic

2.2.1 Definition of Congested Traffic

Free traffic flow (free flow for short) is usually observed, when the vehicle density in traffic is small enough. At small enough vehicle density, interactions between vehicles in free flow are negligible. Therefore, vehicles have an opportunity to move with their desired maximum speeds (if this speed is not restricted by road conditions or traffic regulations).

When the density increases in free flow, the flow rate increases too, however, vehicle interaction cannot be neglected any more. As a result of vehicle interaction in free flow, the average vehicle speed decreases with increase in density.

To illustrate these well-known features [1–3], the flow rate and density, which is calculated with formula (2.8) from the flow rate and average speed measured at a road location, are presented in the flow–density plane (points left of a dashed line FC in Fig. 2.2 (a)). In empirical traffic data, the increase in the flow rate with the density increase in free flow has a limit. At the associated *limit (maximum) point of free flow*, the flow rate and density reach their maximum values denoted by $q_{\max}^{(\text{free, emp})}$ and $\rho_{\max}^{(\text{free, emp})}$, respectively, while the average speed has a minimum value for the free flow:

$$v_{\min}^{(\text{free, emp})} = q_{\max}^{(\text{free, emp})} / \rho_{\max}^{(\text{free, emp})}. \qquad (2.11)$$

These points are well-fitted by a flow–density relationship for free flow, i.e., a certain curve with a positive slope between the flow rate and density associated with averaging of measured data shown left of the dashed line FC in Fig. 2.2 (a) to one average flow rate for each density (curve F in Fig. 2.2 (b)) [1–3, 7, 8]. This flow–density relationship is called the fundamental diagram of free flow. The empirical fundamental diagram of free flow is cut off at the limit point of free flow ($\rho_{\max}^{(\text{free, emp})}$, $q_{\max}^{(\text{free, emp})}$) (Fig. 2.2 (b)) [1–3, 9].

To distinguish free flow points in the flow–density plane, we use in Fig. 2.2 (a, b) the dashed line FC between the origin of the flow–density plane and the limit point of free flow; the slope of the line FC is equal to the minimum speed in free flow $v_{\min}^{(\text{free, emp})}$ (2.11). Thus empirical points of free flow as well as the associated fundamental diagram lie to the left of the dashed line FC in the flow–density plane.

In empirical observations, when density in free flow increases and becomes great enough, the phenomenon of the onset of congestion is observed in this free flow: the average speed decreases abruptly to a lower speed in congested traffic:

- Congested traffic is defined as a state of traffic in which the average speed is *lower* than the minimum average speed that is still possible in free flow (e.g., [3, 10]).

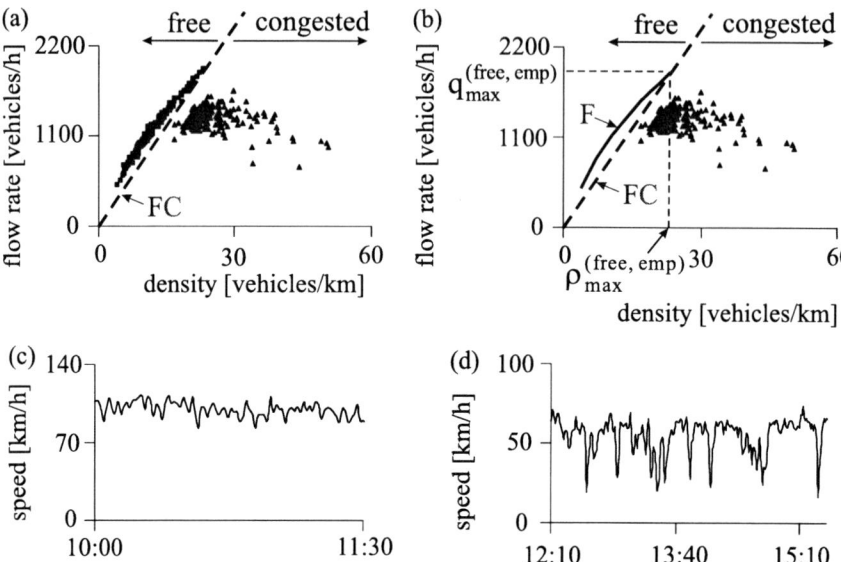

Fig. 2.2 Free flow and congested traffic (e.g., [1–3, 9, 10]). (a) Empirical data for free flow (points left of the dashed line FC) and for congested traffic (points right of the dashed line FC). (b) The fundamental diagram for free flow (curve F) and the same measured data for congested traffic as those in (a). (c, d) Vehicle speed in free flow (c) and congested traffic (d), related to points left and right of the line FC in (a), respectively. 1-min average data measured at a road location

Thus empirical points of congested traffic lie to the right of the dashed line FC in the flow–density plane[3].

Traffic congestion occurs mostly at a highway bottleneck (bottleneck for short). The bottleneck can be a result of road works, on- and off-ramps, a decrease in the number of road lanes, road curves and road gradients, bad weather conditions, accidents, etc. [1–3].

In congested traffic, a great variety of congested traffic patterns are observed [10–16]. A *congested traffic pattern* (congested pattern for short) is defined as follows.

- A congested traffic pattern is a spatiotemporal traffic pattern within which there is congested traffic. The congested pattern is separated from free flow by the downstream and upstream fronts: At the downstream front, vehicles accelerate

---

[3] It must be noted that the definition of congested traffic through the use of the empirical limit point on the fundamental diagram of free flow seems to be easy, however, can lead to an error in measurements of the minimum speed that is possible in free flow. This is because the exact value of this minimum speed that is possible in free flow is associated with the maximum (limit) flow rate $q_{max}^{(free,\ emp)}$ in free flow at which probability of traffic breakdown is equal to one (see explanations of the flow rate dependence of breakdown probability in Sect. 4.2.2). However, it is extremely difficult to find such a free flow in real measured traffic data. This comment is also related to the limit point for free flow shown in Fig. 2.2: the speed $v_{min}^{(free,\ emp)}$ associated with this limit point for free flow gives only an approximate value for the minimum speed that is possible in free flow.

2.2 Free Flow and Congested Traffic

from a lower speed within the pattern to a higher speed in free flow downstream; at the upstream front, vehicles decelerate from a free flow speed to a lower speed within the congested pattern.

In particular, one of the congested traffic patterns is a moving traffic jam [10–16]. A *moving traffic jam* (moving jam for short) is defined as follows:

- A moving jam is a localized congested traffic pattern that moves upstream in traffic flow (Fig. 2.3). Within the moving jam the average vehicle speed is very low (sometimes as low as zero), and the density is very high. The moving jam is spatially restricted by the downstream jam front and upstream jam front. Within the downstream jam front vehicles accelerate from low speed states within the jam to higher speeds in traffic flow downstream of the moving jam. Within the upstream jam front vehicles must slow down to the speed within the jam. Both jam fronts move upstream. Within the jam fronts the vehicle speed, flow rate, and density vary abruptly.

Moving jams have been studied empirically by many authors, in particular, in classic empirical works by Edie *et al.* [11–14], Treiterer *et al.* [15, 16] (Fig. 2.3), and Koshi *et al.* [10].

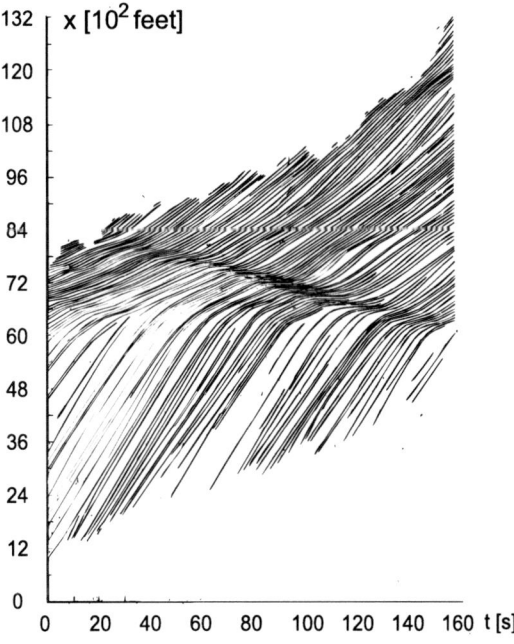

Fig. 2.3 A moving jam: dynamics of vehicle trajectories derived from aerial photography (1 feet is equal to 0.3048 m). Each of the curves in this figure shows a vehicle trajectory in the time–space plane. Taken from Treiterer [16]

2.2.2 Traffic Breakdown

The onset of congestion in an initial free flow is accompanied by a abrupt decrease in average vehicle speed in the free flow to a considerably lower speed in congested traffic (Figs. 2.4 and 2.5). This speed breakdown occurs mostly at highway bottlenecks and is called the breakdown phenomenon or traffic breakdown (see [9, 17–21] and earlier works referred to in the book [1] and in Chap. 2 written by Hall in [3]).

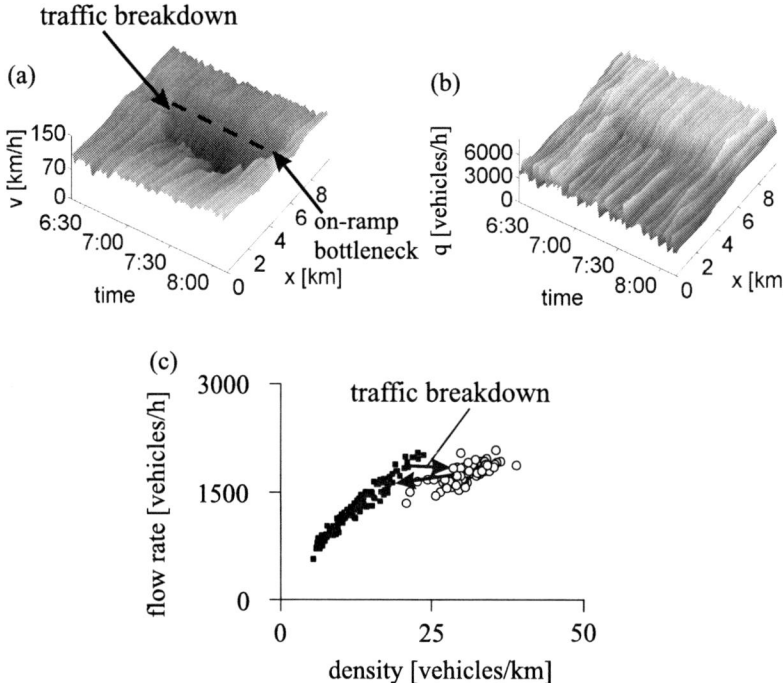

Fig. 2.4 Empirical example of traffic breakdown and hysteresis effect at on-ramp bottleneck: (a, b) Average speed (a) and flow rate (b) on the main road in space and time (note that the flow rate increase downstream of the bottleneck seen in (b) is associated with the on-ramp inflow). (c) Hysteresis effect in the flow–density plane labeled by two arrows representing traffic breakdown and return transition from congested traffic to free flow. 1-min average data. This example of traffic breakdown is qualitatively the same as many other examples observed in various countries (e.g., [9, 17–21])

The flow rate in free flow downstream of a bottleneck measured just before traffic breakdown occurs is called the *pre-discharge flow rate*. The flow rate in free flow downstream of a bottleneck after traffic breakdown has occurred at this bottleneck, i.e., the flow rate in the congested pattern outflow is called the *discharge flow rate* [17].

Hall and Agyemang-Duah have found [17] that

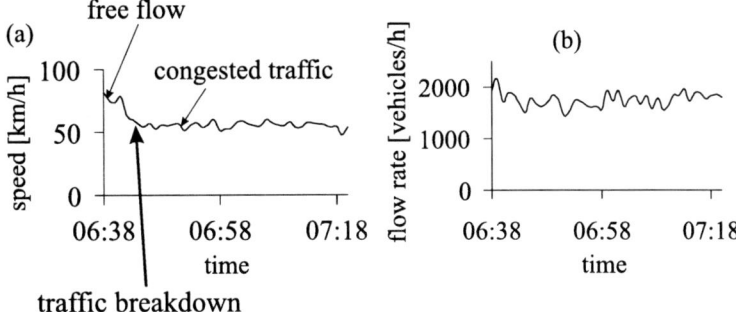

Fig. 2.5 Traffic breakdown at on-ramp bottleneck. Vehicle speed (a) and flow rate downstream of the bottleneck (b) as functions of time related to Fig. 2.4 (e.g., [9, 17–19])

- the discharge flow rate can be as great as the pre-discharge flow rate: in some cases, the discharge flow rate is smaller, however, in other cases it is greater than the pre-discharge flow rate.

2.2.3 Probabilistic Features of Traffic Breakdown

In 1995, Elefteriadou *et al.* found that traffic breakdown at a bottleneck has a probabilistic nature [18]. This means the following: at a given flow rate in free flow downstream of the bottleneck traffic breakdown can occur but it should not necessarily occur. Thus on one day traffic breakdown occurs, however, on another day at the same flow rates traffic breakdown is not observed.

Persaud *et al.* found [19] that empirical probability of traffic breakdown at a bottleneck is an increasing flow rate function (Fig. 2.6). Later such an empirical probability of traffic breakdown was also found on different highways in various countries [22–27].

Another empirical probabilistic characteristic of traffic breakdown is as follows. At given traffic parameters (weather, etc.), the flow rate downstream of an on-ramp bottleneck associated with the empirical maximum flow rate in free flow $q_{\max}^{(\text{free, emp})}$, which was measured on a specific day before congestion occurred, can be greater than the pre-discharge flow rate denoted by $q_{\text{FS}}^{(B)}$ in Fig. 2.7.

After traffic breakdown has occurred, the emergent congested pattern shown in Fig. 2.4 (a, b) exists for about one hour at the bottleneck: at 7:40 free flow occurs at the bottleneck. This restoration of free flow is related to a reverse transition from congested traffic to the free flow at the bottleneck. Traffic breakdown and the reverse transition are accompanied by a well-known *hysteresis effect* and hysteresis loop in the flow–density plane: a congested pattern emerges usually at a greater flow rate downstream of the bottleneck than this flow rate is at which the congested pattern dissolves (see references in [9, 17, 20, 21]) (Fig. 2.4 (c)).

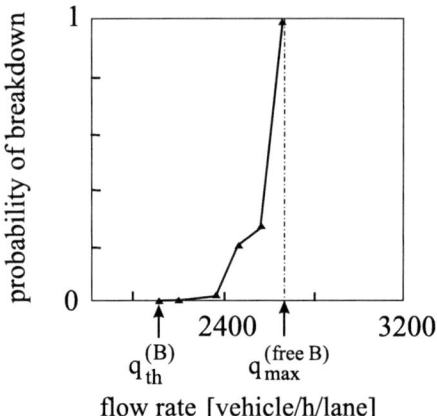

Fig. 2.6 Probability for traffic breakdown at an on-ramp bottleneck for $T_{av} = 10$ min. Taken from Persaud *et al.* [19]. Explanations of characteristic flow rates $q_{th}^{(B)}$ and $q_{max}^{(free\ B)}$ added by the author appears in Sect. 3.3.3

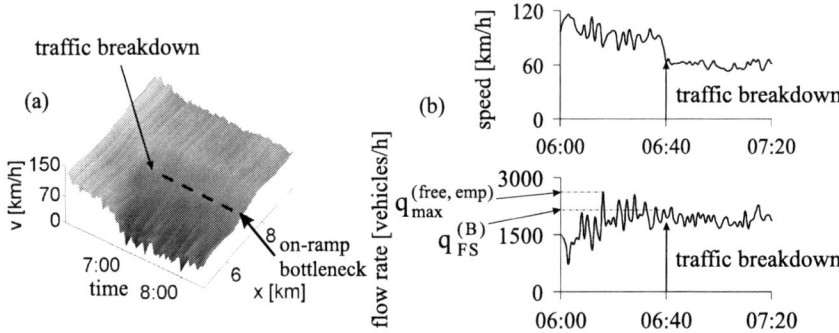

Fig. 2.7 Empirical example in which the flow rate in free flow downstream of on-ramp bottleneck at which traffic breakdown occurs, i.e., the pre-discharge flow rate denoted by $q_{FS}^{(B)}$ is smaller than $q_{max}^{(free,\ emp)}$: (a) Average speed on the main road in space and time. (b) Speed and flow rate (per lane) averaged across the road at location about 100 m downstream of the end of the on-ramp merging region. Up-arrows in (b) show the time instant of the breakdown labeled by "traffic breakdown". The on-ramp bottleneck is the same as that in Fig. 2.4. 1-min average data

2.3 Methodology of Three-Phase Traffic Theory

Three-phase traffic theory introduced by the author [28–36] is a qualitatively theory based on *common* spatiotemporal features of measured (empirical) congested traffic patterns[4]. The methodology of three-phase traffic theory is as follows.

---

[4] A methodology of measurements of congested patterns and their study have been discussed in Sect. 2.4.11 of the book [36].

2.3 Methodology of Three-Phase Traffic Theory

1) Spatiotemporal measurements of traffic flow variables on different highways in various countries over many days and years are collected and analyzed under a variety of traffic parameters.

2) Congested traffic patterns are identified in these measurements.

3) Common qualitative spatiotemporal features of these measured congested traffic patterns are identified [5, 28–40]. There are two main approaches for the classification of the common qualitative spatiotemporal pattern features presented in item 4) and 5) below, respectively.

4) Common pattern features are qualitatively the same *independent of traffic parameters*, i.e., they are qualitatively the same on different highways in various countries over many days and years of measurements *and* they are qualitatively the same for different network infrastructure and highway bottleneck types, weather, percentage of long vehicles, other road conditions, and vehicle technology. Moreover, these common spatiotemporal pattern features must be qualitatively independent of day time, working day or week-end, whether the day is sunny or rainy or else foggy, dry or wet road, or even ice and snow on road, etc.

- Macroscopic and microscopic criteria for traffic phases of three-phase traffic theory (Sects. 2.4–2.6) are some of these common qualitative spatiotemporal features of the measured congested traffic patterns.

5) Common pattern features are qualitatively the same only for a class of congested patterns associated with *a particular set of traffic parameters*. An example of such a class of congested patterns with common qualitative spatiotemporal features are congested patterns at heavy bottlenecks caused by, e.g., bad weather conditions or heavy road works (Sect. 7.2.5).

- An empirical congested traffic pattern classification of Sect. 2.4.6 and Chap. 7 presents some of congested traffic patterns with common qualitative spatiotemporal features[5].

6) The spatiotemporal criteria for the traffic phases and empirical congested traffic pattern classification are the empirical basis for hypotheses of three-phase traffic theory (Chaps. 3–6).

---

[5] A more detailed classification of common qualitative spatiotemporal pattern features can be found in part II of the book [36] and in [38].

2.4 Two Traffic Phases in Congested Traffic: Wide Moving Jam (J) and Synchronized Flow (S)

2.4.1 Empirical Macroscopic Criteria for defining Phases in Congested Traffic

A traffic phase is a traffic state considered in space and time that possesses some unique *empirical spatiotemporal* features. A traffic state is characterized by a particular set of statistical properties of traffic variables.

Three-phase traffic theory assumes that besides the free flow traffic phase there are two other traffic phases in congested traffic: *synchronized flow* and *wide moving jam*. Thus there are three traffic phases in three-phase traffic theory:

Free flow
Synchronized flow
Wide moving jam

The synchronized flow and wide moving jam phases in congested traffic are defined through the use of the following *macroscopic empirical (objective) criteria* for the traffic phases in congested traffic [28–36]:

[J] The wide moving jam phase is defined as follows. A wide moving jam is a moving jam that maintains the mean velocity of the downstream front of the jam as the jam propagates. Vehicles accelerate within the downstream jam front from low speed states (sometimes as low as zero) inside the jam to higher speeds downstream of the jam. A wide moving jam maintains the mean velocity of the downstream jam front, even as it propagates through other different (possibly very complex) traffic states of free flow and synchronized flow or highway bottlenecks. This is a characteristic feature of the wide moving jam phase.

[S] The synchronized flow phase is defined as follows. In contrast to the wide moving jam traffic phase, the downstream front of the synchronized flow phase does *not* maintain the mean velocity of the downstream front. In particular, the downstream front of synchronized flow is often *fixed* at a bottleneck. In other words, the synchronized flow traffic phase does not show the characteristic feature [J] of the wide moving jam phase.

The downstream front of synchronized flow separates synchronized flow upstream from free flow downstream. Within the downstream front of synchronized flow vehicles accelerate from lower speeds in synchronized flow upstream of the front to higher speeds in free flow downstream of the front.

Thus the definitions of the traffic phases in congested traffic are made via the spatiotemporal macroscopic empirical criteria [J] and [S]. The definitions [J] and [S] are associated with dynamic behavior of the downstream front of these phases, while a wide moving jam or synchronized flow propagates through other traffic states or bottlenecks.

2.4 Two Traffic Phases in Congested Traffic

The definitions [J] and [S] of the traffic phases in congested traffic mean that if a congested traffic state is not related to the wide moving jam phase, then the state is associated with the synchronized flow phase. This is because congested traffic can be either within the synchronized flow phase or within the wide moving jam phase. In other words, if in measured data congested traffic states associated with the wide moving jam phase have been identified, then with certainty all remaining congested states in the data set are related to the synchronized flow phase.

2.4.2 Traffic Breakdown at Bottleneck: F→S Transition

Firstly, we apply the phase definitions [S] and [J] to traffic breakdown at the on-ramp bottleneck shown in Fig. 2.5 (a). We see that traffic breakdown leads to congested traffic whose downstream front is fixed at the bottleneck (dashed line in Fig. 2.4 (a)). Thus the congested traffic resulting from traffic breakdown satisfies the definition [S] for the synchronized flow phase.

This is a *common result of all empirical observations of traffic breakdown at any highway bottlenecks*: traffic breakdown is associated with a phase transition from free flow to synchronized flow (F→S transition for short):

- The terms *F→S transition, breakdown phenomenon, traffic breakdown*, and *speed breakdown* are synonyms related to the same effect: the onset of congestion in free flow.

2.4.2.1 Effectual Bottleneck and its Effective Location

Traffic breakdown (F→S transition) occurs usually at the same highway bottlenecks of a road section. These bottlenecks are called *effectual bottlenecks*.

- An effectual bottleneck is a bottleneck at which traffic breakdown most frequently occurs on many different days.

An example of an effectual bottleneck is the on-ramp bottleneck shown in Fig. 2.5 (a). After the breakdown has occurred, synchronized flow is forming at the bottleneck.

- A road location in a neighborhood of the bottleneck at which the downstream front of synchronized flow is fixed is called an *effective location of the effectual bottleneck* (effective location of bottleneck for short).

It must be noted that the effective location of the bottleneck can be different from the location at which traffic breakdown, i.e., F→S transition has occurred leading to congested pattern emergence. Moreover, both the location of traffic breakdown and the effective location of the bottleneck are probabilistic values in real traffic. Even for the same type of congested pattern the effective location of the bottleneck can randomly change over time [36].

2.4.2.2 Flow Rate in Synchronized Flow

The discharge flow rate, i.e., the flow rate in the congested pattern outflow can be as great as the pre-discharge flow rate. This empirical result is associated with a common feature of empirical observations of synchronized flow that during traffic breakdown the flow rate in the emergent synchronized flow at the location of traffic breakdown can be almost as great as the flow rate in free flow has been before traffic breakdown at the bottleneck has occurred. An example can be seen in Fig. 2.5 (b): whereas there is an abrupt decrease in average vehicle speed, the flow rate does not necessarily abruptly decrease during traffic breakdown (F→S transition) and within an emergent synchronized flow at the bottleneck.

In some cases, the flow rate in synchronized flow, which determines in the most degree the discharge flow rate, can be even greater than the pre-discharge flow rate.

- A common empirical feature of synchronized flow that the flow rate within synchronized flow can be as great as in free flow is the important one for feedback on-ramp metering control applications (Sect. 9.2).

2.4.3 Propagation of Wide Moving Jams through Bottlenecks

In Fig. 2.8, there is a sequence of two moving jams propagating upstream on a road section with a bottleneck. These moving jams propagate through different states of traffic flow and through the bottleneck while maintaining the downstream jam front velocity. Thus in accordance with the definition [J], these moving jams are wide moving jams.

There is congested traffic in which speed is much lower than in free flow (compare vehicle speeds in Fig. 2.8 (c, d)). The downstream front of this congested traffic flow, within which vehicles accelerate to free flow, is fixed at the bottleneck (dashed line in Fig. 2.8 (a)). Therefore, in accordance with the definition [S] this congested traffic is synchronized flow.

We can see that whereas in wide moving jams both the speed and flow rate are very low (sometimes as low as zero), in synchronized flow the flow rate is high (compare the flow rates within the wide moving jams and within synchronized flow in Fig. 2.8 (d)). The vehicle speed in synchronized flow is considerably lower than in free flow. However, as abovementioned the flow rates in the free flow and synchronized flow phases can be close to each other (Fig. 2.8 (c, d)).

2.4.4 Explanation of Terms "Synchronized Flow" and "Wide Moving Jam"

The term *synchronized flow* reflects the following features of this traffic phase:

2.4 Two Traffic Phases in Congested Traffic

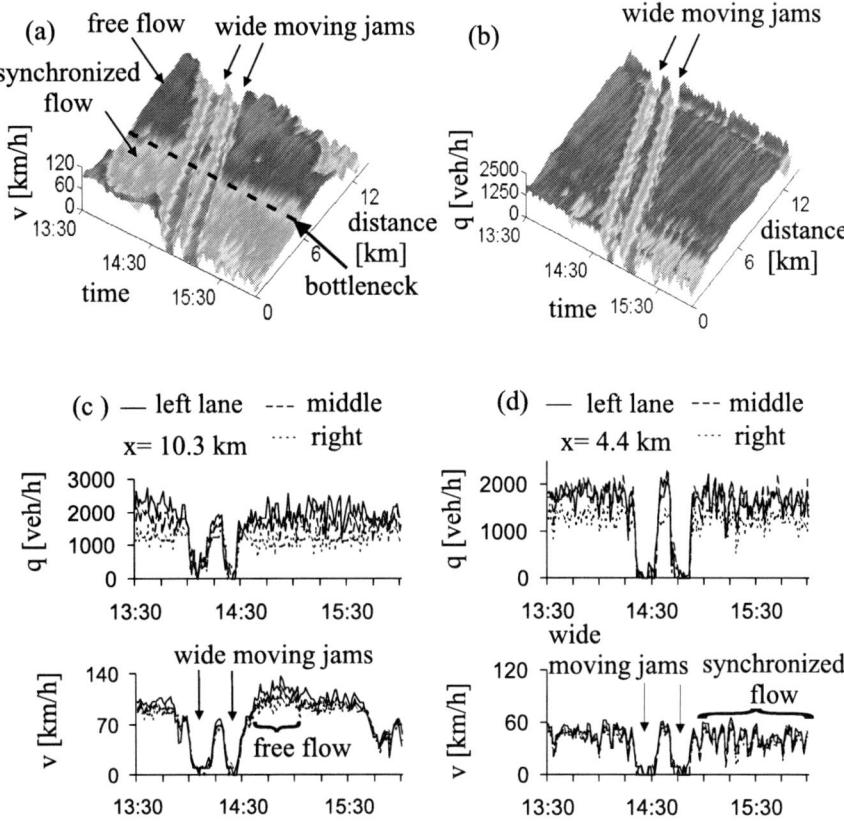

Fig. 2.8 Explanation of the three traffic phases. (a, b) Vehicle speed (a) and flow rate averaged across all road lanes (b) as functions of time and space. (c, d) Flow rate and average vehicle speed at two different freeway locations (c) and (d) in different road lanes. Taken from [36]

(i) It is a continuous traffic flow with no significant stoppage, as often occurs within a wide moving jam. The word *flow* reflects this feature.

(ii) Although in empirical synchronized flow vehicle speeds across different lanes on a multi-lane road should not be necessarily synchronized, there is a *tendency* towards synchronization of these speeds in this flow. In addition, there is a *tendency* towards synchronization of vehicle speeds in each of the road lanes (bunching of vehicles) in synchronized flow. This is due to a relatively small probability of passing in synchronized flow. The word *synchronized* reflects these speed synchronization effects.

The term *wide moving jam* reflects the characteristic feature of the jam to propagate through any other state of traffic flow and through any bottleneck while maintaining the mean velocity of the downstream jam front. The phrase *moving jam* reflects the jam *propagation* as a whole localized structure on a road. To distinguish wide moving jams from narrow moving jams, which do not characteristically maintain the mean velocity of the downstream jam front (see Sect. 2.6.3), we use the term *wide moving jam*. This relates to the fact that if a moving jam has a width (in the longitudinal direction) considerably greater than the widths of the jam fronts, and if the vehicle speed within the jam is zero, the jam always exhibits the characteristic feature of *maintaining* the mean velocity of the downstream jam front[6]. Thus the word *wide* reflects this characteristic jam feature.

2.4.5 Wide Moving Jam Emergence: S→J Transition

Based on an analysis of empirical data measured over many days and years on various freeways it has been found [28, 29] that wide moving jams do not emerge spontaneously in free flow. In contrast, wide moving jams can emerge spontaneously in synchronized flow. Observations show that the greater the density in synchronized flow, the more likely is spontaneous moving jam emergence in that synchronized flow.

A wide moving jam emerges in an initial free flow due to a sequence of two phase transitions [36]:

1) An F→S transition occurs. As a result of this traffic breakdown, synchronized flow emerges.
2) Later and usually at another road location than the location of this F→S transition, a phase transition from synchronized flow to wide moving jam (S→J transition for short) occurs spontaneously leading to wide moving jam emergence. This sequence of phase transitions is called the F→S→J transitions.

This scenario of wide moving jam emergence in real traffic is illustrated in Fig. 2.9 (a): Firstly, an F→S transition occurs (labeled by "traffic breakdown (F→S transition)"). As a result of traffic breakdown, synchronized flow emerges upstream of the bottleneck (labeled by "synchronized flow"). Later and upstream of the location of traffic breakdown, S→J transitions occur in this synchronized flow. Wide moving jams, which have emerged due to these F→S→J transitions, propagate upstream while maintaining the mean velocity of their downstream jam fronts (labeled by "wide moving jams" in Fig. 2.9 (a)).

2.4.6 Definitions of Synchronized Flow and General Congested Patterns

After traffic breakdown, i.e., an F→S transition has occurred at an effectual bottleneck, synchronized flow is formed at the bottleneck. There are two main types of

---

[6] An explanation of this feature of a wide moving jam can be found in Sect. 7.6.5 of [36].

2.4 Two Traffic Phases in Congested Traffic

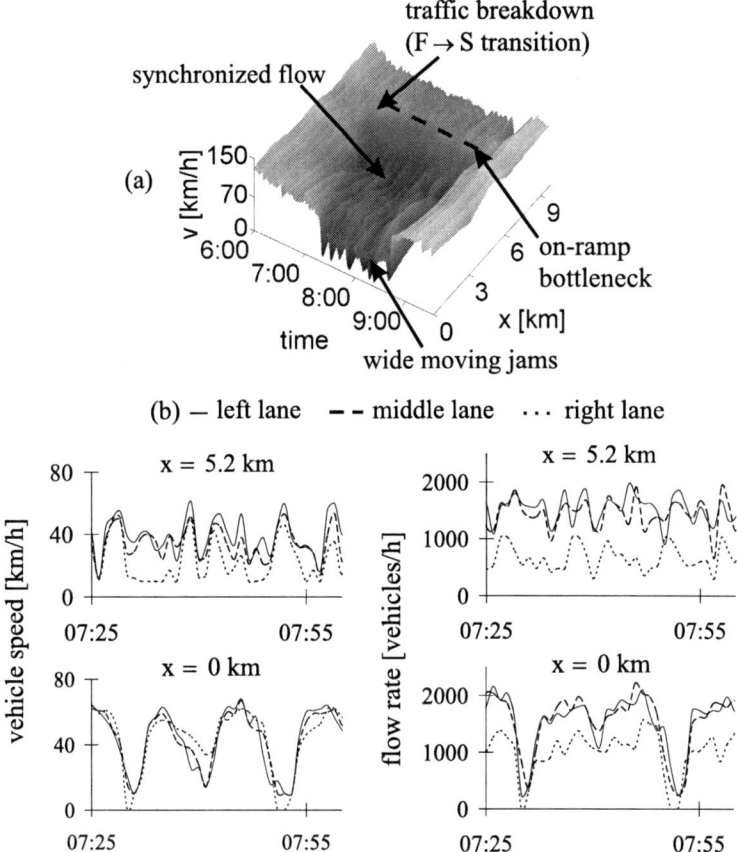

Fig. 2.9 Empirical example of wide moving jam emergence in synchronized flow: (a) Vehicle speed in space and time. (b) Vehicle speed (left) and flow rate (right) in three road lanes within synchronized flow (location 5.2 km) and wide moving jams (location 0 km). 1-min average data. This figure shows the wide moving jam emergence in synchronized flow whose occurrence at the on-ramp bottleneck is shown in Fig. 2.7. Taken from [36]

congested patterns that result from traffic breakdown at the bottleneck: a *synchronized flow traffic pattern* (synchronized flow pattern or SP for short) and a *general congested traffic pattern* (general pattern or GP for short).

The terms *synchronized flow pattern (SP)* and *general pattern (GP)* are defined as follows:

- A synchronized flow pattern (SP) is a spatiotemporal congested traffic pattern in which congested traffic consists only of the synchronized flow traffic phase.

- A general pattern (GP) is a spatiotemporal congested traffic pattern in which congested traffic consists of the synchronized flow and wide moving jam phases[7].

An empirical example of an SP is shown in Fig. 2.4 (a, b). Indeed, this congested traffic pattern consists of only the synchronized flow phase. In contrast, a congested pattern shown in Fig. 2.9 consists of the synchronized flow and wide moving jam phases. Therefore, this pattern is an empirical example of an GP. A more detailed account of SPs and GPs features appears in Chap. 7.

2.5 Characteristic Parameters of Wide Moving Jam Propagation

In a theory of wide moving jam propagation firstly derived in 1994 by Kerner and Konhäuser [41] and in investigations of wide moving jams in real measured traffic data made in [40] have been found that there are characteristic parameters of wide moving jam propagation. The main common feature of the characteristic parameters of wide moving jam propagation is that at given traffic parameters (weather, etc.; see Sect. 2.1) they do not depend on traffic variables in traffic flow upstream of a wide moving jam and they do not change while the jam propagates on a road. These characteristic parameters are the same for different wide moving jams. The jam characteristic parameters are as follows:

(i) The mean velocity of the downstream front of a wide moving jam denoted by v_g. The constancy of v_g while the jam propagates on the road is consistent with the characteristic jam feature [J] of Sect. 2.4.1.

(ii) The flow rate q_{out}, density ρ_{min}, and average vehicle speed v_{max} in the outflow from the wide moving jam. These traffic variables are the characteristic parameters only under condition that free flow is formed in the jam outflow.

(iii) The mean vehicle density within the wide moving jam denoted by ρ_{max} that is also called the jam density.

---

[7] Note an GP is always a general case of a congested traffic pattern. This is because in three-phase traffic theory there are *only* the synchronized flow and wide moving jam phases in congested traffic and, in accordance with the GP definition, any GP consists of the both phases. This is independent of whether an GP appears at an on-ramp bottleneck or the GP is caused by a combination of several on- and off-ramps, etc.: An GP is a *generic* term for many *species*, i.e., subordinate terms of different types of GPs. Examples of these different GP types are an EGP, i.e., an expanded GP whose synchronized flow affects two or more effectual bottlenecks (see Sect. 7.3), a dissolving GP [36], and an GP with a non-regular pinch region (see Sect. 7.2.5).

Respectively, an SP in which congested traffic consists of synchronized flow *only* is the *generic* term for many different types of SPs. Examples of these different SP types are a moving SP, a localized SP, a widening SP (Sect. 7.2.1), and an ESP, i.e., an expanded SP whose synchronized flow affects two or more effectual bottlenecks.

This classification of congested patterns in three-phase traffic theory is made based on *common* empirical features of congested traffic (item 5 of Sect. 2.3) found during many years of observations of empirical congested patterns. Therefore, the classification of congested patterns into different types of GPs and SPs is well predictive and self-contained (Sect. 2.4.10 of [36]).

2.5 Characteristic Parameters of Wide Moving Jam Propagation

However, we should mention that the characteristic parameters can depend considerably on traffic parameters.

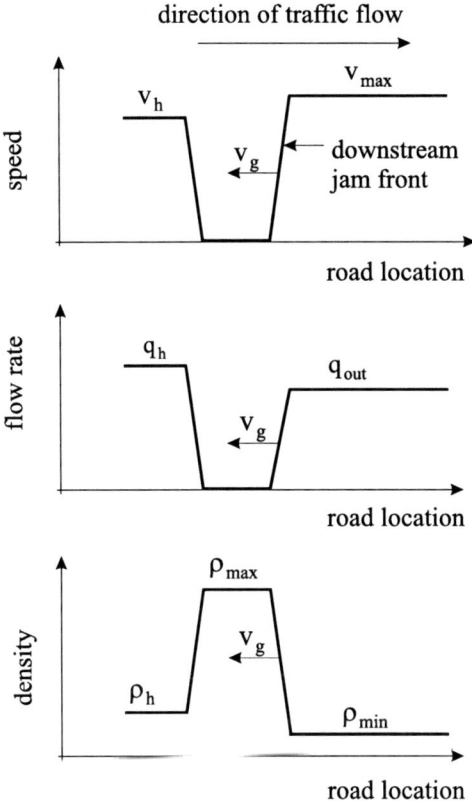

Fig. 2.10 Qualitative illustration of characteristic parameters of wide moving jam propagation [41]. Schematic representation of a wide moving jam at a fixed time instant. Spatial distributions of vehicle speed v, flow rate q, and vehicle density ρ in the wide moving jam, which propagates through a homogeneous state of initial free flow with speed v_h, flow rate q_h, and density ρ_h

The characteristic parameters of wide moving jam propagation are qualitatively illustrated in Fig. 2.10. In this figure, at a given time instant distributions of traffic variables along a road are shown associated with a wide moving jam propagating in an initial homogeneous free flow. The flow rate q_h and density ρ_h in this initial free flow are chosen to be greater, respectively, the average speed v_h to be lower than the related characteristic traffic variables of free flow formed in the jam outflow: $q_h > q_{out}$, $\rho_h > \rho_{min}$, $v_h < v_{max}$. Obviously, upstream of the jam the initial free flow occurs. However, this is not the case downstream of the jam. This is because during the jam propagation vehicles escaping from the jam at the downstream jam front form a new free flow with the flow rate q_{out}, density ρ_{min}, and average speed v_{max}.

2.5.1 Driver Time Delay in Acceleration and Mean Velocity of Downstream Jam Front

Concerning the mean velocity of the downstream front of a wide moving jam v_g, some general assumptions can be made [42]. Each driver within the jam can start to accelerate to either free flow or synchronized flow downstream after two conditions have been satisfied:

(i) The preceding vehicle has already begun to move away from the jam.
(ii) Due to the preceding vehicle motion, after some time the space gap between the two drivers has exceeded a space gap at which a safety condition for driver acceleration is satisfied:

- There is some time delay in vehicle acceleration at the downstream front of the wide moving jam.

The mean time in vehicle acceleration at the downstream front of a wide moving jam will be denoted by $\tau^{(a)}_{\text{del, jam}}$. In empirical data,

$$\tau^{(a)}_{\text{del, jam}} \approx 1.5 - 2 \text{ sec.} \tag{2.12}$$

The motion of the downstream front of a wide moving jam results from acceleration of drivers from the standstill within the jam to flow downstream of the jam. Because the average distance between vehicles inside the jam, including average length of each vehicle, equals $1/\rho_{\text{max}}$, the velocity of the downstream front of the wide moving jam is

$$v_g = -\frac{1}{\rho_{\text{max}} \tau^{(a)}_{\text{del, jam}}}. \tag{2.13}$$

In empirical observations,

$$v_g \sim -15 \text{ km/h}. \tag{2.14}$$

2.5.2 The Line J

The characteristic parameters of wide moving jam propagation and the jam feature [J], which defines the wide moving jam phase in congested traffic (Sect. 2.4.1), can

2.5 Characteristic Parameters of Wide Moving Jam Propagation

be presented by a line in the flow–density plane. This line firstly introduced in [41] and studied in [28–34, 40, 43] is called the "line J" (Figs. 2.11 and 2.12).

The slope of the line J is equal to the characteristic velocity v_g. If in the wide moving jam outflow free flow is formed, then the flow rate q_{out} in this jam outflow and the related density ρ_{min} give the left coordinates (ρ_{min}, q_{out}) of the line J; as abovementioned, q_{out}, ρ_{min}, and the related average vehicle speed v_{max} are characteristic parameters of wide moving jam propagation (Fig. 2.10). The right coordinates $(\rho_{max}, 0)$ of the line J are related to the traffic variables within the jam, the density ρ_{max} and flow rate $q_{min} = \rho_{max} v_{min}$, where the average vehicle speed within the jam v_{min} is here assumed to be zero that results in $q_{min} = 0$ within the jam (Fig. 2.11).

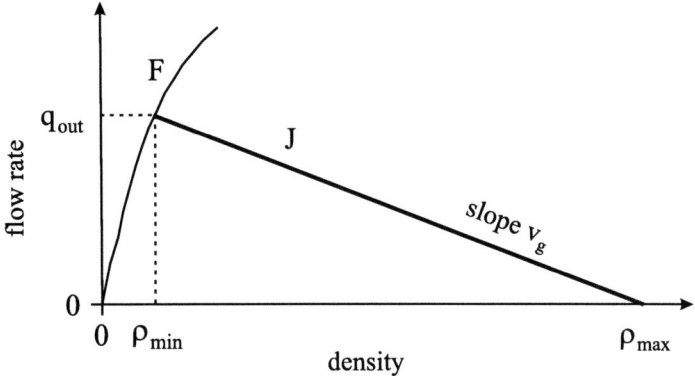

Fig. 2.11 Qualitative representation of the fundamental diagram of free flow (F) together with the line J (line J) whose slope is equal to the mean downstream jam front velocity v_g [41]

Because the slope of the line J is equal to the characteristic jam velocity v_g (2.13), on the line J the derivative

$$\frac{dq}{d\rho} = v_g, \tag{2.15}$$

i.e., in accordance with the line J definition it satisfies the equation

$$q(\rho) = |v_g|(\rho_{max} - \rho) \tag{2.16}$$

that using (2.13) can be rewritten in the equivalent form as

$$q(\rho) = \frac{1}{\tau_{del,\,jam}^{(a)}}\left(1 - \frac{\rho}{\rho_{max}}\right). \tag{2.17}$$

From Eq. (2.16), for the flow rate in free flow formed in the jam outflow we get the formula

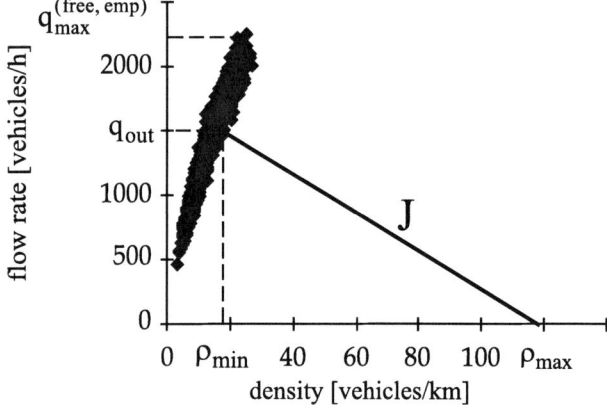

Fig. 2.12 Empirical example of the line J: Representation of the propagation of the downstream front of the downstream jam in the sequence of two wide moving jams in Fig. 2.8 by the line J in the flow–density plane and data for free flow (black diamonds). Taken from [40]

$$q_{\text{out}} = |v_g|(\rho_{\max} - \rho_{\min}), \quad (2.18)$$

which, as follows from (2.17), can also be written as

$$q_{\text{out}} = \frac{1}{\tau_{\text{del, jam}}^{(a)}}\left(1 - \frac{\rho_{\min}}{\rho_{\max}}\right). \quad (2.19)$$

A result of empirical observations of wide moving jam propagation is that the maximum flow rate in free flow $q_{\max}^{(\text{free,emp})}$ is considerably greater than the flow rate in free flow in the wide moving jam outflow q_{out} (Fig. 2.12) [40]:

$$q_{\max}^{(\text{free, emp})} > q_{\text{out}}. \quad (2.20)$$

In particular, for 1-min average measured traffic data has been found that [40]

$$\frac{q_{\max}^{(\text{free, emp})}}{q_{\text{out}}} \approx 1.5. \quad (2.21)$$

2.6 Microscopic Criterion for Traffic Phases in Congested Traffic

In this section, we show that when *microscopic* traffic variables are measured, then the distinguishing the synchronized flow and wide moving jam phases can be possible, even if the microscopic traffic variables are measured at only one road location within a congested pattern.

2.6.1 Traffic Flow Interruption within Wide Moving Jam

The spatiotemporal criteria for a wide moving jam [J], which distinguish the wide moving jam and synchronized flow phases, can be explained by a traffic flow interruption effect within a wide moving jam: traffic flow is interrupted within the wide moving jam. As a result, there is no influence of the inflow into the jam on the jam outflow [36]. A difference between the jam inflow and the jam outflow changes the jam width only. This *traffic flow interruption effect within a wide moving jam* (flow interruption for short) is a general effect for each wide moving jam.

In a hypothetical case, when all vehicles within a moving jam do not move, the criterion for the traffic flow interruption effect within the jam is [5, 39]

$$I = \frac{\tau_J}{\tau_{\text{del, jam}}^{(a)}} \gg 1, \quad (2.22)$$

where τ_J is the duration of a wide moving jam (jam duration for short), i.e., the time interval between the upstream and downstream jam fronts measured when the wide moving jam propagates through a given road location; I is approximately equal to the vehicle number stopped within the jam.

In real traffic, there can be low speed vehicle motion within a wide moving jam (see Sect. 2.6.4). Due to the existence of such a low speed vehicle motion within a wide moving jam instead of hypothetical criterion (2.22) the following *sufficient criterion* for flow interruption within the jam can be used [5, 39]

$$I_s = \frac{\tau_{\max}}{\tau_{\text{del, jam}}^{(a)}} \gg 1, \quad (2.23)$$

where $\tau_{\max}$ is the maximum time headway between two vehicles following each other within the jam ($\tau_{\max} \leq \tau_J$).

We define a flow interruption interval within wide moving jam as follows [5, 39].

- A flow interruption interval within a wide moving jam is a time interval $\tau_{\max}$ for which condition (2.23) is satisfied.

The interruption of traffic flow within a moving jam can also be found in *empirical* single vehicle data. In an example shown in Fig. 2.13, the flow interruption effect occurs two times during upstream jam propagation through a road detector (these time intervals are labeled "flow interruption" in Fig. 2.13 (c)). The values $\tau_{\max}$ for the first flow interruption intervals within the wide moving jam are equal to approximately 50 s in the left lane, 24 s in the middle line, and 80 s in the right line. In accordance with (2.12), these values $\tau_{\max}$ satisfy criterion (2.23). This means that traffic flow is interrupted within the moving jam, i.e., this moving jam can be associated with the wide moving jam phase [5, 39].

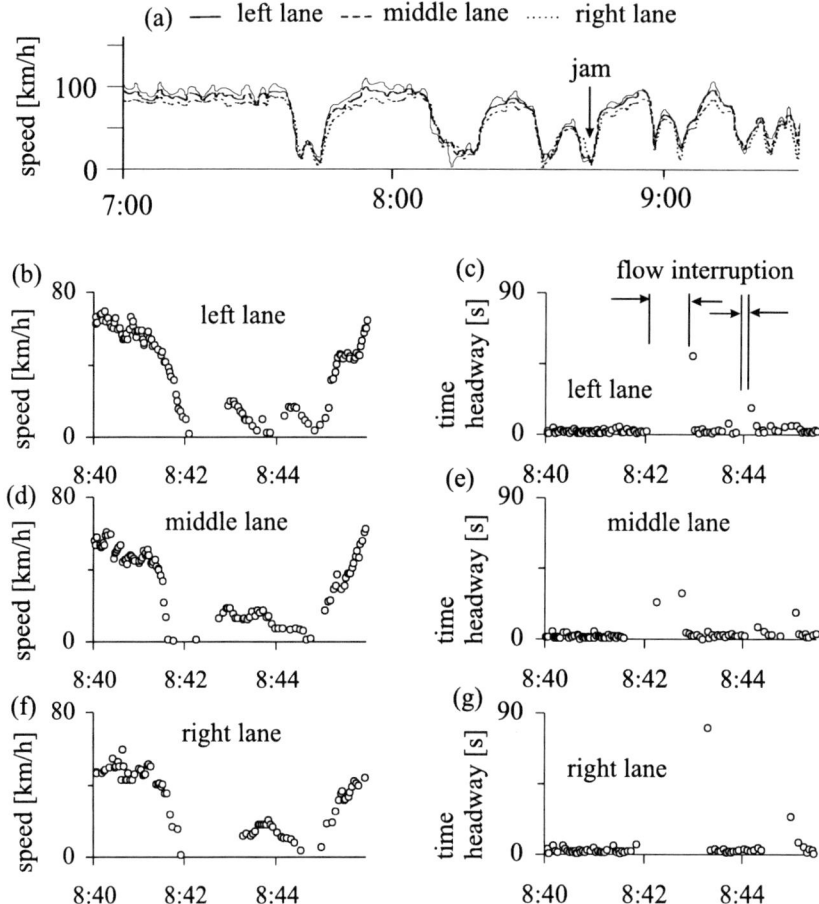

Fig. 2.13 Measured single vehicle data analysis: (a) Overview of measured data (1 min average data). (b–g) Single vehicle data for speed in three freeway lanes for a wide moving jam (left figures) labeled by "jam" in (a) and the related time headways (right figures). Taken from [5]

2.6.2 Flow Interruption Effect and Characteristic Jam Feature [J]

The flow interruption effect is a general effect for each wide moving jam [5, 39]:

- Condition (2.23) can be considered a microscopic criterion for the wide moving jam phase.

This microscopic criterion can be used to distinguish the synchronized flow and wide moving jam phases in single vehicle data. The microscopic criterion for the wide moving jam phase is possible to apply, even if traffic data is measured at a single road location.

This is because there is a deep connection between the flow interruption effect and characteristic feature [J] of wide moving jams: If there is a flow interruption interval within a moving traffic jam, then the jam exhibits the characteristic feature [J] to propagates though any bottleneck while maintaining the mean velocity of the downstream jam front, i.e., this jam is a wide moving jam.

To explain this, we note that during a long enough flow interruption interval τ_{max} in (2.23) there are at least several vehicles within a moving jam that are either in a standstill or they are moving with a negligible low speed in comparison with the speed in the jam inflow and outflow. Because τ_{max} is much longer than any driver time delays, at a given time instant there is a wide (in the longitudinal direction) enough road region associated with the flow interruption interval within which all vehicles can approximately be considered being in the standstill. We call this region within the jam as the flow interruption region. Thus a vehicle, which comes to a stop within the jam at the upstream boundary of the flow interruption region (this boundary coincides often with the upstream jam front), must wait for acceleration at least during the flow interruption interval τ_{max}: During this time interval other vehicles that are stopped downstream of the vehicle within the flow interruption region accelerate one after another from the standstill. This vehicle escaping from the standstill within the flow interruption region of the moving jam is independent of a traffic state in the jam outflow and it does not depend on whether there is a bottleneck or not[8].

This explains also why the downstream front of the wide moving jam propagates through the bottleneck while maintaining the mean velocity of the downstream jam front: this velocity v_g given by formula (2.13) is determined by the successive vehicle acceleration at the downstream front of the moving jam independent of traffic flow characteristics within the bottleneck. Thus under condition (2.23), the moving jam exhibits the characteristic jam feature [J], i.e., the jam is a wide moving jam.

2.6.3 Narrow Moving Jams

In an empirical example shown in Fig. 2.14, there are moving jams (Fig. 2.14 (b, c)). However, rather than wide moving jams these moving jams should be classified as narrow moving jams. In general, a narrow moving traffic jam is defined as follows.

- A narrow moving jam is a moving jam, which does not exhibit the characteristic feature [J] of a wide moving jam; for this reason, narrow moving jams belong to the synchronized flow traffic phase.

---

[8] A more detailed discussion of the link between the flow interruption effect and characteristic jam feature [J] can be found in Sect. 7.6.5 of the book [36].

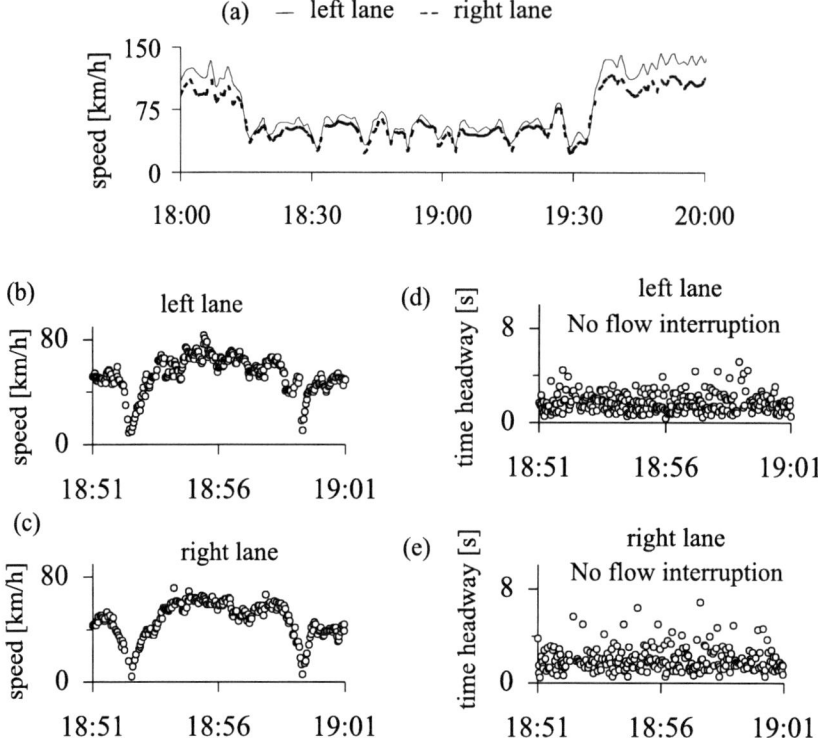

Fig. 2.14 Measured single vehicle data analysis of microscopic structure of narrow moving jams. (a) Overview of measured data (1 min average data). (b–e) Single vehicle data for speed in two freeway lanes for a sequence of narrow moving jams (left figures) and the related time headways (right figures). Taken from [5]

This is because there are no traffic flow interruption intervals within a narrow moving jam. Indeed, in empirical observations upstream and downstream of the jam, as well as within the jam there are many vehicles that traverse the induction loop detector. There is no considerable quantitative difference in time headways for different time intervals associated with these narrow moving jams and in traffic flow upstream or downstream of the jams (Fig. 2.14 (d, e)).

To understand this, we note that even if within a narrow moving jam the speed is equal to zero, then such narrow jam should consist of the jam fronts only: Each vehicle, which meets the narrow moving jam, can nevertheless accelerate later almost without any time delay within the jam. Within the upstream front vehicles must decelerate to a very low speed. However, the vehicles can accelerate almost immediately at the downstream jam front. These assumptions are confirmed by single vehicle data shown in Fig. 2.14 (b–e), in which time intervals between different measurements of time headways for different vehicles exhibit the same behavior outside of and within the jams. Even if within a narrow moving jam there are vehi-

2.6 Microscopic Criterion for Traffic Phases in Congested Traffic

cles that are stopped, the condition

$$l_s = \frac{\tau_{max}}{\tau^{(a)}_{del,\, jam}} \sim 1 \qquad (2.24)$$

can be satisfied. Under this condition, there is no flow interruption interval within this jam. Thus independent of these narrow moving jams, traffic flow is not interrupted, i.e., the narrow moving jams are associated with the synchronized flow phase. This single vehicle analysis enables us to assume that congested traffic in Fig. 2.13 (b–g) is associated with a wide moving jam. In contrast, congested traffic in Fig. 2.14 (b–e) is associated with the synchronized flow phase.

In Sect. 2.6.2, we have shown a deep connection between the characteristic feature [J] and the existence of a flow interruption interval within a wide moving jam. Otherwise, if there is no flow interruption interval within a moving jam, i.e., rather than condition (2.23), the condition (2.24) is satisfied, as this is the case for a narrow moving jam, we can expect that the jam does not exhibit the characteristic feature [J]. This explains both the definition of the narrow moving jam and why the narrow moving jam belongs to the synchronized flow traffic phase.

In Fig. 2.15 (figures left), we support this qualitative explanation of narrow moving jams by numerical simulations in which a moving jam is artificially created in simulations downstream of an effectual on-ramp bottleneck. Before the jam due to its upstream propagation reaches the bottleneck, free flow is at the bottleneck. This means that before the moving jam under consideration reaches the bottleneck the jam is surrounded both downstream and upstream by free flows.

When the moving jam reaches the bottleneck, rather than the jam propagates through the bottleneck while maintaining the mean velocity of the downstream jam front, the jam causing traffic breakdown is caught at the bottleneck. In other words, the jam transforms into qualitatively another congested pattern whose downstream front is fixed at the bottleneck. Thus the characteristic feature [J] is not satisfied for the jam. We can conclude that this moving jam is a narrow moving jam, i.e., the jam belongs to the synchronized flow phase. The same conclusion can be made from a consideration of time headways within this moving jam presented in Fig. 2.15 (c, e) (figures left). We see that these time headways do not satisfy the microscopic criterion (2.23) for a wide moving jam: there is no traffic flow interval within this jam.

- A moving jam, within which there is no flow interruption interval, does not also exhibit the characteristic feature [J], i.e., such a moving jam is a narrow moving jam.

A qualitatively different case is presented in Fig. 2.15 (figures right). In this case, time headways within a moving jam satisfies the microscopic criterion (2.23) for a wide moving jam: there are traffic flow intervals within this jam both in the left and right road lanes. As expected, in the case the jam propagates through the bottleneck while maintaining the mean velocity of the downstream jam front (characteristic feature [J]), i.e., this jam is a wide moving jam.

The results of numerical simulations (Fig. 2.15) allow us to conclude that

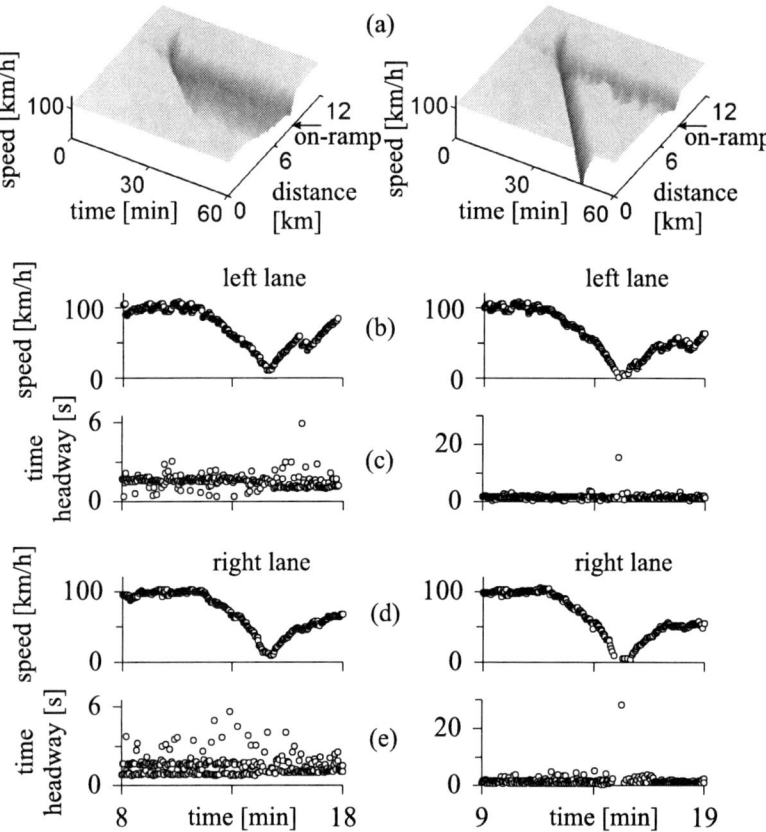

Fig. 2.15 Comparison of microscopic criterion (2.23) with macroscopic spatiotemporal objective criteria [J] and [S] for the phases in congested traffic of identical vehicles (results of model simulations). Left figures are related to a narrow moving jam. Right figures are related to a wide moving jam. (a) Vehicle speed on the main road in space and time that is averaged across two lanes. (b, d) Single vehicle data for speed in the left (b) and in right lanes (d). (c, e) Time headways associated with (b) and (d), respectively. Data in (b–e) is related to a road location 50 m downstream of the end of the on-ramp merging region. Taken from [39]

- the microscopic criterion (2.23) enables us indeed to distinguish the wide moving jam and synchronized flow phases in single vehicle data measured at one road location[9].

---

[9] It must be stressed that as we see from numerical simulations presented in Fig. 2.15, for an empirical proof of the microscopic criterion (2.23) for the traffic phases, we should compare results of phase identification based on this criterion in measured single vehicle data with the traffic phase definitions, i.e., with the macroscopic criteria for the traffic phases [J] and [S] (Sect. 2.4.1). However, such an empirical proof is possible, if the single vehicle data associated with a moving jam in the neighborhood of an effectual bottleneck is measured at many roads locations includ-

2.6.4 Moving Blanks within Wide Moving Jam

Between flow interruption intervals within the wide moving jam shown in Fig. 2.13 (b–g), vehicles within the jam exhibit time headways about 1.5–7 sec (Fig. 2.16). These time headways are considerably shorter than flow interruption intervals in Fig. 2.13 (c, e, g). The time headways are related to low speed states measured at detectors within the jam (Fig. 2.13 (b, d, f)).

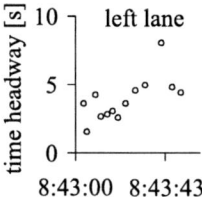

Fig. 2.16 Measured time headways associated with moving blanks in the left lane within a wide moving jam shown in Fig. 2.13 (b, c). Taken from [5]

To understand the effect of these low speed states, note that when vehicles meet the wide moving jam, firstly they decelerate usually to a standstill at the upstream jam front. As a result, the first flow interruption interval in all road lanes appears (Fig. 2.13 (c, e, g)). Space gaps between these vehicles can be very different and the mean space gap can exceed a safe space gap considerably, i.e., regions with no vehicles called as *blanks* can appear within the jam. A blank within the jam is defined as follows.

- A blank within a wide moving jam is a region with no vehicles.

Later vehicles move covering these blanks. This low speed vehicle motion is responsible for low speed states mentioned above (Fig. 2.13 (b, d, f)). Consequently, due to this low speed vehicle motion new blanks between vehicles occur upstream, i.e., the blanks move upstream within the jam. Then other vehicles within the jam that are upstream of these blanks begin also to move covering the latter blanks. This leads to moving blanks that propagate upstream within the jam.

We define a *moving blank* as follows [5, 36, 39].

- A moving blank within a wide moving jam is a region without vehicles, which moves upstream due to vehicle motion within the jam.

Thus we see that in a general case a microscopic structure of wide moving jam is as follows [5]:

---

ing locations upstream and downstream of this bottleneck. In addition, before the jam reaches the bottleneck, free flow should be at the bottleneck. This is because only in this case we can prove whether the downstream front of the congested pattern exhibits the characteristic jam feature [J] or not. Unfortunately, we do not have such measured single vehicle data.

- A microscopic structure of wide moving jam consists of complex spatiotemporal alternations of flow interruption intervals and moving blanks.

2.7 Motivation for Traffic Phase Definitions

In some *averaged*, i.e., macroscopic traffic data of congested traffic, in particular which is measured outside of effectual bottlenecks, the traffic phase definitions [S] and [J] (Sect. 2.4.1) *cannot* be applied to perform an accurate identification of traffic phases. Thus a question arises:

- What is the sense of the traffic phase definitions [S] and [J], if the definitions cannot be used for the identification of traffic phases in some real macroscopic data of congested traffic?

A response on this question is as follows.

- Rather than to distinguish traffic phases in real measured macroscopic traffic data, the main *sense* of the phase definitions [S] and [J] is that the definitions result from and, therefore, distinguish qualitatively different *common* spatiotemporal characteristics of congested patterns (see item 4 of Sect. 2.3).
- These *common* spatiotemporal characteristics are associated with the behavior of the *downstream front* of a congested pattern at an effectual bottleneck at which traffic breakdown is possible: (i) If the pattern propagates through the bottleneck while maintaining the mean velocity of the downstream pattern front, then the congested pattern is a wide moving jam. (ii) In contrast with the wide moving jam, the downstream front of synchronized flow does not maintain the mean front velocity, in particular, the downstream front of synchronized flow is often fixed at the bottleneck.
- In turn, these common spatiotemporal characteristics of congested traffic patterns are the *origin* of hypotheses of three-phase traffic theory[10].
- This theory explains traffic breakdown and resulting congested traffic patterns in all known real measured traffic data [36].

Thus we can make the following conclusion about the motivation for the traffic phase definitions [S] and [J] made in three-phase traffic theory:

- The motivation for the traffic phase definitions [S] and [J] is the understanding of vehicular traffic congestion with the objective to derive new effective and reliable methods for traffic control and managements.

Additionally, the traffic phase definitions [S] and [J] are important for the validation of traffic flow models used for traffic control and dynamic traffic management.

The identification of the traffic phases in single vehicle data of congested traffic can be made with the microscopic criterion (2.23) for a wide moving jam. As

---

[10] Explanations of why the phase definitions [S] and [J] are the origin of hypotheses of three-phase traffic theory appear in Sect. 6.1.

explained in [5, 39], after downstream and upstream congested pattern fronts have been identified in averaged measured traffic data, this criterion can be used for the traffic phase identification, even if the single vehicle data is measured at one road location. Nevertheless, the identified traffic phases are valid only for the road location at which the single vehicle data is measured. The reason for this is as follows.

- A traffic phase that has just been identified with the microscopic criterion (2.23) at a road location can transform into another traffic phase at an adjacent road location outside of the location at which the single vehicle data is measured.
- Thus to perform an accurate traffic phase identification in congested traffic, single vehicle data measured simultaneously in space and time are needed.

Because the accurate traffic phase identification in congested traffic is important to overcome the problems of traffic control, single vehicle data measured simultaneously in time and space should be available in the future. This can be achieved through the use of new intelligent transportation systems, like car-to-car and car-to-infrastructure communication.

References

1. A.D. May, *Traffic Flow Fundamentals*, (Prentice-Hall, Inc., New Jersey, 1990)
2. *Highway Capacity Manual 2000*, (National Research Council, Transportation Research Board, Washington, D.C., 2000)
3. N.H. Gartner, C.J. Messer, A. Rathi (editors), *Traffic Flow Theory: A State-of-the-Art Report*, (Transportation Research Board, Washington DC, 2001)
4. Next Generation Simulation Programs, http://ngsim.camsys.com/
5. B.S. Kerner, S.L. Klenov, A. Hiller, H. Rehborn, Phys. Rev. E **73**, 046107 (2006)
6. B.S. Kerner, in *Encyclopedia of Complexity and System Science*, ed. by R.A. Meyers. (Springer, Berlin, 2009), pp. 9302–9355
7. I. Prigogine, R. Herman, *Kinetic Theory of Vehicular Traffic*, (American Elsevier, New York, 1971)
8. W. Leutzbach, *Introduction to the Theory of Traffic Flow*, (Springer, Berlin, 1988)
9. F.L. Hall, V.F. Hurdle, J.H. Banks, Transp. Res. Rec. **1365** 12–18 (1992)
10. M. Koshi, M. Iwasaki, I. Ohkura, in *Proc. 8th International Symposium on Transportation and Traffic Theory*, ed. by V.F. Hurdle. (University of Toronto Press, Toronto, Ontario, 1983), p. 403
11. L.C. Edie, R.S. Foote, Highway Res. Board Proc. Ann. Meeting **37**, 334–344 (1958)
12. L.C. Edie, R.S. Foote, in *Highway Research Board Proceedings*, (39. HRB, National Research Council, Washington, D.C., 1960), pp. 492–505
13. L.C. Edie, Oper. Res. **9**, 66–77 (1961)
14. L.C. Edie, P. Herman, T.N. Lam, Transp. Science **14**, 55–76 (1980)
15. J. Treiterer, J.A. Myers, in *Procs. 6th International Symposium on Transportation and Traffic Theory*, ed. by D.J. Buckley. (A.H. & AW Reed, London, 1974), pp. 13–38
16. J. Treiterer, *Investigation of Traffic Dynamics by Aerial Photogrammetry Techniques*, (Ohio State University Technical Report PB 246 094, Columbus, Ohio, 1975)
17. F.L. Hall, K. Agyemang-Duah, Trans. Res. Rec. **1320**, 91–98 (1991)
18. L. Elefteriadou, R.P. Roess, W.R. McShane, Transp. Res. Rec. **1484**, 80–89 (1995)
19. B.N. Persaud, S. Yagar, R. Brownlee, Trans. Res. Rec. **1634**, 64–69 (1998)
20. F.L. Hall, M.A. Gunter, Trans. Res. Rec. **1091** 1–9 (1986)

21. F.L. Hall, Trans. Res. A **21** 191–201 (1987)
22. H. Okamura, S. Watanabe, T. Watanabe, Trans. Res. Cir. **E-C018** (2000)
23. M. Lorenz, L. Elefteriadou, Trans. Res. Cir. **E-C018** 84–95 (2000)
24. H. Zurlinden, Ganzjahresanalyse des Verkehrsflusses auf Straßen. *Schriftenreihe des Lehrstuhls für Verkehrswesen der Ruhr-Universität Bochum*, Heft 26 (Ruhr-Universität Bochum, Bochum, 2003)
25. W. Brilon, J. Geistefeld, M. Regler, in *Transportation and Traffic Theory*, ed. by H.S. Mahmassani. Proceedings of the 16th Inter. Sym. on Transportation and Traffic Theory, (Elsevier, Amsterdam, 2005), pp. 125–144
26. W. Brilon, H. Zurlinden, Straßenverkehrstechnik, Heft 4, 164 (2004)
27. W. Brilon, M. Regler, J. Geistefeld, Straßenverkehrstechnik, Heft 3, 136 (2005)
28. B.S. Kerner, in *Proceedings of the 3rd Symposium on Highway Capacity and Level of Service*, ed. by R. Rysgaard. Vol 2 (Road Directorate, Ministry of Transport – Denmark, 1998), pp. 621–642
29. B.S. Kerner, Phys. Rev. Lett. **81**, 3797–3400 (1998)
30. B.S. Kerner, in *Traffic and Granular Flow' 97*, ed. by M. Schreckenberg, D.E. Wolf. (Springer, Singapore, 1998), pp. 239–267
31. B.S. Kerner, Trans. Res. Rec. **1678**, 160–167 (1999)
32. B.S. Kerner, in *Transportation and Traffic Theory*, ed. by A. Ceder. (Elsevier Science, Amsterdam 1999), pp. 147–171
33. B.S. Kerner, Physics World **12**, 25–30 (August 1999)
34. B.S. Kerner, J. Phys. A: Math. Gen. **33**, L221-L228 (2000); Networks and Spatial Economics, **1**, 35 (2001); in *Progress in Industrial Mathematics at ECMI 2000*, (Springer, Berlin, 2001), pp. 286–292; Mathematical and Computer Modelling, **35**, 481–508 (2002)
35. B.S. Kerner, Trans. Res. Rec. **1710**, 136 (2000)
36. B.S. Kerner, *The Physics of Traffic*, (Springer, Berlin, New York, 2004)
37. B.S. Kerner, Phys. Rev. E **65**, 046138 (2002)
38. B.S. Kerner, J. Phys. A: Math. Theor. **41**, 215101 (2008); 369801 (2008)
39. B.S. Kerner, S.L. Klenov, A. Hiller, J. Phys. A: Math. Gen. **39**, 2001–2020 (2006)
40. B.S. Kerner, H. Rehborn, Phys. Rev. E **53**, R1297–R1300 (1996); R4275–R4278 (1996); Phys. Rev. Lett. **79**, 4030–4033 (1997)
41. B.S. Kerner, P. Konhäuser, Phys. Rev. E **50**, 54–83 (1994)
42. B.S. Kerner, in *Transportation Systems 1997*, ed. by M. Papageorgiou, A. Pouliezos. (Elsevier Science Ltd., London, 1997), pp. 765–770
43. B.S. Kerner, S.L. Klenov, P. Konhäuser, Phys. Rev. E **56**, 4200–4216 (1997)

Chapter 3
Nature of Traffic Breakdown at Bottleneck

For the understanding of the nature of traffic breakdown, probably the most important fundamental *empirical feature* of traffic breakdown is the possibility of both *spontaneous* and *induced* traffic breakdowns at the same bottleneck. Indeed, this feature leads to the conclusions about the nucleation nature of traffic breakdown and the existence of the infinite number of highway capacities of free flow at the bottleneck [1].

3.1 Induced Traffic Breakdown

3.1.1 Definitions of Spontaneous and Induced Traffic Breakdowns at Bottleneck

In observations of traffic breakdown, i.e., an F→S transition at a bottleneck, we distinguish two different cases: (i) *spontaneous* traffic breakdown (spontaneous F→S transition) and (ii) induced traffic breakdown (induced F→S transition) [1]:

- If before traffic breakdown occurs at the bottleneck, there is *free flow* at the bottleneck as well as upstream and downstream in a neighborhood of the bottleneck, then traffic breakdown at the bottleneck is called spontaneous traffic breakdown.
- An induced traffic breakdown at the bottleneck is traffic breakdown induced by the propagation of a spatiotemporal *congested* traffic pattern. This congested pattern has occurred *earlier* than the time instant of traffic breakdown at the bottleneck and at a *different* road location (for example at another bottleneck) than the bottleneck location. When this congested pattern reaches the bottleneck, the pattern induces traffic breakdown at the bottleneck.

Empirical examples of spontaneous traffic breakdowns are presented in Figs. 2.4 (a) and 2.7 (a). Indeed, in each of these cases before traffic breakdown occurs at a bottleneck, there is free flow at the bottleneck as well as upstream and downstream in a neighborhood of the bottleneck.

3.1.2 Empirical Induced Traffic Breakdown and Catch Effect

Traffic breakdown can be induced by a wide moving jam propagating upstream through a bottleneck. Such an empirical example is shown in Fig. 3.1: a moving jam propagating through an on-ramp bottleneck (labeled by "on-ramp bottleneck 1" in Fig. 3.1 (a)) *induces* the synchronized flow phase at this bottleneck (the induced traffic breakdown is labeled by "induced F→S transition"). Indeed, the downstream front of resulting congested traffic is fixed at the bottleneck, i.e., in accordance with the definition [S] (Sect. 2.4.1) this congested traffic is associated with the synchronized flow phase. Synchronized flow is self-sustaining for a very long time (more than an hour) upstream of the bottleneck.

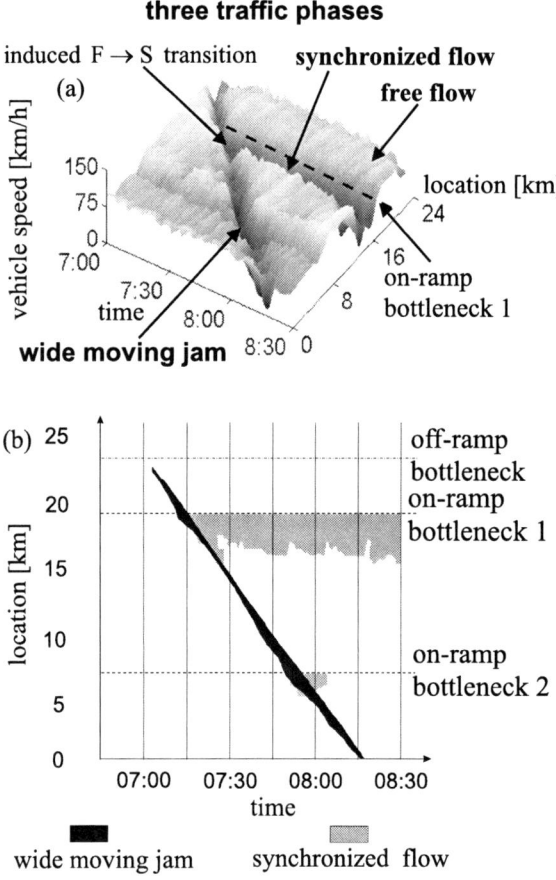

Fig. 3.1 Empirical example of induced traffic breakdown caused by wide moving jam propagation through on-ramp bottleneck. (a) Vehicle speed in space and time. (b) A graph of (a) with the free flow phase, the synchronized flow phase, and the wide moving jam phase. Taken from [1]

3.1 Induced Traffic Breakdown

In contrast with synchronized flow, the moving jam propagates through this bottleneck with the mean velocity of the downstream jam front remaining unchanged (Fig. 3.1). Thus in accordance with the definition [J] (Sect. 2.4.1), this moving jam is associated with the wide moving jam phase.

An induced traffic breakdown at a bottleneck can also occur when a region of synchronized flow first occurs downstream of this bottleneck, and the region later reaches the bottleneck due to the upstream propagation of synchronized flow. However, in contrast to the above case of the induced traffic breakdown caused by wide moving jam propagation, the initial synchronized flow is usually caught at the bottleneck [1]. The latter effect is called the *catch effect*.

An example of an induced traffic breakdown at an on-ramp bottleneck with the associated catch effect is shown in Fig. 3.2. In this case, a moving synchronized flow pattern (MSP), i.e., a moving congested pattern consisting of only the synchronized flow phase propagates upstream of the on-ramp bottleneck (labeled by "MSP" in Fig. 3.2). The MSP has initially earlier occurred at an off-ramp bottleneck that is downstream of the on-ramp bottleneck. Due to the MSP upstream propagation, the MSP approaches the on-ramp bottleneck. Before the MSP reaches the bottleneck, there is free flow at the on-ramp bottleneck (Fig. 3.2 (a, c)). After the MSP has reached the on-ramp bottleneck, the MSP induces traffic breakdown at the on-ramp bottleneck. As a result of this induced traffic breakdown, the MSP is caught at this bottleneck (labeled by "catch effect and induced F→S transition" in Fig. 3.2). This catch effect is inconsistent with the traffic phase definition [J]. In other words, the MSP satisfies the traffic phase definition [S], i.e., the MSP is indeed associated with the synchronized flow phase. The downstream front of a congested pattern resulting from the induced traffic breakdown is fixed at the bottleneck, i.e., the resulting congested pattern is also associated with the synchronized flow phase.

3.1.3 Fundamental Empirical Features of Traffic Breakdown

Real measured traffic data discussed in this section and in Chap. 2, allow us to conclude that traffic breakdown at a bottleneck exhibits the following fundamental empirical features:

A Traffic breakdown is an F→S transition (Sect. 2.4.2).
B At the same bottleneck, traffic breakdown can be either spontaneous or induced.
C Onset and dissolution of congestion are accompanied by a hysteresis effect (Sect. 2.2.2).
D Traffic breakdown exhibits a probabilistic nature (Sect. 2.2.3).

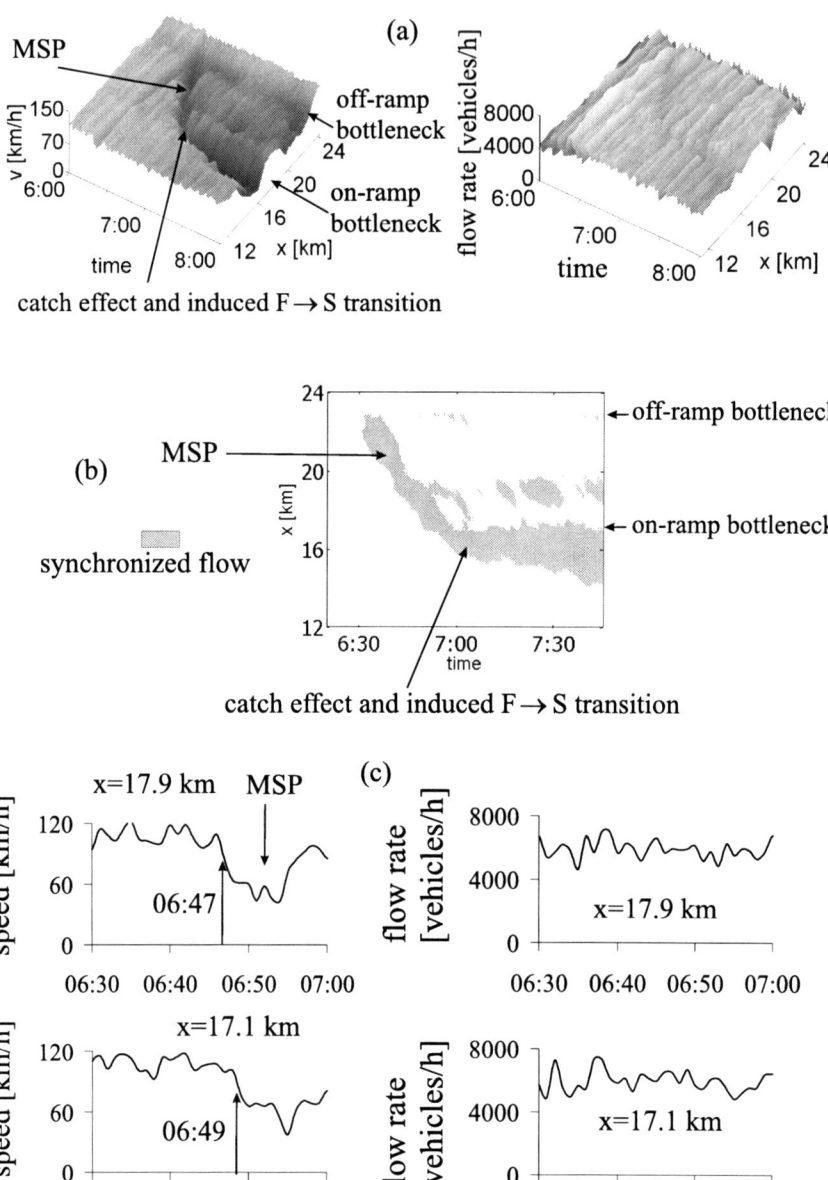

Fig. 3.2 Empirical example of induced traffic breakdown at on-ramp bottleneck. (a) Vehicle speed (left) and flow rate (right) in space and time. (b) A graph of (a) with the free flow phase (white) and the synchronized flow phase (gray). (c) Average (across the road) speed (left) and flow rate (right) within the MSP at two road locations; location $x = 17.1$ km is about 100 m downstream of the end of the merging region of the on-ramp bottleneck. 1-min average data. Taken from [1]

3.2 Explanation of Nature of Traffic Breakdown at Bottleneck through Fundamental Hypothesis of Three-Phase Traffic Theory

In three-phase traffic theory, the nature of traffic breakdown is explained by a competition between two opposing tendencies occurring within a random local disturbance in which the speed is lower and vehicle density is greater than in an initial free flow [1]:

(i) a tendency towards synchronized flow due to vehicle deceleration associated with a *speed adaptation effect*;
(ii) a tendency towards the initial free flow due to vehicle acceleration associated with an *over-acceleration effect*.

3.2.1 Steady States of Synchronized Flow and Fundamental Hypothesis of Three-Phase Traffic Theory

For a qualitative consideration of the speed adaptation and over-acceleration effects as well as their competition, we should introduce the term a *steady state of synchronized flow*:

- A steady state of synchronized flow is a *hypothetical* state of synchronized flow of identical vehicles and drivers in which all vehicles move with the same time-independent speed and have the same space gaps (a space gap is a net distance between two following each other vehicles), i.e., this synchronized flow is homogeneous in time and space.

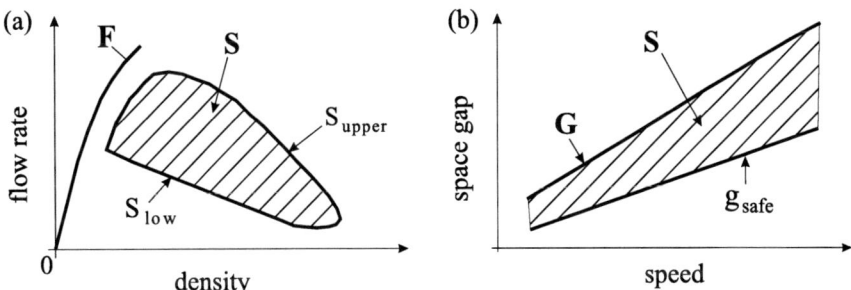

Fig. 3.3 Fundamental hypothesis of three-phase traffic theory [2–7]: (a) Qualitative representation of free flow states (F) and 2D steady states of synchronized flow (dashed region S) on a multi-lane road in the flow–density plane. (b) A part of the 2D steady states of synchronized flow shown in (a) in the space-gap–speed plane (dashed region S)

A hypothesis about steady states of synchronized flow, which is the fundamental hypothesis of three-phase traffic theory, is as follows [2–8]:

- Steady states of synchronized flow cover a two-dimensional (2D) region in the flow–density plane (Fig. 3.3 (a)). The multitudes of free flow states overlap steady states of synchronized flow in the vehicle density. The free flow states on a multi-lane road and steady states of synchronized flow are separated by a gap in the flow rate and, therefore, by a gap in the speed at a given density: at each given density the synchronized flow speed is lower than the free flow speed[1].

The basic driver behavioral assumptions of the fundamental hypothesis of three-phase traffic theory are as follows:

(i) At relatively small space gaps, which occur in synchronized flow, a driver can recognize whether the space gap to the preceding vehicle in car-following increases or decreases over time, i.e., the driver recognizes whether she/he is slower or faster than the preceding vehicles. This is independent of the speed difference $\Delta v = v_\ell - v$ between the speed of the preceding vehicle v_ℓ and the vehicle speed v. In other words, the basic driver behavioral assumption is true even if the speed difference Δv in car-following is negligible.

(ii) At a given speed in synchronized flow, the driver can make an *arbitrary choice* in the space gap to the preceding vehicle within a finite range of space gaps associated with the 2D region of steady states of synchronized flow (Fig. 3.3 (b)): the driver accepts different space gaps at different times and does not control a fixed space gap to the preceding vehicle[2].

In any steady state of synchronized flow the vehicle speed $v > 0$. Therefore, in a steady state of synchronized flow, the space gap g and time headway τ to the preceding vehicle are connected with the vehicle speed v by the formula

---

[1] It must be noted that a well-known wide scattering of empirical data for congested traffic in the flow–density plane *cannot* be considered a proof of the fundamental hypothesis of three-phase traffic theory: a 2D-region for steady states of synchronized flow (Fig. 3.3 (a)) is a *theoretical* assumption only. Indeed, the wide scattering of measured data for congested traffic in the flow–density plane can be associated for example with a variety of driver characteristics and vehicle parameters as well as with speed and density disturbances, which are always present in real traffic flow. As a proof of the fundamental hypothesis, mathematical results of traffic flow models can be considered in which this and other hypotheses of three-phase traffic theory have been used; this is because these results explain and predict empirical features of traffic breakdown and resulting congested patterns [1].

[2] These driver behavioral assumptions leading to the fundamental hypothesis of three-phase traffic theory contradict driver behavioral assumptions made in earlier traffic flow theories, in particular, in the General Motors traffic flow model by Herman, Gazis *et al.* (Sect. 10.3.1), optimal velocity (OV) traffic flow models of Newell, Bando *et al.*, Treiber *et al.* intelligent driver model (IDM) (Sect. 10.3.2), Payne's and Aw-Rascle macroscopic traffic flow models (Sect. 10.3.3), Wiedemann's psychophysical traffic flow model (Sect. 10.3.4) as well as in the Nagel-Schreckenberg cellular automata (CA) traffic flow model (Sect. 10.3.5) and a stochastic traffic flow model of Krauß (Sect. 10.3.6). Traffic breakdown in these traffic flow models is associated with wide moving jam formation in free flow (see Fig. 10.6 and related explanations in Sect. 10.3.7). This is inconsistent with the fundamental empirical features of traffic breakdown (Sect. 3.1.3).

3.2 Traffic Breakdown and Fundamental Hypothesis of Three-Phase Traffic Theory

$$g = v\tau. \tag{3.1}$$

The upper boundary of the 2D region of the steady states of synchronized flow (labeled by S_{upper} in Fig. 3.3 (a)) is determined by a *safe space gap* denoted by g_{safe} associated with a safe time headway denoted by τ_{safe}. In accordance with (3.1), we get

$$g_{\text{safe}}(v) = v\tau_{\text{safe}}(v). \tag{3.2}$$

In (3.2), we have taken into account that τ_{safe} can be a function of the speed.

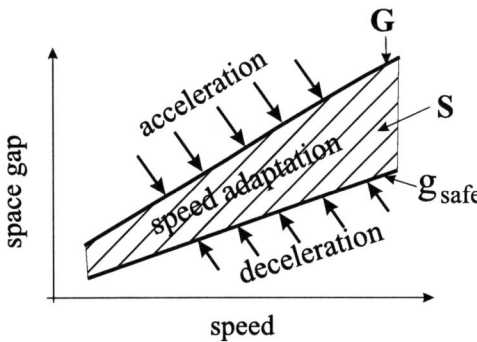

Fig. 3.4 Qualitative explanation of car-following in three-phase traffic theory [2–7]: A vehicle accelerates at $g > G$ and decelerates at $g < g_{\text{safe}}$, whereas under condition (3.12) the vehicle adapts its speed to the speed of the preceding vehicle without caring what the precise space gap is. A dashed region of synchronized flow is taken from Fig. 3.3 (b)

In steady states of synchronized flow, the space gap g and, therefore, the time headway τ should be not smaller than the safe space gap, respectively, the safe time gap. Otherwise, when

$$g < g_{\text{safe}} \tag{3.3}$$

that is equivalent to

$$\tau < \tau_{\text{safe}}, \tag{3.4}$$

the vehicle decelerates (labeled by "deceleration" in Fig. 3.4), i.e., there are no steady states of synchronized flow.

The lower boundary of the 2D-region for steady states of synchronized flow in the flow–density plane (labeled by S_{low}) is determined by a *synchronization space gap* between vehicles denoted by G associated with a synchronization time headway denoted by τ_G. In accordance with (3.1), we get

$$G(v) = v\tau_G(v). \tag{3.5}$$

In (3.5), we have taken into account that τ_G can be a function of the speed. In steady states of synchronized flow, the space gap g and therefore the time headway τ should be not greater than the synchronization space gap, respectively, the synchronization time headway. Otherwise, when

$$g > G \qquad (3.6)$$

that is equivalent to

$$\tau > \tau_G, \qquad (3.7)$$

the vehicle accelerates (labeled by "acceleration" in Fig. 3.4), i.e., there are no steady states of synchronized flow.

The synchronization space gap G is defined through the condition that in steady states of synchronized flow there is a vehicle speed range within which at each given steady speed v the synchronization space gap is greater than the safe space gap (Fig. 3.3 (b)):

$$G(v) > g_{\text{safe}}(v). \qquad (3.8)$$

In accordance with (3.1), the inequality (3.8) is equivalent to

$$\tau_G(v) > \tau_{\text{safe}}(v). \qquad (3.9)$$

Under condition (3.8), at a given steady speed v in synchronized flow there are the *infinite number* of space gaps g within the range

$$g_{\text{safe}} \leq g \leq G \qquad (3.10)$$

at which a driver can move with this steady, i.e., time-independent speed v. Conditions (3.10) are equivalent to

$$\tau_{\text{safe}} \leq \tau \leq \tau_G. \qquad (3.11)$$

The sense of the infinite number of space gaps g (3.10) and time headways τ (3.11) in the steady states of synchronized flow is as follows[3]. If the preceding vehicle moves at a time-independent synchronized flow speed v_ℓ, the following driver can move with the same speed $v = v_\ell$ at any space gap from the space gap range (3.10), i.e., at any time headway τ from the time headway range (3.11). This is associated with the driver behavioral assumption made above that at a given speed in steady states of synchronized flow, the driver can make an arbitrary choice in the space gap to the preceding vehicle within the gap range (3.10). Thus under condition

---

[3] Steady states of synchronized flow are only some *hypothetical flow states*, which cannot be exactly found and, therefore, measured in real synchronized flow. This is at least because of fluctuations, which are always present in real traffic flow. Moreover, even if we neglect the fluctuations, we should mention that moving during a long enough time interval the driver can arbitrarily change the space gap within the range (3.10): she/he moves usually at a particular space gap only during a finite time interval changing to the moving at another space gap within the range (3.10) during another time interval, and so on. In other words, even at a given speed in synchronized flow there are always dynamic transitions over time between different synchronized flow states; these dynamic synchronized flow states correspond only very approximately to the hypothetical steady states of synchronized flow. However, in three-phase traffic theory non-linear features of the dynamic synchronized flow states are associated in a great degree with the features of steady states of synchronized flow. This emphasizes the importance of a consideration of steady states of synchronized flow for the understanding of the nature of traffic breakdown as well as other traffic flow phenomena.

3.2 Traffic Breakdown and Fundamental Hypothesis of Three-Phase Traffic Theory

(3.10) drivers do not try to reach a particular space gap (particular time headway) to the preceding vehicle, but adapt the speed while keeping the gap in a range.

3.2.2 Speed Adaptation Effect

When a driver approaches a slower moving preceding vehicle and the driver cannot pass it, the driver decelerates within the synchronization space gap G adapting the speed to the speed of the preceding vehicle (Fig. 3.5 (a)). This speed adaptation within the synchronization gap is associated with the 2D-region of steady states of synchronized flow given by conditions (3.10) (Fig. 3.3 (b)).

In a general case of car-following, speeds of the vehicle and the preceding vehicle are time-dependent. Then we should consider a *dynamic* synchronization space gap $G(v, v_\ell)$ and a dynamic safe space gap $g_{\text{safe}}(v, v_\ell)$, which can depend on the vehicle speed *and* the speed[4] of the preceding vehicle v_ℓ.

As abovementioned, at $g > G(v, v_\ell)$ a vehicle accelerates whereas at $g < g_{\text{safe}}(v, v_\ell)$, the vehicle decelerates (Fig. 3.4). We define the dynamic synchronization space gap $G(v, v_\ell)$ as a space gap g between the vehicle and the preceding vehicle within which the vehicle tends to adapt the speed to the speed of the preceding vehicle without caring, what the precise space gap is, as long as this space gap is not smaller than the safe gap g_{safe}. This space gap g can be any space gap from the space gaps within the space gap range

$$g_{\text{safe}}(v, v_\ell) \leq g \leq G(v, v_\ell). \tag{3.12}$$

In general, the speed adaptation effect is defined as follows:

- The speed adaptation effect is the adaptation of the vehicle speed to the speed of the preceding vehicle at any space gap within the space gap range (3.12), i.e., without caring what the precise space gap to the preceding vehicle is.

Under condition (3.12), a vehicle accelerates when it is slower than the preceding vehicle, and decelerates when it is faster than the preceding vehicle.

- Vehicle motion occurring under condition (3.12) will be called car-following in the framework of three-phase traffic theory (car-following for short) (Fig. 3.4).

3.2.3 Over-Acceleration Effect

When condition (3.12) is satisfied, a vehicle accelerates when it is slower than the preceding vehicle, and decelerates when it is faster than the preceding vehicle. However, under condition (3.12) the vehicle can also accelerate, even if the vehicle is *not*

---

[4] In a general case, the dynamic synchronization and safe space gaps can also depend on the acceleration (deceleration) of the preceding vehicle a_ℓ: $G = G(v, v_\ell, a_\ell)$, $g_{\text{safe}} = g_{\text{safe}}(v, v_\ell, a_\ell)$.

slower than the preceding vehicle and the preceding vehicle does *not* accelerate. This vehicle acceleration can be considered *vehicle over-acceleration* [1]. In general, the over-acceleration effect is defined as follows:

- The over-acceleration effect is driver maneuver[5] leading to a higher speed from initial car-following at a lower speed occurring under condition (3.12).

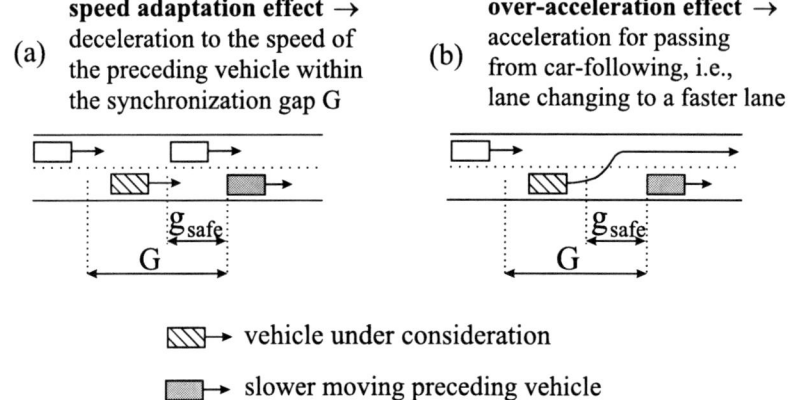

Fig. 3.5 Qualitative explanation of vehicle deceleration due to speed adaptation (a) and vehicle acceleration due to over-acceleration (b) [1, 5–7, 9]

To understand the sense of the term *over-acceleration effect*, we consider a scenario in which a vehicle that moves in free flow on a multi-lane road approaches a slower moving preceding vehicle. If firstly the vehicle cannot pass the preceding vehicle, then the vehicle decelerates within the synchronization space gap to the speed of the preceding vehicle, i.e., the speed adaptation effect is realized leading to car-following of the slow preceding vehicle (Fig. 3.5 (a)). We assume in this scenario that later the vehicle can pass this slow moving preceding vehicle. To pass the preceding vehicle, the vehicle should change lane and accelerate. The vehicle acceleration takes place, even if the vehicle is *not* currently *slower* than the preceding vehicle and the preceding vehicle does *not* accelerate. In the case under consideration, the over-acceleration effect includes vehicle acceleration for passing from car-following, i.e., lane changing to a faster lane (Fig. 3.5 (b)); as a result, the probability of the over-acceleration effect denoted by P_{OA} (probability of over-acceleration for short) is equal to the passing probability from car-following[6] denoted by P:

---

[5] It must be noted that in real traffic flow the speed adaptation and over-acceleration effects appear usually in their dynamic competition within a local disturbance in free flow at a bottleneck. For this reason, a separate consideration of the over-acceleration effect made here is a simplification of the reality. This can be seen from a discussion of a dual role of lane changing in free flow at the bottleneck that appears in Sect. 3.4.

[6] To estimate passing probability, we introduce a *probe vehicle* in traffic flow. The probe vehicle moves with a time-independent speed that is lower than vehicle speed in a neighboring lane(s)

3.2 Traffic Breakdown and Fundamental Hypothesis of Three-Phase Traffic Theory

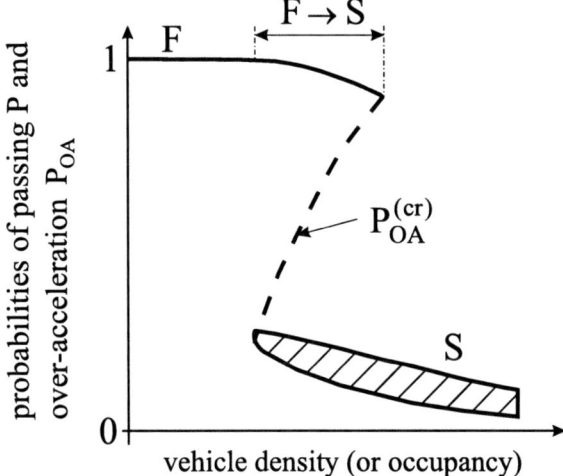

Fig. 3.6 Qualitative Z-shaped density function of passing probability and probability of over-acceleration [1, 5–7, 9]. F – free flow, S – synchronized flow, dashed curve between states F and S is related to a density function of critical probabilities for passing and over-acceleration (the latter is denoted by $P_{OA}^{(cr)}$). Traffic breakdown is possible at any density in initial free flow that is within a density range labeled by "F→S". 2D-region for synchronized flow (dashed region) is associated with the 2D-region for steady states of synchronized flow shown in Fig. 3.3 (a)

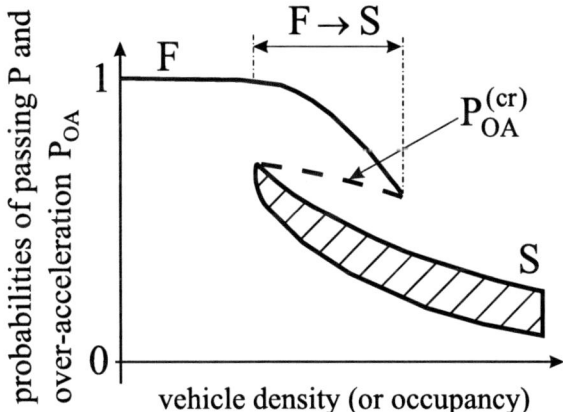

Fig. 3.7 Qualitative Z-shaped density function for passing probability and probability of over-acceleration for a case, when in synchronized flow of smaller densities passing probability is greater than in free flow of greater densities. F – free flow, S – synchronized flow, dashed curve between states F and S is related to a density function of critical probabilities for passing and over-acceleration $P_{OA}^{(cr)}$

$$P_{\text{OA}} = P. \tag{3.15}$$

At a given vehicle density, passing probability P is greater in free flow than in synchronized flow. This assumption together with the fundamental hypothesis of three-phase traffic theory, specifically that at each given density there is a speed gap between states of free flow and synchronized flow lead to the following hypothesis [1, 5–7]:

- Passing probability P and, therefore, over-acceleration probability P_{OA} exhibit a *discontinuous* character; in particular, these probabilities are Z-shaped density functions (Fig. 3.6): At a given density, there is a *drop* in passing and over-acceleration probabilities, when free flow transforms into synchronized flow.

At a given density in free flow and synchronized flow, a discontinuity in passing probability can also be realized, when in synchronized flow of smaller densities passing probability is greater than in free flow of greater densities[7]. This possible case is shown in Fig. 3.7.

Now based on the Z-shape density function for probabilities of vehicle passing and over-acceleration we discuss qualitatively a *competition* between two opposing tendencies – the tendency towards free flow due to the over-acceleration effect and the tendency towards synchronized flow due to the speed adaptation effect. This competition occurs within a random local disturbance in which the speed is lower and density is greater than in an initial free flow. This competition is qualitatively illustrated by arrows between states of free flow and synchronized flow labeled by "over-acceleration" and "speed adaptation" in Fig. 3.8 (a); regions in the space-gap–speed, speed–density, and flow–density planes in which this competition occurs are qualitatively shown by gray regions in Fig. 3.8.

---

of a multi-lane road. We assume that all following vehicles try to pass the slower moving probe vehicle during a given long enough time interval denoted by $T_{\text{follow}}^{(\text{max})}$. This time interval is chosen to be long enough considering passing probability as being zero, if no vehicles can pass the probe vehicle during this time interval (for example, $T_{\text{follow}}^{(\text{max})} = 10$ min). Furthermore, we consider the *average* time interval denoted by $T_{\text{follow}}^{(\text{av})}$ within which vehicles under condition (3.12) have to follow the probe vehicle before they can pass it. Clear is that $T_{\text{follow}}^{(\text{av})} \leq T_{\text{follow}}^{(\text{max})}$ and the shorter $T_{\text{follow}}^{(\text{av})}$ in comparison with $T_{\text{follow}}^{(\text{max})}$, the greater the passing probability. In other words, the relation

$$\frac{T_{\text{follow}}^{(\text{av})}}{T_{\text{follow}}^{(\text{max})}} \tag{3.13}$$

can be considered probability of car-following of a slow preceding vehicle before passing. Then passing probability P can be defined by the formula

$$P = 1 - \frac{T_{\text{follow}}^{(\text{av})}}{T_{\text{follow}}^{(\text{max})}}. \tag{3.14}$$

[7] Some other possible driver maneuver that lead to a discontinuous character of over-acceleration will be discussed in Sect. 11.3.3.

3.2 Traffic Breakdown and Fundamental Hypothesis of Three-Phase Traffic Theory   53

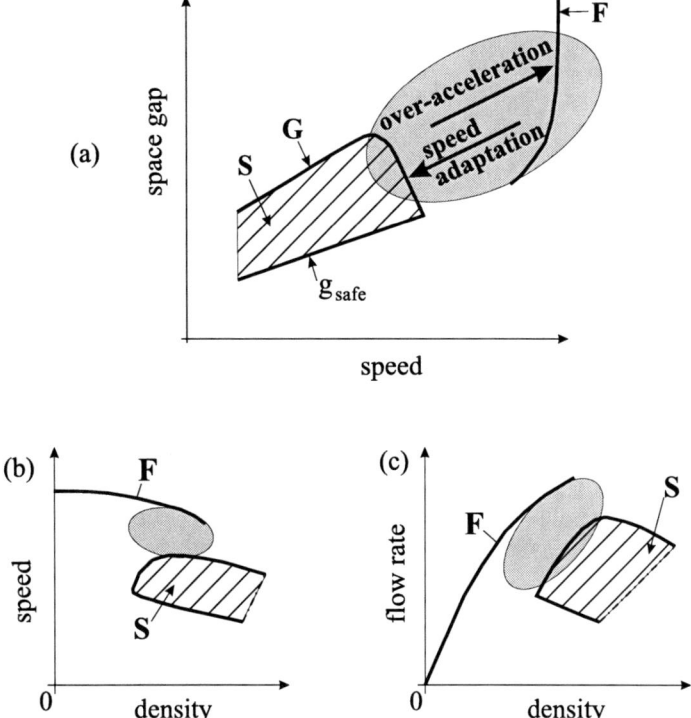

Fig. 3.8 Qualitative illustration of regions in the space-gap–speed (a), speed–density (b), and flow–density (c) planes (gray regions) in which a competition occurs between the tendency towards free flow due to the over-acceleration effect (arrow labeled by "over-acceleration" in (a)) and the tendency towards synchronized flow due to the speed adaptation effect (arrow labeled by "speed adaptation" in (a)). F – free flow, S – a part of synchronized flow states taken from Fig. 3.3 (a)

The Z-shape density function for probability of vehicle passing (Figs. 3.6 and 3.7) explains traffic breakdown as follows [1]. Free flow remains on a multi-lane road as long as the over-acceleration effect, which describes the tendency towards free flow, is stronger than the speed adaptation effect that describes the tendency towards synchronized flow. However, the greater the density and, therefore, the lower the average speed in free flow, the smaller the probability of over-acceleration P_{OA} (3.15), i.e., the weaker the over-acceleration effect.

There is a critical probability of over-acceleration denoted by $P_{OA}^{(cr)}$ that depends on the density (dashed curves between probability of over-acceleration in states F and S in Figs. 3.6 and 3.7). To understand the critical probability of over-acceleration, we consider an initial hypothetical homogeneous free flow at a given density. We assume that in this free flow a random local disturbance occurs within which the speed is lower and density is greater than these traffic variables are outside of the disturbance in the initial free flow. The critical probability of over-acceleration is defined as follows:

- Probability of over-acceleration within a local disturbance in free flow is equal to the critical probability of over-acceleration, i.e.,

$$P_{\rm OA} = P_{\rm OA}^{\rm (cr)}, \qquad (3.16)$$

when at some density and speed within the disturbance the tendency towards the initial free flow due to over-acceleration is on average as strong as the tendency towards synchronized flow due to speed adaptation.

The greater the density and the lower the speed within a local disturbance in the initial free flow under consideration, the smaller the probability of over-acceleration within the disturbance. For this reason, if the *density within the disturbance* increases (speed within the disturbance decreases) in comparison with the density and speed at which the condition (3.16) is satisfied, then within this disturbance over-acceleration probability becomes smaller than the critical probability of over-acceleration:

$$P_{\rm OA} < P_{\rm OA}^{\rm (cr)}. \qquad (3.17)$$

In this case, within the disturbance the over-acceleration effect is weaker than the speed adaptation effect, i.e., the tendency towards synchronized flow due to speed adaptation overcomes the tendency towards free flow due to over-acceleration. This results in the growth of the disturbance leading to traffic breakdown. Traffic breakdown is possible at any density in the initial free flow that is within a density range labeled by "F→S" in Figs. 3.6 and 3.7.

On the contrary, at the same density in the initial free flow outside of the disturbance, if the *density within the disturbance* decreases (speed within the disturbance increases) in comparison with the density and speed at which the condition (3.16) is satisfied, then within this disturbance over-acceleration probability becomes greater than the critical probability of over-acceleration:

$$P_{\rm OA} > P_{\rm OA}^{\rm (cr)}. \qquad (3.18)$$

In this case, within the disturbance the over-acceleration effect is stronger than the speed adaptation effect: the tendency towards free flow due to over-acceleration overcomes the tendency towards synchronized flow due to speed adaptation. As a result, the disturbance decays, i.e., no traffic breakdown occurs and free flow remains.

3.2.4 Critical Speed and Density in Free Flow required for Traffic Breakdown

In free flow, there are always local disturbances within which the speed is lower and the density is greater than outside of the disturbances. Such local disturbances in free flow can be associated with lane changing causing deceleration of following vehicles, vehicle merging from other roads, fluctuations in upstream flow rates, slow moving vehicles, etc.

As explained in Sect. 3.2.3, at a high enough speed and small density in an initial free flow the over-acceleration effect is stronger than the speed adaptation effect, therefore no traffic breakdown is possible. However, if a local disturbance occurs within which the vehicle density is greater and the average speed is lower than in the initial free flow, then within the disturbance the over-acceleration effect becomes weaker than this effect is in the initial free flow outside of the disturbance.

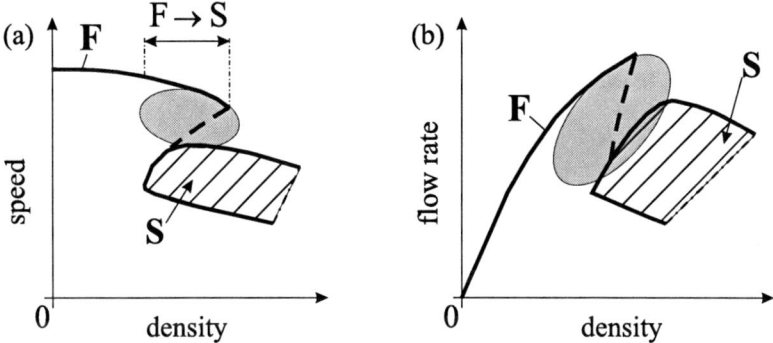

Fig. 3.9 Qualitative illustration of critical density and speed required for traffic breakdown by dashed curves between states of free flow F and synchronized flow S in the speed–density (a) and flow–density (b) planes. States F and S as well as gray regions in which the competition between the over-acceleration and speed adaptation effects occurs are taken from Fig. 3.8 (b, c)

Thus at each given density in the initial free flow that is within a density range labeled by "F→S" in Figs. 3.6, 3.7, and 3.9 (a) there must be a *critical speed* and associated *critical density* within the disturbance required for traffic breakdown defined as follows:

- Critical speed and critical density required for traffic breakdown are respectively the vehicle speed and density within a local disturbance occurring in free flow at which the tendency towards free flow due to the over-acceleration effect is on average as strong as the tendency towards synchronized flow due to the speed adaptation effect.
- It is equivalent to consider either the critical speed or critical density required for traffic breakdown.

The critical density and speed required for traffic breakdown are qualitatively illustrated by dashed curves between states of free flow F and synchronized flow S in the speed–density and flow–density planes shown in Fig. 3.9.

These definitions mean that if within the disturbance the speed is higher than the critical speed (density is smaller than the critical density), the vehicle over-acceleration becomes stronger than the speed adaptation effect; as a result, traffic breakdown cannot occur.

In contrast, when a local disturbance occurs within which the speed is lower than the critical speed (density is greater than the critical density), the speed adaptation effect becomes stronger than the over-acceleration; as a result, traffic breakdown must occur. This local disturbance can be considered a *nucleus* for traffic breakdown[8]. The term *nucleus required for traffic breakdown* is defined as follows[9].

- A nucleus required for traffic breakdown is a local disturbance in free flow within which the vehicle speed is equal to or lower than the critical speed required for traffic breakdown, respectively, the vehicle density is equal to or greater than the critical density.

When a nucleus appears in free flow, traffic breakdown occurs. In other words, traffic breakdown exhibits a *nucleation nature* [2–7, 10].

3.2.5 Linking of Critical Speed, Critical Density, Nucleus required for Traffic Breakdown, and Over-Acceleration Probability

The nucleation nature of traffic breakdown is associated with the discontinuous character of over-acceleration probability P_{OA} (Figs. 3.6 and 3.7), specifically with the existence of the critical over-acceleration probability $P_{OA}^{(cr)}$:

- the speed and density within a local disturbance in free flow at which over-acceleration probability P_{OA} is equal to the critical over-acceleration probability $P_{OA}^{(cr)}$, i.e., the condition (3.16) is satisfied are the critical speed and critical density, respectively.

In other words, a local disturbance, within which the vehicle density is greater and the average speed is lower than these traffic variables are an initial free flow, is a nucleus required for traffic breakdown, when within this disturbance over-acceleration probability is equal to or smaller than the critical over-acceleration

---

[8] It should be noted that the use of the term *nucleus* is associated with a huge number of qualitatively other nucleation phenomena observed in a variety of systems of natural science like fluid dynamics, semiconductors, gas plasma, optical systems, chemical reactions, and biological systems [17–26]; see also Sect. 6.2.4.

[9] In real traffic flow, a nucleus required for traffic breakdown exhibits two attributes: (i) critical speed (critical density) and (ii) spatial distributions of traffic variables within the nucleus. However, in a qualitative discussion of the nature of traffic breakdown made here we ignore spatial distributions of traffic variables within the nucleus.

3.3 Nucleation Features of Traffic Breakdown at Bottleneck

probability, i.e.,

$$P_{\rm OA} \leq P_{\rm OA}^{\rm (cr)}. \tag{3.19}$$

Under this condition, within the disturbance the tendency towards free flow due to over-acceleration is on average as strong as or weaker than the tendency towards synchronized flow due to speed adaptation.

3.3 Nucleation Features of Traffic Breakdown at Bottleneck

3.3.1 Permanent Speed Disturbance at Bottleneck

To explain why traffic breakdown is mostly observed at a highway bottleneck, in three-phase traffic theory is assumed that in free flow at the bottleneck there is a *permanent* and on average motionless local speed disturbance in which the speed is lower and the vehicle density is greater than outside of the bottleneck [1, 10, 12–15]. This permanent disturbance called also a *"deterministic" disturbance* at the bottleneck is a permanent non-homogeneity in free flow localized in a neighborhood of the bottleneck. Naturally, the deterministic disturbance in free flow at the bottleneck occurs only, when flow rates upstream of the bottleneck are great enough.

Due to the existence of the deterministic disturbance, the critical speed required for traffic breakdown is more probable to occur within this deterministic disturbance at the bottleneck. For this reason, traffic breakdown is much more probable to occur at the bottleneck rather than outside of the bottleneck [10]. Thus three-phase traffic theory explains the empirical evidence that traffic breakdown is observed mostly at highway bottlenecks by the following hypothesis:

- Probability of the occurrence of a nucleus required for traffic breakdown is much greater at the bottleneck than outside of it.

For a simplification of a qualitative analysis of traffic breakdown at a bottleneck, we assume that as long as free flow is at the bottleneck, the flow rate in free flow downstream of the bottleneck denoted by $q_{\rm sum}$ is equal to the sum of the flow rates in free flow(s) just upstream of the bottleneck. However, this condition is not satisfied, after traffic breakdown has occurred and a congested pattern is forming at the bottleneck. For this reason, we discuss here conditions for traffic breakdown only, rather than features of resulting congested patterns (see Chap. 7).

3.3.1.1 On-Ramp Bottleneck

A deterministic disturbance in free flow at an on-ramp bottleneck is caused by two flows, which merge at the bottleneck: (i) An on-ramp inflow with the rate $q_{\rm on}$. (ii) A flow on the main road upstream of the bottleneck with the rate $q_{\rm in}$ (Fig. 3.10 (a)). At given high enough flow rates $q_{\rm on}$ and $q_{\rm in}$, vehicles that merge from the on-ramp onto the main road force the vehicles on the main road to decelerate in the vicinity of an on-ramp merging region (Fig. 3.11). This leads to a local decrease in speed and consequently to a local increase in density in free flow in the vicinity of the bottleneck. This flow merging occurs permanent and within the same road region associated with the on-ramp merging region. For this reason, the non-homogeneity of free flow at the bottleneck, i.e., the deterministic disturbance is on average motionless and permanent (solid curves in Fig. 3.10 (b, c)).

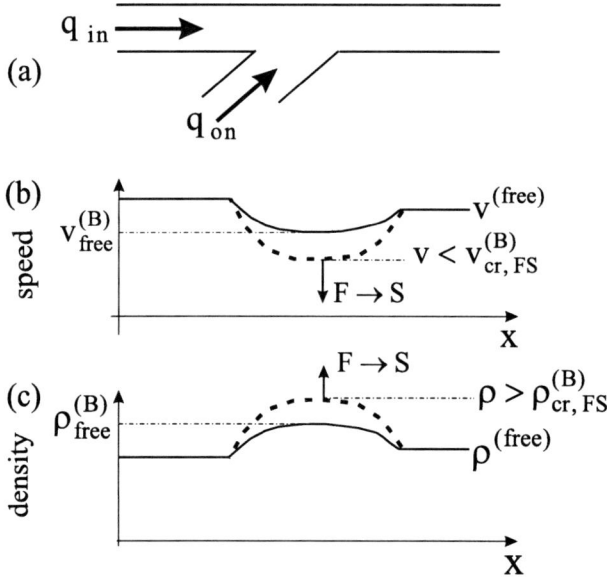

Fig. 3.10 Qualitative explanation of traffic breakdown at on-ramp bottleneck: (a) Sketch of an on-ramp bottleneck. (b, c) Qualitative spatial speed (b) and density (c) distributions in free flow within a local disturbance at the bottleneck; solid curves show a deterministic disturbance; dashed curves show the local disturbance at a fixed time instant for the case when fluctuations in free flow lead to a decrease in the speed and an increase in the density in comparison, respectively, with the speed $v_{\rm free}^{(B)}$ and density $\rho_{\rm free}^{(B)}$ within the deterministic disturbance. Arrows labeled F→S symbolize traffic breakdown that occurs, when a nucleus for traffic breakdown appears at the bottleneck

The average speed $v_{\rm free}^{(B)}$ and density $\rho_{\rm free}^{(B)}$ in free flow within the deterministic disturbance (Fig. 3.10 (b, c)) correspond to the conditions

$$v_{\rm free}^{(B)} < v^{\rm (free)}, \; \rho_{\rm free}^{(B)} > \rho^{\rm (free)}, \tag{3.20}$$

where $v^{\rm (free)}$ and $\rho^{\rm (free)}$ are the average vehicle speed and density in free flow on the main road downstream of the bottleneck (Fig. 3.10 (b, c)). Because the deterministic disturbance is assumed to be permanent and motionless, at given $q_{\rm in}$ and $q_{\rm on}$, the

3.3 Nucleation Features of Traffic Breakdown at Bottleneck

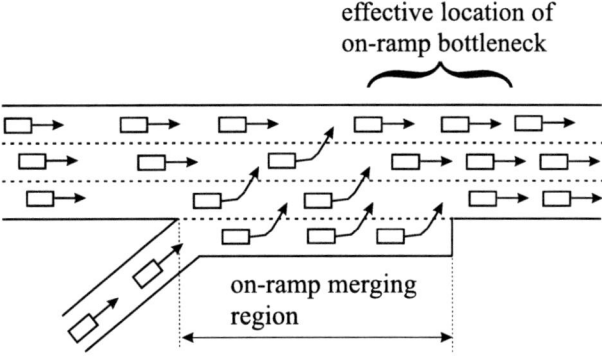

Fig. 3.11 Sketch of on-ramp bottleneck

flow rate[10]

$$q_{\text{sum}} = q_{\text{in}} + q_{\text{on}} \qquad (3.21)$$

does not depend on the spatial co-ordinate within the disturbance. For this reason, the speed and density on the main road satisfy the following condition:

$$q_{\text{sum}} = v^{(\text{free})} \rho^{(\text{free})} = v_{\text{free}}^{(B)} \rho_{\text{free}}^{(B)}. \qquad (3.22)$$

Within the deterministic disturbance a nucleus required for traffic breakdown has a greater probability to occur spontaneously in comparison with other road locations outside of the bottleneck. Indeed, the deterministic disturbance represents a real local disturbance at the bottleneck only on the average. Due to random fluctuations caused for example by a random process of vehicle merging from on-ramp onto the main road, the speed and density within the local disturbance at the bottleneck change randomly over time in the vicinity of the speed $v_{\text{free}}^{(B)}$ and density $\rho_{\text{free}}^{(B)}$ within the deterministic disturbance, respectively. In other words, the local disturbance at the bottleneck (dashed curves in Fig. 3.10 (b, c)) can be considered consisting of two components:

(a) the deterministic disturbance and
(b) a random disturbance component.

---

[10] In a neighborhood of a bottleneck the number of road lanes, which includes the number of lanes on the main road and the number of on- and off-ramp lanes, depends usually on the road location. For this reason, we assume here and below that the flow rate q_{sum} and densities $\rho^{(\text{free})}$, $\rho_{\text{free}}^{(B)}$ are found per road lane at each road location in the neighborhood of the bottleneck as follows: Firstly, the *total* flow rate and densities are calculated across all lanes of the main road *and* of on- and off-ramps at the location. Then regardless of this location, these total flow rate and densities are divided by the number of lanes at some location downstream of the bottleneck. Accordingly, $v^{(\text{free})} = q_{\text{sum}}/\rho^{(\text{free})}$ and $v_{\text{free}}^{(B)} = q_{\text{sum}}/\rho_{\text{free}}^{(B)}$.

We consider a random disturbance component, which leads to a speed decrease and density increase in comparison, respectively, with the speed $v^{(B)}_{\text{free}}$ and density $\rho^{(B)}_{\text{free}}$ within the deterministic disturbance. In accordance with the definition of a nucleus required for traffic breakdown (Sect. 3.2.4), when due to this random speed decrease the speed v within the disturbance becomes equal to or lower than a critical speed denoted by $v^{(B)}_{\text{cr, FS}}$, respectively, the density ρ within the disturbance becomes equal to or greater than a critical density denoted by $\rho^{(B)}_{\text{cr, FS}}$, i.e.,

$$v \leq v^{(B)}_{\text{cr, FS}} \quad (\rho \geq \rho^{(B)}_{\text{cr, FS}}), \tag{3.23}$$

the local disturbance is a nucleus required for traffic breakdown: this disturbance grows leading to traffic breakdown at the bottleneck. When the opposite condition

$$v > v^{(B)}_{\text{cr, FS}} \quad (\rho < \rho^{(B)}_{\text{cr, FS}}), \tag{3.24}$$

is satisfied, no traffic breakdown occurs at the bottleneck.

We can see that

- due to the existence of the deterministic disturbance localized at the bottleneck in which the speed $v^{(B)}_{\text{free}}$ is lower and the density $\rho^{(B)}_{\text{free}}$ is greater than, respectively, the speed $v^{(\text{free})}$ and density $\rho^{(\text{free})}$ outside of the disturbance, a nucleus required for traffic breakdown occurs more probably within this disturbance than outside of it.
- Therefore, probability of traffic breakdown is much greater within the deterministic disturbance localized at the bottleneck than outside of the disturbance.
- This explains why traffic breakdown is observed much more frequently at highway bottlenecks than outside of them.

3.3.1.2 Off-Ramp Bottleneck

A deterministic disturbance at an off-ramp bottleneck is localized usually *upstream* of the off-ramp merging region within which vehicles change from the main road onto an off-ramp lane(s) (Figs. 3.12 and 3.13). A reason for this is as follows.

If the flow rate q_{in} upstream of the bottleneck and the percentage of vehicles going to the off-ramp denoted by η are great enough, then in free flow at some distance upstream of the beginning of the off-ramp merging region the vehicle density in the right lane (in general, in the lane that is the neighboring one to the off-ramp lane(s)) is greater and the associated speed is lower than in other lanes of the main road. Moreover, due to the lower speed in the right lane many vehicles going to the off-ramp try to move in the other lanes as long as possible, i.e., they change to the right lane at a relatively small distance upstream of the beginning of the off-ramp merging region. This lane changing leads to an increase in the density in the right lane upstream of the beginning of the off-ramp merging region as well as to great disturbances in free flow.

3.3 Nucleation Features of Traffic Breakdown at Bottleneck

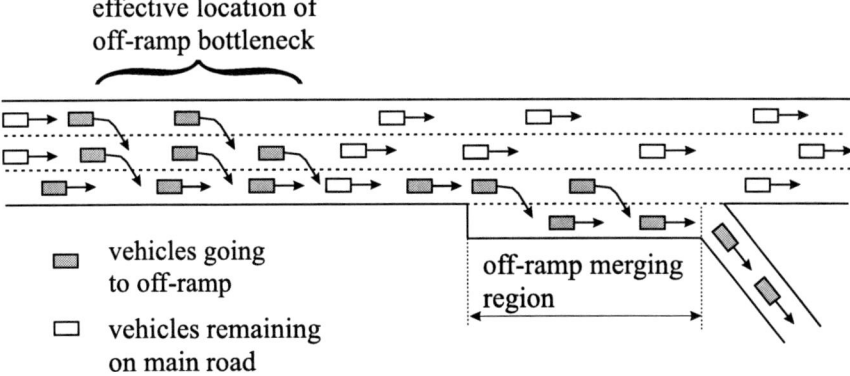

Fig. 3.12 Sketch of off-ramp bottleneck

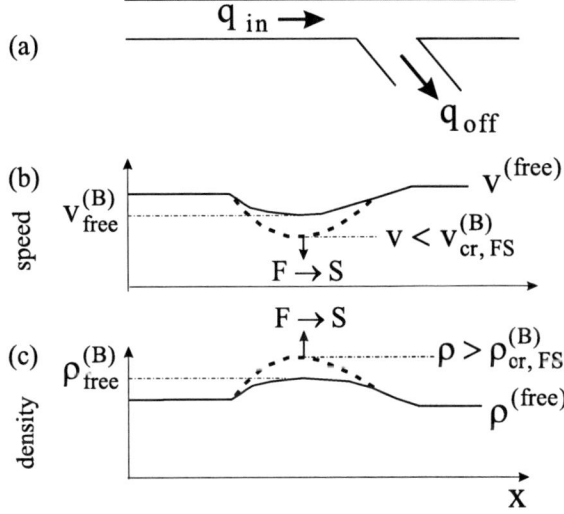

Fig. 3.13 Qualitative explanation of traffic breakdown at off-ramp bottleneck: (a) Sketch of an off-ramp bottleneck. (b, c) Qualitative spatial speed (b) and density (c) distributions in free flow within a local disturbance at the bottleneck; solid curves show a deterministic disturbance; dashed curves show the local disturbance at a fixed time instant for the case when fluctuations in free flow lead to a decrease in the speed and an increase in the density in comparison, respectively, with the speed $v^{(B)}_{\text{free}}$ and density $\rho^{(B)}_{\text{free}}$ within the deterministic disturbance. Arrows labeled F→S symbolize traffic breakdown that occurs, when a nucleus for traffic breakdown appears at the bottleneck

For an off-ramp bottleneck, $q_{sum} = q_{in}$ (Fig. 3.13) and the speed $v_{free}^{(B)}$ and density $\rho_{free}^{(B)}$ within the deterministic disturbance satisfy formulae (3.20) and (3.22). Thus if due to a random speed decrease in comparison with the speed $v_{free}^{(B)}$, the speed within the disturbance becomes equal to or lower than the critical speed required for breakdown, i.e., condition (3.23) is satisfied, the local disturbance grows and traffic breakdown occurs at the bottleneck. Otherwise, i.e., under condition (3.24) no traffic breakdown is realized at the bottleneck.

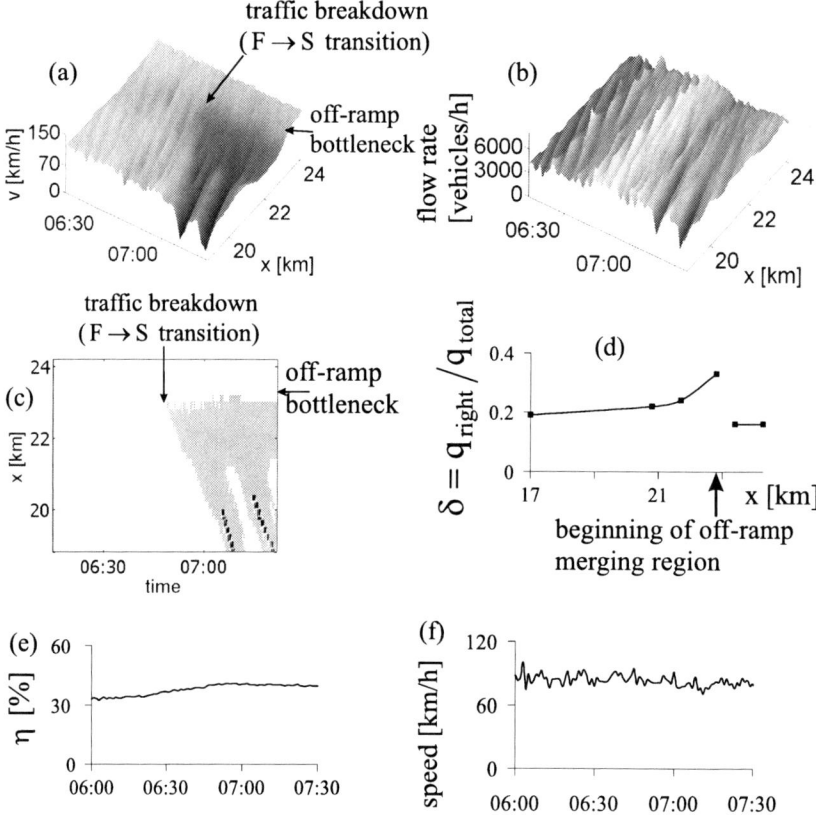

Fig. 3.14 Empirical example of traffic breakdown at off-ramp bottleneck: (a, b) Average speed (a) and total flow rate (b) on the main road in space and time. (c) Overview of traffic phases in space and time associated with (a, b). (d) Ratio $\delta = q_{right}/q_{total}$ as a function of road location. (e) Time-dependence of percentage of vehicles going to the off-ramp η. (f) Average speed across off-ramp lanes measured about 600 m downstream of the end of the off-ramp merging region. 1-min average data. Taken from [1]

An empirical example of traffic breakdown at an off-ramp bottleneck is shown in Fig. 3.14. We can see that the share of vehicles moving in the right lane, i.e., the ratio $\delta = q_{right}/q_{total}$ (where q_{right} and q_{total} are respectively the flow rate in the right lane and total flow rate across the main road) increases strongly, when the distance to the beginning of the off-ramp merging region decreases (Fig. 3.14 (d)). This shows that many vehicles going to the off-ramp try indeed to move in the other lanes as long as possible before they have to change to the right lane with the goal to leave the main road to the off-ramp. A drop in $\delta = q_{\mathrm{right}}/q_{\mathrm{total}}$ downstream of the end of the merging region of the off-ramp (Fig. 3.14 (d)) is associated with vehicles that left the main road to the off-ramp. A relatively great value of the percentage of vehicles going to the off-ramp η (Fig. 3.14 (e)) and an increase in the flow rate q_{in} upstream of the bottleneck (Fig. 3.14 (b)) together with this lane changing can explain that traffic breakdown and the downstream front of the resulting synchronized flow appear upstream of the beginning of the off-ramp merging region (Fig. 3.14).

3.3.1.3 Effective Location of Bottleneck

After traffic breakdown has occurred at a bottleneck, synchronized flow is formed at the bottleneck. The downstream front of the synchronized flow is fixed at the bottleneck. Within the downstream front of synchronized flow vehicles accelerate from synchronized flow upstream of the front to free flow downstream of the front. As mentioned in Sect. 2.4.2, a road location in the neighborhood of the bottleneck at which this downstream front of synchronized flow is fixed is an effective location of bottleneck.

- The effective location of an on-ramp bottleneck is usually within the on-ramp merging region (Fig. 3.11) or at a small distance downstream of the on-ramp merging region. The reason for this is the same as that for the occurrence of the deterministic disturbance at the on-ramp bottleneck explained above.
- In contrast with the on-ramp bottleneck, the effective location of an off-ramp bottleneck is usually upstream of the beginning of the off-ramp merging region within which vehicles change from the main road onto an off-ramp lane(s) (Fig. 3.12). The reason for this is the same as that for the upstream occurrence of the deterministic disturbance upstream of the off-ramp.

3.3.2 Z-Shaped Speed–Flow and S-Shaped Density–Flow Characteristics for Traffic Breakdown

The critical speed $v_{\mathrm{cr,\,FS}}^{(B)}$ is an increasing function of the flow rate in free flow downstream of the bottleneck q_{sum} (Fig. 3.15), respectively, the associated critical density $\rho_{\mathrm{cr,\,FS}}^{(B)}$ is a decreasing function of q_{sum} (Fig. 3.16).

To explain this general conclusion, we consider an on-ramp bottleneck at a given great enough flow rate to the on-ramp q_{on} (Fig. 3.10). The greater the flow rate q_{sum} (due to an increase in the flow rate q_{in} on the main road upstream of the bottleneck),

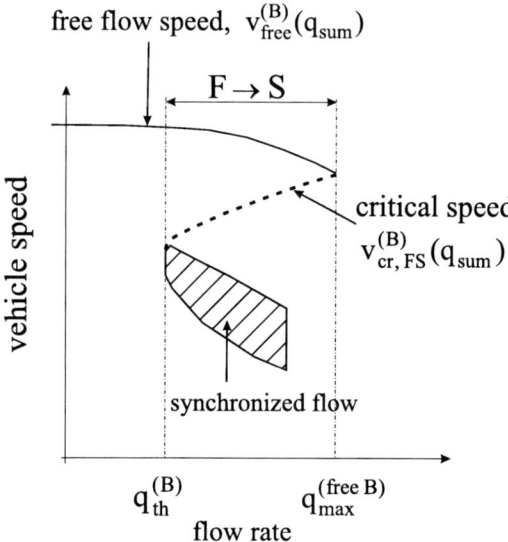

Fig. 3.15 Qualitative Z-characteristic for traffic breakdown at a bottleneck in the speed–flow plane: Vehicle speed at the location of the deterministic disturbance at the bottleneck as a function of the flow rate in free flow downstream of the bottleneck shown in a limited flow rate range. The Z-characteristic consists of free flow speed (curve $v_{\text{free}}^{(B)}$), critical speed in free flow required for traffic breakdown (dashed curve $v_{\text{cr, FS}}^{(B)}$), and a part of synchronized flow states (dashed region). Characteristic flow rates $q_{\text{th}}^{(B)}$ and $q_{\text{max}}^{(\text{free B})}$ will be explained in Sect. 3.3.3. Taken from [1]

the lower the speed $v_{\text{free}}^{(B)}$ (Fig. 3.15) and the greater the density $\rho_{\text{free}}^{(B)}$ within the deterministic disturbance at the bottleneck (Fig. 3.16):

- The greater the flow rate q_{sum}, the smaller the probability of over-acceleration, i.e., the weaker the over-acceleration effect within the deterministic disturbance at the bottleneck.

Thus the greater the flow rate q_{sum}, the higher the critical speed and the smaller the critical density at which the speed adaptation effect becomes on average as strong as the over-acceleration effect. This explains why $v_{\text{cr, FS}}^{(B)}(q_{\text{sum}})$ is an increasing flow function (Fig. 3.15), respectively, $\rho_{\text{cr, FS}}^{(B)}(q_{\text{sum}})$ is a decreasing flow function (Fig. 3.16).

This qualitative consideration leads to the conclusion that the average vehicle speed at the location of the deterministic disturbance at the bottleneck as a function of the flow rate in free flow downstream of the bottleneck is a Z-shaped characteristic in the speed–flow plane (Fig. 3.15). The Z-shaped speed–flow characteristic consists of the flow rate dependence of free flow speed $v_{\text{free}}^{(B)}(q_{\text{sum}})$ within the deterministic disturbance localized at the bottleneck, the flow rate dependence of critical speed within the disturbance $v_{\text{cr, FS}}^{(B)}(q_{\text{sum}})$, and synchronized flow states.

3.3 Nucleation Features of Traffic Breakdown at Bottleneck

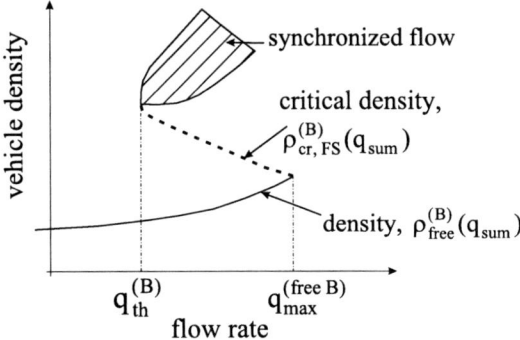

Fig. 3.16 Qualitative S-characteristic for traffic breakdown in the density–flow plane associated with the Z-characteristic shown in Fig. 3.15: Vehicle density at the location of the deterministic disturbance at the bottleneck as a function of the flow rate in free flow downstream of the bottleneck shown in a limited flow rate range. The S-characteristic consists of free flow density (curve $\rho_{\rm free}^{(B)}$), critical density in free flow required for traffic breakdown (dashed curve $\rho_{\rm cr,\,FS}^{(B)}$), and a part of synchronized flow states (dashed region)

In accordance with this Z-characteristic, the vehicle density at the location of the deterministic disturbance at the bottleneck as a function of the flow rate in free flow downstream of the bottleneck should be an S-shaped characteristic in the density–flow plane (Fig. 3.16). As follows from conclusions of Sect. 3.2.5, the Z- and S-characteristics for traffic breakdown result from the hypothesis of three-phase traffic theory about a Z-shaped density function of probability of vehicle over-acceleration (Figs. 3.6 and 3.7).

3.3.3 Maximum and Threshold Flow Rates in Free Flow at Bottleneck

From Z-shaped speed–flow and S-shaped density–flow characteristics (Figs. 3.15 and 3.16) we can see that for each of the flow rate in free flow $q_{\rm sum}$, which is within a finite flow rate range

$$q_{\rm th}^{(B)} \leq q_{\rm sum} \leq q_{\rm max}^{(\text{free B})}, \tag{3.25}$$

traffic breakdown is possible (this flow rate range is labeled by "F→S" in Fig. 3.15).

When initially free flow is at the bottleneck and the flow rate in the free flow satisfies conditions (3.25), then traffic breakdown is realized, if a random great enough speed decrease (density increase) (3.23) within a localized disturbance at the bottleneck occurs associated with the occurrence of a nucleus required for traffic breakdown.

The characteristic flow rate $q_{\text{max}}^{(\text{free B})}$ is the maximum flow rate in free flow: Due to the occurrence of traffic breakdown at the bottleneck the flow rate $q_{\text{sum}} > q_{\text{max}}^{(\text{free B})}$ cannot be observed in free flow.

The characteristic flow rate $q_{\text{th}}^{(\text{B})}$ is a threshold flow rate for traffic breakdown[11]: At the flow rate

$$q_{\text{sum}} < q_{\text{th}}^{(\text{B})} \qquad (3.26)$$

no traffic breakdown is possible at the bottleneck. In other words, under condition (3.26) independent of how great a time-limited speed decrease and density increase within a local disturbance in free flow are, traffic breakdown cannot occur at the bottleneck.

The threshold flow rate is smaller than the maximum flow rate in free flow:

$$q_{\text{th}}^{(\text{B})} < q_{\text{max}}^{(\text{free B})}. \qquad (3.27)$$

- This means that there are the *infinite number* of the flow rates q_{sum} in free flow downstream of the bottleneck satisfying (3.25) at which traffic breakdown at the bottleneck is possible.

Thus we can conclude that traffic breakdown occurs, if the following two conditions are satisfied:

(i) The flow rate in free flow downstream of a bottleneck satisfies conditions (3.25).
(ii) A nucleus required for traffic breakdown appears in free flow at the bottleneck, i.e., condition (3.23) is satisfied.

Item (i) is associated with a Z-shaped density function of vehicle over-acceleration probability (Figs. 3.6 and 3.7 in Sect. 3.2.3) resulting in the nucleation nature of traffic breakdown. Item (ii) means that to initiate traffic breakdown, a local disturbance in free flow should appear within which the speed is equal to or lower than the critical speed, respectively, the density is equal to or greater than the critical density.

Note that we can consider all free flow states at the bottleneck within the flow rate range

$$q_{\text{th}}^{(\text{B})} \leq q_{\text{sum}} < q_{\text{max}}^{(\text{free B})} \qquad (3.28)$$

as *metastable* free flow states with respect to traffic breakdown (F→S transition)[12]. The term *metastable* free flow with respect to traffic breakdown is defined as follows.

---

[11] It must be noted that the maximum flow rate in free flow $q_{\text{max}}^{(\text{free B})}$ and the threshold flow rate in free flow $q_{\text{th}}^{(\text{B})}$ depend on the averaging time interval for traffic variables T_{av}. The reason for dependences $q_{\text{max}}^{(\text{free B})}(T_{\text{av}})$ and $q_{\text{th}}^{(\text{B})}(T_{\text{av}})$ will be discussed in Chap. 4.

[12] Indeed, in accordance with the wide-accepted terminology used in the natural science (see references in [17–26]), the nucleation feature of traffic breakdown allows us to conclude that traffic breakdown is a first-order local F→S transition and under conditions (3.28) free flow is in a *metastable state* with respect to this F→S transition.

- Metastable free flow with respect to traffic breakdown is free flow with the flow rate downstream of a bottleneck that satisfies conditions (3.28). This metastable free flow is stable with respect to small enough local disturbances in this flow, i.e., when condition (3.24) is satisfied; however, traffic breakdown occurs in the free flow, if a nucleus required for traffic breakdown appears at the bottleneck, i.e., when condition (3.23) is satisfied.

3.3.4 Linking Empirical Induced and Spontaneous Traffic Breakdowns with Over-Acceleration, Critical Speed, and Critical Density

Three-phase traffic theory explains the nature of traffic breakdown by a Z-shaped density function of probability of vehicle over-acceleration and the associated competition between the speed adaptation and over-acceleration effects within a local disturbance in an initial free flow. The sense of this hypothesis about a Z-shaped density function of over-acceleration probability (Figs. 3.6 and 3.7 in Sect. 3.2.3) is that the application of this hypothesis leads to the conclusion about the nucleation nature of traffic breakdown.

This in turn explains the fundamental empirical feature of traffic breakdown about the possibility of either spontaneous or induced traffic breakdown (Sect. 3.1.3) as follows. Traffic breakdown occurs, if a nucleus required for traffic breakdown appears at the bottleneck. There are qualitatively two *different sources* of the occurrence of this nucleus:

(i) A random nucleus occurrence within an initial free flow at the bottleneck due to for example lane changing, fluctuations in the upstream flow rates, vehicle merging from other roads, an unexpected vehicle deceleration, etc. In this case, a spontaneous traffic breakdown occurs at the bottleneck. Thus the term *spontaneous* means that the nucleus required for traffic breakdown appears due to some random disturbances within the *initial free flow* in a neighborhood of the bottleneck.

(ii) The role of a nucleus required for traffic breakdown exhibits a congested traffic pattern that propagates on the road; this pattern has occurred *earlier* than the instant of traffic breakdown at the bottleneck and at *another road location* than the bottleneck location. The speed within this congested pattern is considerably lower and the density is considerably greater than the critical speed and density required for traffic breakdown, respectively. For this reason, when this pattern reaches the bottleneck at which free flow has been before that satisfies conditions (3.25), the pattern must induce traffic breakdown at the bottleneck. This explains the term *induced* traffic breakdown.

- Thus rather than the nature of traffic breakdown, the terms *spontaneous* and *induced* traffic breakdowns at a bottleneck distinguish different *sources* of the occurrence of a nucleus required for the initiating of traffic breakdown.

3.3.5 Simulations of Traffic Breakdown at Bottlenecks

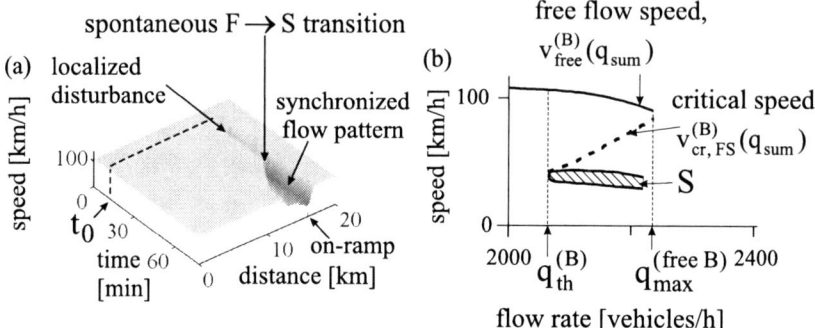

Fig. 3.17 Simulations of nucleation features of traffic breakdown at on-ramp bottleneck with three-phase traffic flow model: (a) Localized speed disturbance with spontaneous traffic breakdown at on-ramp bottleneck; at $t = t_0$ on-ramp inflow is switched on. (b) Z-characteristic for traffic breakdown. Taken from [16]

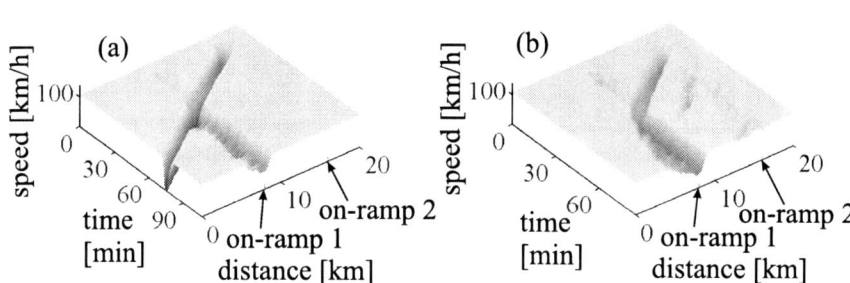

Fig. 3.18 Simulations of induced traffic breakdowns at upstream on-ramp bottleneck labeled "on-ramp 1" with three-phase traffic flow model: (a) Wide moving jam, which propagates through the bottleneck while maintaining the jam downstream front velocity, induces traffic breakdown. (b) Moving synchronized flow pattern is caught at bottleneck inducing traffic breakdown. Taken from [16]

Three-phase traffic flow models can show and predict all fundamental empirical features of traffic breakdown. In particular, a simulated example of a spontaneous traffic breakdown at an on-ramp bottleneck exhibits an initial disturbance localized at the bottleneck and the spontaneous traffic breakdown at the location of the disturbance with the subsequent synchronized flow emergence (Fig. 3.17 (a)); this traffic breakdown is associated with an Z-shaped speed–flow characteristic found in simulations (Fig. 3.17 (b)) as discussed above based on three-phase traffic theory.

3.4 Dual Role of Lane Changing in Free Flow 69

Simulated examples of an induced traffic breakdown due to upstream propagation of a wide moving jam through an upstream on-ramp bottleneck (Fig. 3.18 (a)) and an induced traffic breakdown at the same on-ramp with the catch effect due to upstream propagation of a moving synchronized flow pattern (MSP) are qualitatively the same ones as those in empirical Figs. 3.1 and 3.2, respectively.

3.4 Dual Role of Lane Changing in Free Flow: Maintenance of Free Flow or Traffic Breakdown

Lane changing can play a dual role in free flow at a bottleneck: lane changing can maintain free flow or in contrast lead to traffic breakdown at the bottleneck. The reason for this lane changing effect on traffic flow is as follows.

(i) Lane changing to a faster lane is responsible for the over-acceleration effect that describes the tendency towards free flow at a bottleneck. This maintains free flow at the bottleneck. However, this over-acceleration effect can maintain free flow only if due to lane changing a vehicle passes the preceding vehicle and this vehicle passing does *not* force the following vehicle in the target lane to decelerate strongly.

(ii) In contrast, lane changing can lead to the occurrence of a nucleus required for traffic breakdown at a bottleneck. This can occur, if lane changing forces the following vehicles in the target lane to decelerate strongly.

Thus the over-acceleration effect due to lane changing to a faster lane can lead to free flow only in the case, when the following vehicles in the target lane should not decelerate strongly, specifically, no nucleus required for traffic breakdown occurs due to this lane changing.

Otherwise, lane changing to a faster lane leads to traffic breakdown at the bottleneck. This emphasizes that in real traffic flow the speed adaptation and over-acceleration effects appear usually in their dynamic competition in free flow at the bottleneck.

To illustrate this dual role of lane changing for traffic breakdown, we consider results of simulations of the spontaneous emergence of a moving synchronized flow pattern (MSP) at an on-ramp bottleneck (Fig. 3.19 (a)).

The dual role of lane changing for traffic breakdown can clearly be seen in vehicle trajectories, if they are presented in the co-ordinate system that moves with a positive velocity v_{system}. In this moving co-ordinate system, the merging region of the on-ramp within which vehicles merge from the on-ramp onto the right lane of the main road is moving at a negative velocity $-v_{system}$ (dashed lines in Fig. 3.19 (b, c)).

The decay of a local disturbance at the bottleneck through lane changing to the faster lane, i.e., through the over-acceleration effect is shown in Fig. 3.19 (b). This effect of lane changing is labeled by "dissolving disturbance" in Figs. 3.19 (a, b). The initial disturbance occurs only in the right lane due to the merging of a vehi-

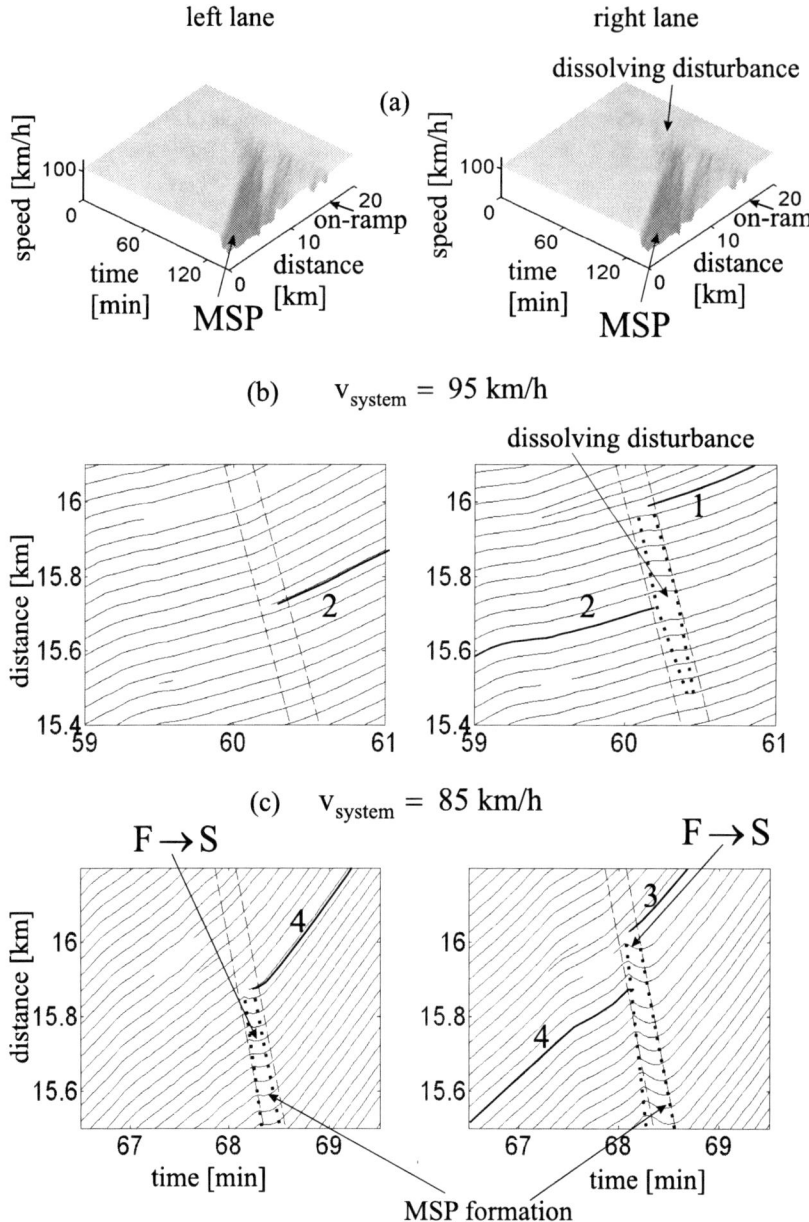

Fig. 3.19 Simulation of MSP emergence at on-ramp bottleneck: (a) Speed in time and space. (b, c) Vehicle trajectories associated with MSP emergence at on-ramp bottleneck shown in (a). Vehicle trajectories are shown in the co-ordinate systems moving at the velocity $v_{\rm system} = 95$ km/h (b) and 85 km/h (c); in these moving co-ordinate systems, dashed lines show the merging region of the on-ramp of the length 300 m. Left and right figures are related to the left and right road lanes, respectively. Taken from [28]

3.4 Dual Role of Lane Changing in Free Flow

cle from the on-ramp (fat trajectory of this merging vehicle and the region of the disturbance are labeled, respectively, by number 1 and dotted curves in Fig. 3.19 (b), right). The disturbance is localized within the on-ramp merging region (dashed lines in Fig. 3.19 (b)). Each following vehicle moving in the right lane of the main road and reaching the disturbance must slow down within the disturbance. However, later one of the vehicles in the right lane that approaches the disturbance upstream on the main road (fat trajectory of this vehicle is labeled by number 2 in Fig. 3.19 (b), right) can pass the slow moving preceding vehicle before the vehicle reaches this disturbance. This occurs through lane changing to the faster left lane of the main road (trajectory 2 in Fig. 3.19 (b), left). As a result of this vehicle passing, the initial disturbance in the right lane dissolves over time (dotted curves in Fig. 3.19 (b), right). In other words, this lane changing to the faster left lane is associated with the over-acceleration effect maintaining free flow.

The opposite effect of lane changing from the right lane to the left lane leading to an F→S transition in the left lane on the main road is shown in Fig. 3.19 (c), left. In this case, we use $v_{system} = 85$ km/h. This is because the vehicle speed about 85 km/h is close to the maximum synchronized flow speed in the stochastic microscopic traffic flow model used for simulations presented in Fig. 3.19. Thus traffic breakdown can be identified through the occurrence of vehicle trajectories with a negative slope associated with the vehicle speed $v < 85$ km/h, i.e., with synchronized flow.

A scenario of this traffic breakdown in the left lane due to lane changing is as follows. Firstly, as in the case shown in Fig. 3.19 (b), right, due to the merging of a vehicle from the on-ramp an initial disturbance occurs only in the right lane (fat trajectory of this merging vehicle and the region of the disturbance are labeled, respectively, by number 3 and dotted curves in Fig. 3.19 (c), right). However, in contrast with the first case the disturbance grows leading to an F→S transition in the right lane (labeled by arrow F→S in Fig. 3.19 (c), right). This traffic breakdown can be seen due to the occurrence of vehicle trajectories with a negative slope within the disturbance associated with the vehicle speed $v < 85$ km/h, i.e., with synchronized flow. At the time instant of this traffic breakdown in the right lane, free flow is still observed in the left lane (Fig. 3.19 (c), left).

Later, one of the vehicles in the right lane that reaches the disturbance upstream on the main road (trajectory of this vehicle is labeled by number 4 in Fig. 3.19 (c), right) can pass the slow moving preceding vehicle. This occurs through lane changing to the faster left lane of the main road (trajectory 4 in Fig. 3.19 (c), left). However, in contrast with the dissolution of the disturbance at the bottleneck as this has been observed in Fig. 3.19 (b), right, this vehicle passing leads to the deceleration of the following vehicles in the left lane that causes the emergence of synchronized flow, i.e., traffic breakdown in the left lane (labeled by arrow F→S in Fig. 3.19 (c), left). As a result, an MSP appears in both lanes propagating upstream of the bottleneck (labeled by "MSP" and "MSP formation" in Figs. 3.19 (a, c), respectively).

References

1. B.S. Kerner, *The Physics of Traffic*, (Springer, Berlin, New York, 2004)
2. B.S. Kerner, in *Proceedings of the 3rd Symposium on Highway Capacity and Level of Service*, ed. by R. Rysgaard. Vol 2 (Road Directorate, Ministry of Transport – Denmark, 1998), pp. 621–642
3. B.S. Kerner, Phys. Rev. Lett. **81**, 3797–3400 (1998)
4. B.S. Kerner, in *Traffic and Granular Flow' 97*, ed. by M. Schreckenberg, D.E. Wolf. (Springer, Singapore, 1998), pp. 239–267
5. B.S. Kerner, Trans. Res. Rec. **1678**, 160–167 (1999)
6. B.S. Kerner, in *Transportation and Traffic Theory*, ed. by A. Ceder. (Elsevier Science, Amsterdam 1999), pp. 147–171
7. B.S. Kerner, Physics World **12**, 25–30 (August 1999)
8. B.S. Kerner, J. Phys. A: Math. Gen. **33**, L221-L228 (2000); Networks and Spatial Economics, **1**, 35 (2001); in *Progress in Industrial Mathematics at ECMI 2000*, (Springer, Berlin, 2001), pp. 286–292; Mathematical and Computer Modelling, **35**, 481–508 (2002)
9. B.S. Kerner, in *Encyclopedia of Complexity and System Science*, ed. by R.A. Meyers. (Springer, Berlin, 2009), pp. 9302–9355
10. B.S. Kerner, Trans. Res. Rec. **1710**, 136–144 (2000)
11. B.S. Kerner, J. Phys. A: Math. Theor. **41**, 215101 (2008); 369801 (2008)
12. B.S. Kerner, S.L. Klenov, cond-mat/0502281, e-print in http://arxiv.org/abs/cond-mat/0502281 (2005)
13. B.S. Kerner, S.L. Klenov, Physica A **364**, 473–492 (2006)
14. B.S. Kerner, S.L. Klenov, Transp. Res. Rec. **1965**, 70–78 (2006)
15. B.S. Kerner, S.L. Klenov, in *Encyclopedia of Complexity and System Science*, ed. by R.A. Meyers. (Springer, Berlin, 2009), pp. 9282–9302
16. B.S. Kerner, S.L. Klenov, Phys. Rev. E **68**, 036130 (2003)
17. S. Chandrasekhar, *Hydrodynamic and Hydromagnetic Stability*, (Oxford University Press, Oxford, 1961)
18. C.W. Gardiner, *Handbook of Stochastic Methods*, (Springer, Berlin, 1994)
19. H. Haken, *Synergetics*, (Springer, Berlin 1977)
20. H. Haken, *Advanced Synergetics*, (Springer, Berlin 1983)
21. H. Haken, *Information and Self-Organization*, (Springer, Berlin 1988)
22. V.A. Vasil'ev, Yu.M. Romanovskii, D.S. Chernavskii, V.G. Yakhno, *Autowave Processes in Kinetic Systems*, (Springer, Berlin 1990)
23. A.S. Mikhailov, *Foundations of Synergetics Vol. I*, (Springer, Berlin 1994), 2nd ed.
24. A.S. Mikhailov, A.Yu. Loskutov, *Foundation of Synergetics II. Complex patterns*, (Springer, Berlin 1991)
25. B.S. Kerner, V.V. Osipov, *Autosolitons: A New Approach to Problems of Self-Organization and Turbulence*, (Kluwer, Dordrecht, Boston, London 1994); Sov. Phys. Usp. **32**, 101–138 (1989); Sov. Phys. Usp. **33**, 679–719 (1990)
26. N. N. Rosanov, *Spatial Hysteresis and Optical Patterns*, (Springer, Berlin, 2002)
27. B.S. Kerner, S.L. Klenov, D.E. Wolf, J. Phys. A: Math. Gen. **35**, 9971–10013 (2002)
28. B.S. Kerner, S.L. Klenov, unpublished

Chapter 4
Infinite Number of Highway Capacities of Free Flow at Bottleneck

Highway capacity of free flow at a bottleneck (called also as bottleneck capacity) is limited by traffic breakdown at the bottleneck [1–19].

However, there are the infinite number of the flow rates in free flow downstream of the bottleneck $q_{\rm sum}$ (3.25) at which traffic breakdown can occur at the bottleneck. In three-phase traffic theory, the following main statement has been made about the highway capacity [20]:

- All flow rates $q_{\rm sum}$ in free flow downstream of the bottleneck, which satisfy conditions (3.25), i.e., the *infinite number* of these flow rates are highway capacities of free flow at the bottleneck that we denote by $q_{\rm C}^{(\rm B)}$.
- In accordance with (3.25), the infinite number of highway capacities $q_{\rm C}^{(\rm B)}$ of free flow at the bottleneck satisfy conditions

$$q_{\rm th}^{(\rm B)} \leq q_{\rm C}^{(\rm B)} \leq q_{\rm max}^{(\rm free\ B)}. \tag{4.1}$$

To explain this statement, note that if the flow rate in free flow downstream of the bottleneck $q_{\rm sum}$ is smaller than the threshold flow rate for traffic breakdown $q_{\rm th}^{(\rm B)}$ (3.26), then no traffic breakdown is possible at the bottleneck. This means that all flow rates in free flow downstream of the bottleneck that satisfy condition (3.26) are smaller than highway capacity of free flow at the bottleneck. Traffic breakdown at the bottleneck can occur at the infinite number of the flow rates in free flow downstream of the bottleneck $q_{\rm sum}$ that satisfy conditions (3.25). For this reason, there are the infinite number of highway capacities of free flow at the bottleneck given by conditions (4.1).

The conclusion about the infinite number of highway capacities of free flow at the bottleneck made in three-phase traffic theory [20] contradicts assumptions about a particular (fixed or stochastic) highway capacity made in other traffic flow theories [1–19]. The understanding of the infinite number of highway capacities of three-phase traffic theory contradicts also a basic assumption of most existing traffic operation methods like dynamic traffic assignment models, on-ramp metering, speed limit control, which assume the existence of a particular highway capacity and use the highway capacity as a control parameter (see Sect. 10.6).

As explained in Chap. 3, the key empirical evidence for the nucleation nature of traffic breakdown leading to conditions (3.25) and, therefore, to the conclusion about the infinite number of highway capacities of free flow at the bottleneck (4.1) is the possibility of both spontaneous and induced traffic breakdowns at the same bottleneck observed in real measured traffic data.

4.1 Definition of Highway Capacity of Free Flow at Bottleneck

For a given averaging time interval for traffic variables T_{av}, traffic breakdown can occur with some probability denoted by $P_{\mathrm{FS}}^{(\mathrm{B})}$. For this reason, an attribute of highway capacity at the bottleneck is probability

$$P_{\mathrm{C}}^{(\mathrm{B})} = 1 - P_{\mathrm{FS}}^{(\mathrm{B})} \qquad (4.2)$$

that free flow remains at the bottleneck during the time interval T_{av}. Highway capacity is reached when

$$P_{\mathrm{C}}^{(\mathrm{B})} < 1. \qquad (4.3)$$

Because highway capacity of free flow at the bottleneck is limited by traffic breakdown at the bottleneck, the fundamental empirical features of traffic breakdown of Sect. 3.1.3 are also the fundamental empirical features of highway capacity. The definition of highway capacity of free flow at the bottleneck made in three-phase traffic theory, which satisfies these fundamental empirical features of highway capacities, reads as follows [20]:

- Highway capacity of free flow at a bottleneck is equal to the flow rate downstream of the bottleneck at which free flow remains at the bottleneck with the probability $P_{\mathrm{C}}^{(\mathrm{B})} < 1$ (4.3) during a given averaging time interval for traffic variables T_{av}.

The infinite number of the flow rates q_{sum} (3.25) satisfy condition (4.3). For this reason, the capacity definition is consistent with the infinite number of highway capacities (4.1). Each of the highway capacities (4.1) has two attributes:

(1) The probability $P_{\mathrm{C}}^{(\mathrm{B})}$ (4.3) that free flow remains at the bottleneck during a given averaging time interval for traffic variables T_{av}.
(2) The time interval T_{av}.

To understand the sense of the capacity attributes, we should consider dependences of breakdown probability $P_{\mathrm{FS}}^{(\mathrm{B})}$ on the flow rate q_{sum} and time interval T_{av}.

4.2 Characteristics of Highway Capacities

4.2.1 Minimum and Maximum Highway Capacities

Under condition

$$q_{\text{sum}} < q_{\text{th}}^{(B)}, \qquad (4.4)$$

where the threshold flow rate $q_{\text{th}}^{(B)}$ can be a function of T_{av}, no traffic breakdown is possible at the bottleneck and, therefore, breakdown probability[1] is

$$P_{\text{FS}}^{(B)}(q_{\text{sum}}) \big|_{q_{\text{sum}} < q_{\text{th}}^{(B)}} = 0. \qquad (4.5)$$

This conclusion is also true if a moving congested traffic pattern, which has initially occurred at a different road location than the bottleneck location, reaches the bottleneck: after the pattern has passed the bottleneck, free flow returns at the bottleneck, if condition (4.4) is satisfied. In other words, neither spontaneous nor induced traffic breakdown is possible under condition (4.4).

In accordance with (4.2), condition (4.5) is equivalent to

$$P_{\text{C}}^{(B)} = 1 \qquad (4.6)$$

Thus the flow rate in free flow downstream of the bottleneck q_{sum} that is smaller than any of the capacities satisfies condition (4.5):

- The threshold flow rate $q_{\text{th}}^{(B)}$ is the *minimum highway capacity* (Fig. 4.1).

In contrast, under condition

$$q_{\text{sum}} \geq q_{\text{max}}^{(\text{free B})} \qquad (4.7)$$

already small local disturbances in free flow at the bottleneck lead to traffic breakdown. In other words, at $q_{\text{sum}} = q_{\text{max}}^{(\text{free B})}$ associated with at a given T_{av} traffic breakdown occurs at the bottleneck with probability[2]

---

[1] Probability of traffic breakdown $P_{\text{FS}}^{(B)}$ at a bottleneck is defined as follows. A large number N_{FS} of different realizations (days) is studied in which an initial free flow at the bottleneck is observed. Each of these realizations refers to the same flow rate q_{sum}, the same value of T_{av}, and the same traffic parameters (weather, percentage of long vehicles, etc.). Let us assume that in n_{FS} of these N_{FS} realizations spontaneous traffic breakdown has been observed at the bottleneck. Then the breakdown probability associated with the flow rate q_{sum} and time interval T_{av} is equal to

$$P_{\text{FS}}^{(B)} = \frac{n_{\text{FS}}}{N_{\text{FS}}}.$$

[2] We should note that the maximum flow rate $q_{\text{max}}^{(\text{free B})}$ in free flow is associated with an empirical limit flow rate $q_{\text{max}}^{(\text{free. emp})}$ in free flow that determines the limit point of free flow discussed in Sect. 2.2.1. In accordance with (4.8), probability of traffic breakdown at this flow rate is equal to

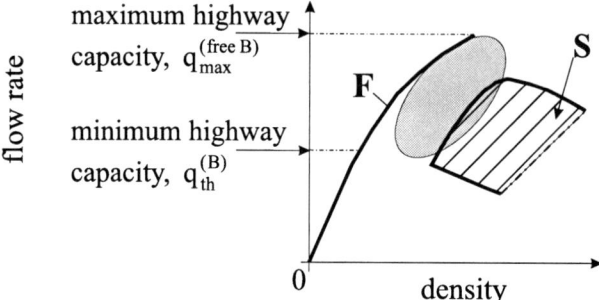

Fig. 4.1 Qualitative illustration of minimum and maximum highway capacities in the flow–density plane. In a gray region, a competition occurs between a tendency towards free flow due to the over-acceleration effect and a tendency towards synchronized flow due to the speed adaptation effect. F – free flow, S – a part of synchronized flow states taken from Fig. 3.3 (a)

$$P_{\text{FS}}^{(\text{B})}(q_{\text{sum}}) \big|_{q_{\text{sum}} = q_{\text{max}}^{(\text{free B})}} = 1. \tag{4.8}$$

In accordance with (4.2), condition (4.8) is equivalent to the condition

$$P_{\text{C}}^{(\text{B})} = 0. \tag{4.9}$$

This means that

- in accordance with (4.8), the maximum flow rate $q_{\text{max}}^{(\text{free B})}$ is the *maximum highway capacity* (Fig. 4.1).

4.2.2 Dependence of Breakdown Probability on Flow Rate

The maximum flow rate in free flow at a bottleneck, i.e., the maximum highway capacity $q_{\text{max}}^{(\text{free B})}$ is related to the flow for which breakdown probability reaches 1 (4.8). The threshold flow rate, i.e., the minimum highway capacity $q_{\text{th}}^{(\text{B})}$ characterizes the flow rate in free flow downstream of the bottleneck at which traffic breakdown is still possible: under condition (4.4) breakdown probability is 0 (4.5). Thus if the flow rate increases from $q_{\text{sum}} = q_{\text{th}}^{(\text{B})}$ to $q_{\text{sum}} = q_{\text{max}}^{(\text{free B})}$, breakdown probability must increase, i.e., the probability is an increasing function of the flow rate q_{sum} as found in empirical data (Fig. 2.6).

To explain this behavior of breakdown probability, let us consider the flow rate range (3.28) within which free flow is in a metastable state with respect to traffic breakdown. In the metastable free flow, the probability of the nucleus occurrence is equal to breakdown probability: if the nucleus occurs, traffic breakdown must occur

---

one. This explains why the exact values of the empirical limit flow rate and minimum vehicle speed that is possible in free flow are very difficult to find in real measured traffic data.

4.2 Characteristics of Highway Capacities

in this metastable free flow. The greater the flow rate q_{sum}, the higher the critical speed, respectively, the smaller the critical density within a local disturbance in free flow required for traffic breakdown (Sect. 3.3.2).

Thus the greater the flow rate q_{sum}, the smaller the speed decrease within a local disturbance at the bottleneck should be to initiate traffic breakdown in metastable free flow. Obviously, probability of the random occurrence of a small local disturbance in free flow is considerably greater than probability for the random occurrence of a great local disturbance. This explains why probability of traffic breakdown is an increasing function of the flow rate q_{sum}.

Theoretical probability for traffic breakdown at an on-ramp bottleneck found in [21] (Fig. 4.2) shows qualitatively the same features as empirical breakdown probability firstly found by Persaud *et al.* (Fig. 2.6) [6]; theoretical probability $P_{FS}^{(B)}$ as a function of the flow rate q_{sum} at a given flow rate to the on-ramp q_{on} can be approximated by the formula [21]

$$P_{FS}^{(B)} = \frac{1}{1 + \exp[\alpha(q_P - q_{sum})]}, \qquad (4.10)$$

where α and q_P are functions of the on-ramp inflow rate q_{on} and a time interval within which traffic breakdown is studied.

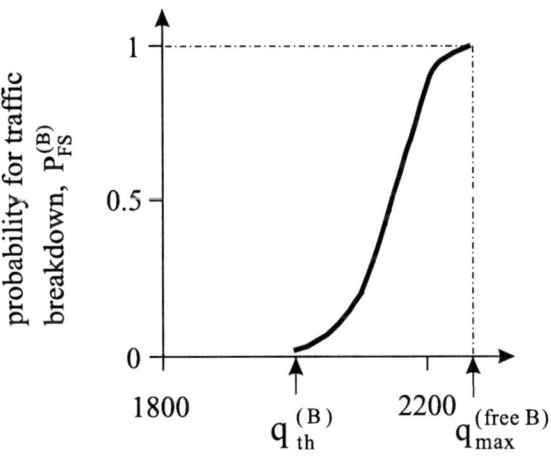

Fig. 4.2 Probability $P_{FS}^{(B)}(q_{sum})$ that an F→S transition occurs at on-ramp bottleneck within time interval 15 min after the on-ramp inflow was switched on as a function of the flow rate in free flow downstream of the bottleneck. Taken from [21]

4.2.3 Dependence of Minimum and Maximum Capacities on Averaging Time Interval for Traffic Variables

At a given flow rate $q_{\rm sum}$, breakdown probability depends on the averaging time interval for traffic variables $T_{\rm av}$. This explains why $T_{\rm av}$ is an important attribute of highway capacity.

The longer the averaging time interval $T_{\rm av}$ for the flow rate $q_{\rm sum}$, the more probably the occurrence of a nucleus in free flow at a bottleneck during the time interval $T_{\rm av}$. However, under condition (4.1) the greater the probability for the nucleus occurrence, the greater the breakdown probability. Thus the longer $T_{\rm av}$, the greater the breakdown probability at the same flow rate $q_{\rm sum}$ satisfying (3.25). This conclusion has been confirmed in empirical and numerical studies of traffic breakdown probability [6, 21]. This is also valid for each of the flow rates within the flow rate range (3.25), i.e., the maximum and minimum capacities $q_{\rm max}^{\rm (free\ B)}$ and $q_{\rm th}^{\rm (B)}$ are decreasing functions of the time interval $T_{\rm av}$:

- Within a limited range[3] of $T_{\rm av}$, the longer the time interval $T_{\rm av}$, the smaller the maximum and minimum capacities $q_{\rm max}^{\rm (free\ B)}$ and $q_{\rm th}^{\rm (B)}$.

The maximum and minimum capacities can depend considerably on traffic parameters (weather, percentage of long vehicles, etc.)[4]. In empirical observations, the maximum and minimum capacities are found from a study of a flow rate dependence of breakdown probability $P_{\rm FS}^{\rm (B)}(q_{\rm sum})$ (Fig. 2.6). In other words, $q_{\rm max}^{\rm (free\ B)}$ and $q_{\rm th}^{\rm (B)}$ are mean values found in many different realizations (days) at which traffic breakdowns have occurred.

Summarizing results of this chapter, we can conclude that the infinite number of highway capacities (4.1) and the capacity definition made in three-phase traffic theory can be explained as follows. At each flow rate $q_{\rm sum}$ downstream of a bottleneck, which satisfies conditions (3.25), traffic breakdown at the bottleneck is possible: traffic breakdown occurs either spontaneously with a probability $P_{\rm FS}^{\rm (B)} > 0$ or traffic breakdown can be induced. Thus each of these flow rates is highway capacity. Breakdown probability $P_{\rm FS}^{\rm (B)}$ depends strongly on $q_{\rm sum}$. Thus probability $P_{\rm C}^{\rm (B)}$ (4.2) that traffic breakdown does not occur spontaneously, i.e., that free flow remains at the bottleneck is an attribute of highway capacity. This capacity attribute distinguishes different highway capacities. The probability function $P_{\rm C}^{\rm (B)}(q_{\rm sum})$ depends on $T_{\rm av}$. Thus $T_{\rm av}$ is another attribute of highway capacity. Traffic breakdown is not

---

[3] On the one hand, $T_{\rm av}$ should be appreciably longer than a time duration of traffic breakdown, i.e., the transition time from free flow to synchronized flow after a nucleus for traffic breakdown has occurred. In empirical observation, this transition time is about 1 min. On the other hand, $T_{\rm av}$ should be considerably shorter than the duration of the period of high travel demand in a traffic network when traffic breakdown is possible. The latter period is usually about 30 – 180 min.

[4] Note that for an on-ramp bottleneck, $q_{\rm max}^{\rm (free\ B)}$ and $q_{\rm th}^{\rm (B)}$ depend on the flow rate to the on-ramp $q_{\rm on}$ and the flow rate in free flow upstream of the bottleneck $q_{\rm in}$; for this reason, there are also infinite maximum and infinite minimum capacities associated with different values $q_{\rm max}^{\rm (free\ B)}(q_{\rm on}, q_{\rm in})$ and $q_{\rm th}^{\rm (B)}(q_{\rm on}, q_{\rm in})$, respectively. For more detail, see the book [20].

possible at the bottleneck only, when the flow rate $q_{\rm sum}$ satisfies condition (4.4). Thus only the flow rates that satisfy condition (4.4) are smaller than any highway capacity.

References

1. A.D. May, *Traffic Flow Fundamentals*, (Prentice-Hall, Inc., New Jersey, 1990)
2. *Highway Capacity Manual 2000*, (National Research Council, Transportation Research Board, Washington, D.C., 2000)
3. N.H. Gartner, C.J. Messer, A. Rathi (editors), *Traffic Flow Theory: A State-of-the-Art Report*, (Transportation Research Board, Washington DC, 2001)
4. F.L. Hall, K. Agyemang-Duah, Trans. Res. Rec. **1320**, 91–98 (1991)
5. L. Elefteriadou, R.P. Roess, W.R. McShane, Transp. Res. Rec. **1484**, 80–89 (1995)
6. B.N. Persaud, S. Yagar, R. Brownlee, Trans. Res. Rec. **1634**, 64–69 (1998)
7. W. Brilon, J. Geistefeld, M. Regler, in *Transportation and Traffic Theory* ed. by H.S. Mahmassani. Proceedings of the 16th Inter. Sym. on Transportation and Traffic Theory, (Elsevier, Amsterdam, 2005), pp. 125–144
8. I. Prigogine, R. Herman, *Kinetic Theory of Vehicular Traffic*, (American Elsevier, New York, 1971)
9. W. Leutzbach, *Introduction to the Theory of Traffic Flow*, (Springer, Berlin, 1988)
10. R. Wiedemann, *Simulation des Verkehrsflusses*, (University of Karlsruhe, Karlsruhe, 1974)
11. G.F. Newell, *Applications of Queuing Theory*, (Chapman Hall, London, 1982)
12. C.F. Daganzo, *Fundamentals of Transportation and Traffic Operations*, (Elsevier Science Inc., New York, 1997)
13. M. Papageorgiou, *Application of Automatic Control Concepts in Traffic Flow Modeling and Control*, (Springer, Berlin, New York, 1983)
14. D.C. Gazis, *Traffic Theory*, (Springer, Berlin, 2002)
15. D. Chowdhury, L. Santen, A. Schadschneider, Physics Reports **329**, 199 (2000)
16. D. Helbing, Rev. Mod. Phys. **73**, 1067–1141 (2001)
17. R. Mahnke, J. Kaupužs, I. Lubashevsky, Phys. Rep. **408**, 1–130 (2005)
18. T. Nagatani, Rep. Prog. Phys. **65**, 1331–1386 (2002)
19. K. Nagel, P. Wagner, R. Woesler, Oper. Res. **51**, 681–716 (2003)
20. B.S. Kerner, *The Physics of Traffic*, (Springer, Berlin, New York, 2004); Physica A **333**, 379–440 (2004)
21. B.S. Kerner, S.L. Klenov, D.E. Wolf, J. Phys. A: Math. Gen. **35**, 9971–10013 (2002)

Chapter 5
Nature of Moving Jam Emergence

5.1 Pinch Effect in Synchronized Flow

Wide moving jams do not emerge spontaneously in free flow: no spontaneous phase transition from the free flow phase to the wide moving jam phase (F→J transition for short) has been observed in real measured traffic data (Sect. 2.4.5). Wide moving jams can emerge *spontaneously only* in the synchronized flow phase (S→J transition) [1–4].

In synchronized flow at higher speeds, wide moving jams should not necessarily emerge spontaneously. Observations show that the greater the density in synchronized flow, the more likely is spontaneous moving jam emergence in that synchronized flow. Thus a wide moving jam emerges in an initial free flow due to a sequence of two phase transitions: Firstly, an F→S transition occurs and synchronized flow emerges. Later and usually at another road location than the location of the F→S transition, an S→J transition occurs spontaneously leading to wide moving jam emergence. This sequence of phase transitions is called the F→S→J transitions [1–4].

In empirical observations, S→J transition development is associated with a *pinch effect* in synchronized flow occurring within the associated pinch region of synchronized flow [1–4]. The pinch effect and pinch region of synchronized flow are defined as follows.

- The pinch effect in synchronized flow is the spontaneous emergence of growing narrow moving jams in the synchronized flow.
- The pinch region of synchronized flow is a region of synchronized flow within which the pinch effect occurs.

To explain the term *the pinch effect in synchronized flow*, note that in most of the empirical observations growing narrow moving jams emerge spontaneously in synchronized flow with a *self-compression* of this synchronized flow. This self-compression of synchronized flow means that the average density increases and average speed decreases considerably in the synchronized flow. Therefore, in the related pinch region of synchronized flow, the density is great and the speed is low, however, the average flow rate can be great.

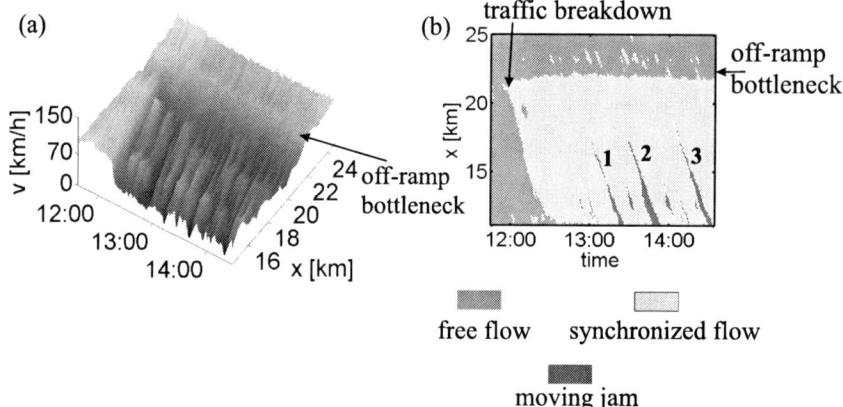

Fig. 5.1 Empirical example of GP emergence at an off-ramp bottleneck. (a) Vehicle speed in space and time. (b) A graph of (a) with the free flow phase, the synchronized flow phase, and wide moving jam phase (wide moving jams labeled 1, 2 and 3 in (b)). Taken from [3]

In an empirical example shown in Fig. 5.1, firstly traffic breakdown occurs at an off-ramp bottleneck leading to the emergence of synchronized flow upstream of the bottleneck. The downstream front of synchronized flow is fixed at the bottleneck, while the upstream front of synchronized flow propagates upstream. Later, the density in synchronized flow increases. In the associated pinch region of synchronized flow, narrow moving jams emerge spontaneously. Some of these narrow moving jams grow over time propagating upstream. Finally, growing narrow moving jams transform into wide moving jams (labeled by 1, 2 and 3 in Fig. 5.1), i.e., S→J transitions occur spontaneously. Locations of these S→J transitions, which can vary considerably for different wide moving jams, are the locations of the upstream boundary of the pinch region. This means that the pinch region width (in the longitudinal direction) can be a complicated time-function. Due to the F→S→J transitions a general pattern (GP) occurs at the off-ramp bottleneck that consists of synchronized flow upstream of the bottleneck and wide moving jams that emerge spontaneously in that synchronized flow.

5.2 Nucleation Features of Wide Moving Jam Emergence in Synchronized Flow

5.2.1 Explanations of S→J Transitions through Three-Phase Traffic Theory

Empirical S→J transitions are explained in three-phase traffic theory by the following hypotheses [1, 2, 5–8]:

(a) All states of traffic flow (infinite number of states) in the flow–density plane that lie on the line J are threshold states for wide moving jam existence and emergence (Fig. 5.2)[1].

(b) The line J intersects the 2D-region of steady states of synchronized flow in the flow–density plane, i.e., there are steady states of synchronized flow below and above the line J (Fig. 5.2).

(c) The line J separates all steady states of synchronized flow in the flow–density plane into two qualitatively different classes (Fig. 5.2):

- All steady states of synchronized flow below the line J are stable with respect to wide moving jam emergence, i.e., no S→J transitions are possible within these states of *homogeneous (in space and time) synchronized flow* (homogeneous synchronized flow for short). In particular, among these states there are homogeneous synchronized flow of a very great density (Fig. 5.2)[2].

- In all synchronized flow states on and above the line J a wide moving jam can emerge and exist.

(d) The steady states of synchronized flow in the flow–density plane that lie on and above the line J are metastable states with respect to wide moving jam emergence:

- A metastable state of synchronized flow (metastable synchronized flow for short) with respect to wide moving jam emergence is the state of synchronized flow, which is stable with respect to small enough local disturbances in this flow: the small disturbances do not grow in this flow. However, if a great enough local disturbance in this flow state appears, the disturbance grows leading to the formation of a wide moving jam.

(e) A feature of metastable states of synchronized flow is as follows. At a given synchronized flow speed, the greater the density in synchronized flow, the greater the probability of an S→J transition.

---

[1] The definition of the line J has been discussed in Sect. 2.5.2.

[2] However, such great density synchronized flow that is homogeneous in space and time has not been observed up to now in real measured traffic data. In contrast, congested traffic of a great density that occurs in the data at heavy bottlenecks is very non-homogeneous in space and time (see Sect. 7.2.5). For a more detailed discussion of homogeneous synchronized flow of a great density, see Sect. 6.3.3 of the book [3].

(f) The maximum flow rate in the jam outflow q_{out} is smaller than the maximum flow rate in free flow $q_{\text{max}}^{(\text{free B})}$ [9, 10]:

$$q_{\text{out}} < q_{\text{max}}^{(\text{free B})}. \qquad (5.1)$$

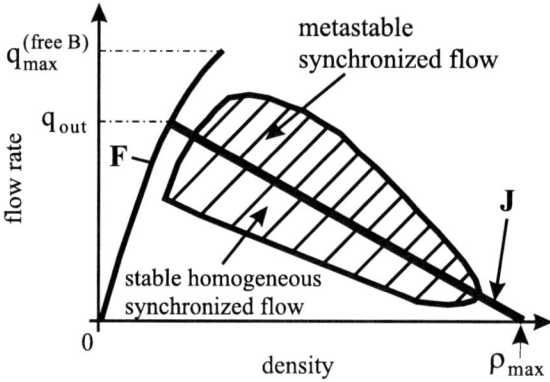

Fig. 5.2 Three-phase traffic theory in the flow–density plane [1,2,5–8]: Free flow (F), steady states of synchronized flow (2D dashed region) taken from Fig. 3.3 (a), and the line J

To explain the hypothesis (c), let us assume that a state of synchronized flow directly upstream of a wide moving jam (Fig. 5.3 (a)) is associated with a point k in the flow–density plane. This point is below the line J (Fig. 5.3 (b)). Because the velocity of the upstream front of the wide moving jam $v_g^{(\text{up})}$ equals the slope of the line K (from a point k in synchronized flow to the point $(\rho_{\text{max}}, 0)$), the absolute value $|v_g^{(\text{up})}|$ is always less than that of the downstream front $|v_g|$ determined by the slope of the line J, i.e., the formula

$$|v_g^{(\text{up})}| < |v_g| \qquad (5.2)$$

is valid. Therefore, the width of the wide moving jam gradually decreases and the jam dissolves. This means that no wide moving jams can persist continuously, i.e., all steady states of synchronized flow below the line J are stable with respect to wide moving jam emergence.

In contrast, assume that a state of synchronized flow upstream of another wide moving jam (Fig. 5.3 (c)) is associated with a point n in the flow–density plane. This state is above the line J (Fig. 5.3 (d)). In this case, the velocity of the upstream front of the wide moving jam $v_g^{(\text{up})}$ equals the slope of the line N (from a point n in synchronized flow to the point $(\rho_{\text{max}}, 0)$), i.e., the absolute value $|v_g^{(\text{up})}|$ is always greater than that of the downstream front $|v_g|$, i.e., the formula

5.2 Nucleation Features of Wide Moving Jam Emergence in Synchronized Flow

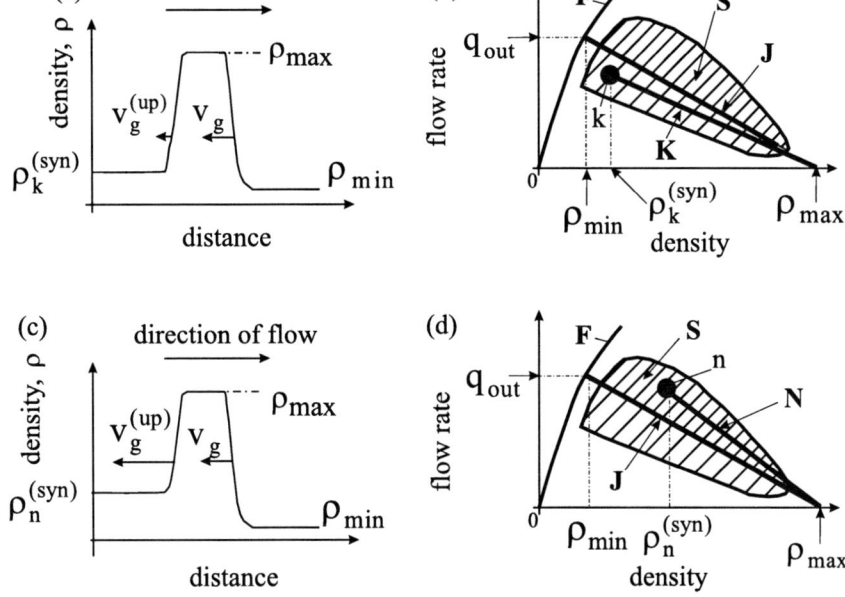

Fig. 5.3 Explanations of moving jam emergence in synchronized flow [1, 2, 5–8]: (a, c) Qualitative distributions of the density within wide moving jams at a fixed time instant for two different densities in synchronized flow upstream of the jams, $\rho^{(\mathrm{syn})} = \rho_k^{(\mathrm{syn})}$ (a) and $\rho^{(\mathrm{syn})} = \rho_n^{(\mathrm{syn})}$ (c). (b, d) Representation of the line J and upstream wide moving jam fronts (lines K (b) and N (d)) in the flow–density plane for the wide moving jams in (a) and (c), respectively. In (b, d) states for free flow (curve F), steady states of synchronized flow (dashed region) are taken from Fig. 5.2

$$|v_g^{(\mathrm{up})}| > |v_g| \qquad (5.3)$$

is valid. Therefore, the width of the wide moving jam in Fig. 5.3 (c) should gradually increase. For these reasons, wide moving jams can be formed and persist in states of synchronized flow that lie on or above the line J. Thus there should be some time-limited disturbances in the states of synchronized flow that can lead to the emergence of wide moving jams, i.e., the states are metastable ones with respect to wide moving jam emergence. This explains the hypothesis (d).

In synchronized flow states that lie on the line J, the velocities of the downstream and upstream fronts of a wide moving jam are equal to each other; this explains the hypothesis (a) that these states are threshold ones for S→J transitions (see for more detail Sect. 6.3.1 of [3]).

To explain the hypothesis (b), note that if a steady state of synchronized flow with a given speed is formed in the outflow of a wide moving jam, then this state is on the line J in the flow–density plane. Therefore, the space gap between vehicles in this synchronized flow state is smaller than the synchronization gap G and it is greater than the safe space gap g_{safe}. Taking into account formula (3.9) for the synchronization time headway τ_G and safe time headway τ_{safe} associated with, respectively, the synchronization space gap G and safe space gap g_{safe} as well as the equation for the line J (2.17), we get

$$\tau_G > \tau^{(a)}_{del,\,jam} > \tau_{safe}, \tag{5.4}$$

where $\tau^{(a)}_{del,\,jam}$ is the time delay in vehicle acceleration at the downstream front of a wide moving jam (Sect. 2.5.1). Condition (5.4) results from (3.9), i.e., from the fundamental hypothesis of three-phase traffic theory and the hypothesis (b).

5.2.2 Competition Between Over-Deceleration and Speed Adaptation Effects

Three-phase traffic theory [3] explains hypotheses (d) and (e) of Sect. 5.2.1 about metastable synchronized flow states with respect to S→J transitions and their features by a competition between two following effects:

(i) the speed adaptation effect (Sect. 3.2.2), which describes a tendency towards synchronized flow and
(ii) an over-deceleration effect, which describes a tendency towards a wide moving jam.

This competition occurs within a local disturbance in an initial synchronized flow; we consider only such a local disturbance within which the vehicle speed is lower and density is greater than, respectively, the speed and density are outside of the disturbance in synchronized flow.

The over-deceleration effect is as follows[3]: If a vehicle begins to decelerate unexpectedly, then due to a driver reaction time the following vehicle starts deceleration with a time delay; as a result, when the time delay is long enough to avoid collisions the driver decelerates longer that leads to a lower speed than the speed of the preceding vehicle.

To explain the competition between the speed adaptation and over-deceleration effects, we assume that a vehicle moving in metastable synchronized flow (a state of synchronized flow that is above the line J in Fig. 5.3 (d)) decelerates. If the following vehicle cannot pass this preceding vehicle, there are two possibilities:

---

[3] Over-deceleration was introduced by Herman *et al.* [13, 14] as well as Kometani and Sasaki [15, 16]. However, in earlier traffic flow models and theories the over-deceleration effect causes an instability of free flow beginning at a critical density [14] (see references in [17–22]). This free flow instability should explain traffic breakdown. In contrast, as stressed in Sect. 3.2.2, rather than this model instability of free flow associated with the over-deceleration effect, in three-phase traffic theory the nucleation nature of traffic breakdown associated with a competition between the speed adaptation and over-acceleration effects explains the fundamental empirical features of traffic breakdown. A more detailed critical discussion of earlier traffic flow theories and models with this free flow instability due to the over-deceleration effect appears in Sect. 10.3.

1) The following vehicle decelerates and due to the speed adaptation effect is able to adapt the speed to the speed of the preceding vehicle. Then rather than an S→J transition, a new synchronized flow state is formed.

2) Due to over-deceleration, the following vehicle decelerates stronger than it would be needed for speed adaptation, i.e., its speed becomes lower than the speed of the preceding vehicle. If each of the following vehicles decelerates also to lower speed than the associated preceding vehicle, then finally the speed upstream decreases to zero leading to an S→J transition, i.e., to the emergence of a wide moving jam in the synchronized flow.

5.2.3 Nucleus for Wide Moving Jam Emergence in Metastable Synchronized Flow

To understand the competition between the over-deceleration and speed adaptation effects, firstly we assume that within a local disturbance that occurs in metastable synchronized flow the decrease in speed is *small enough*. Then vehicles approaching the disturbance can decelerate within the synchronization space gap. This occurs even after a time delay in vehicle deceleration: After the time delay, each of the following vehicles has enough time to adapt the speed to a lower speed within the disturbance. In this case, all following vehicles decelerate within the synchronization space gap to the lower speed within the disturbance, i.e., conditions (3.12) for the speed adaptation effect remain satisfied. This means that due to this speed adaptation effect rather than an S→J transition, a new synchronized flow state is formed. Thus we see that within small enough local disturbances in synchronized flow the tendency towards synchronized flow due to speed adaptation is on average stronger than the tendency towards a wide moving jam due to over-deceleration.

In contrast, if the decrease in speed within a local disturbance occurring in metastable synchronized flow is *great enough*, then due to the time delay in vehicle deceleration each of the following vehicles approaching the disturbance decelerates to a lower speed than the speed of the associated preceding vehicle within the disturbance. As a result, the speed within the disturbance decreases over time up to zero. Thus within great enough local disturbances in synchronized flow the tendency towards a wide moving jam due to over-deceleration should be on average stronger than the tendency towards synchronized flow due to speed adaptation.

Thus there must be some critical density and critical speed within the disturbance required for an S→J transition in metastable synchronized flow defined as follows:

- The critical speed and critical density required for an S→J transition in an initial metastable synchronized flow are, respectively, the vehicle speed and density within a local disturbance occurring in this flow at which the tendency towards a wide moving jam due to the over-deceleration effect is on average as strong as the tendency towards synchronized flow due to the speed adaptation effect.

Thus it is equivalent to consider either the critical speed or critical density required for the S→J transition, i.e. for wide moving jam emergence in synchronized flow. We denote the critical speed and critical density required for an S→J transition in metastable synchronized flow by $v_{\text{cr}}^{(SJ)}$ and $\rho_{\text{cr}}^{(SJ)}$, respectively. We can conclude that

- if within a local disturbance in metastable synchronized flow the speed is higher than the critical speed (density is smaller than the critical density), the over-deceleration effect is on average weaker than the speed adaptation effect; as a result, the disturbance decays over time and no wide moving jam emerges.
- In contrast, when a local disturbance occurs in metastable synchronized flow within which the speed is lower than the critical speed (density is greater than the critical density), the over-deceleration effect becomes on average stronger than the speed adaptation effect; as a result, the disturbance grows leading to the emergence of a wide moving jam in the synchronized flow. This local disturbance can be considered a *nucleus* required for an S→J transition, i.e., for wide moving jam emergence in synchronized flow.

The term *nucleus required for an S→J transition* is defined as follows[4].

- A nucleus required for an S→J transition is a local disturbance in a metastable state of synchronized flow within which the vehicle speed is equal to or lower than the critical speed (respectively, the density is equal to or greater than the critical density) required for the S→J transition, i.e.,

$$v \leq v_{\text{cr}}^{(SJ)} \quad (\rho \geq \rho_{\text{cr}}^{(SJ)}). \tag{5.5}$$

In other words, an S→J transition exhibits a *nucleation nature* [1,2,5–8].

In empirical observations, a nucleus that leads to an S→J transition in synchronized flow has a form of a growing narrow moving jam (Sects. 2.6.3 and 5.1):

- A growing narrow moving jam in synchronized flow is a synonym of a nucleus required for an S→J transition.

To explain the hypothesis (e) of Sect. 5.2.1, we consider two different metastable steady states of synchronized flow associated with a given vehicle speed denoted by v_{syn} that is the same in both states labeled by black circles 1 and 2 in Fig. 5.4 (a). Both synchronized flow states 1 and 2 lie above the line J, i.e., as explained above they are metastable states with respect to an S→J transition. The states 1 and 2 satisfy conditions (3.12) associated with car-following within the synchronization gap. However, the space gap in the state 1 is greater than the gap in the state 2, respectively, the density in the state 1 is smaller than the density in the state 2. To cause the emergence of a wide moving jam in the synchronized flow state 1 with a greater space gap, there should be a greater decrease in speed within a local disturbance in comparison with the one in the state 2. This is because the greater the

---

[4] In real metastable synchronized flow, a nucleus required for an S→J transition exhibits two attributes: (i) a critical speed (critical density) and (ii) spatial distributions of traffic variables within the nucleus. However, in a qualitative discussion of the nature of moving jam emergence made here we ignore spatial distributions of traffic variables within the nucleus.

5.2 Nucleation Features of Wide Moving Jam Emergence in Synchronized Flow

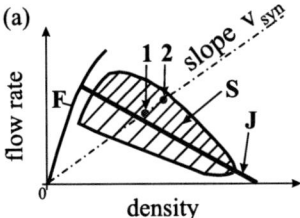 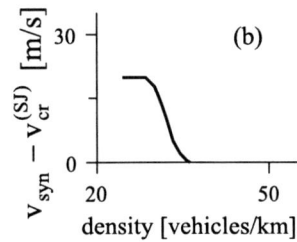

Fig. 5.4 Explanations of wide moving jam emergence in metastable synchronized flow [1,2,5–8]: (a) Free flow (F), steady states of synchronized flow (2D dashed region) and the line J taken from Fig. 5.2. (b) Numerical simulations [23] of difference $v_{\rm syn} - v_{\rm cr}^{(SJ)}$ as a density function in metastable steady states of synchronized flow at a given synchronized flow speed.

space gap, the longer the time that the vehicle has for speed adaptation within the disturbance. Therefore, the critical speed $v_{\rm cr}^{(SJ)}$ within a local disturbance required for an S→J transition in the state 1 should be lower than the one in the state 2:

- At a given synchronized flow speed, the greater the density in a metastable synchronized flow state, the higher the critical speed, respectively, the smaller the critical density required for an S→J transition.

These qualitative explanations of the hypotheses (d) and (e) (Fig. 5.4 (a)) are confirmed by numerical simulations of a three-phase traffic flow model shown in Fig. 5.4 (b) [23]. We can see that for different metastable synchronized flow states associated with a given speed $v_{\rm syn}$ that is the same for all these states, the greater the density of metastable synchronized flow, the higher the critical speed $v_{\rm cr}^{(SJ)}$ and, therefore, the smaller the critical density required for an S→J transition.

5.2.4 Spontaneous and Induced Wide Moving Jam Emergence in Synchronized Flow

We define the term *spontaneous* wide moving jam emergence in synchronized flow, i.e., a spontaneous S→J transition as follows [1–3, 5–8]:

- Spontaneous wide moving jam emergence in metastable synchronized flow is a phase transition from the synchronized flow phase to wide moving jam phase occurring due to the growth of a nucleus required for moving jam emergence that appears *within the synchronized flow*.

In an initial metastable synchronized flow, there are various *sources* for the occurrence of a nucleus required for an S→J transition, for example, unexpected braking of a vehicle in synchronized flow, lane changing and vehicle merging from other roads (e.g., at bottlenecks) that cause vehicle deceleration, fluctuations in upstream flow rates, slow moving vehicles, etc.

Empirical spontaneous S→J transitions are shown in Figs. 2.9 and 5.1. In both cases, growing narrow moving jams, i.e., nuclei for S→J transitions emerge *within initial synchronized flows*. Another empirical example, in which the spontaneous occurrence of narrow moving jams growing in synchronized flow has been observed, has been studied in [24]; in this case, the source for the occurrence of nuclei required for the spontaneous emergence of moving jams has been lane changing that causes deceleration of the following vehicles in synchronized flow.

In simulations of spontaneous S→J transitions based on three-phase traffic flow models, it has been found [25] that vehicle merging from other roads and lane changing in a neighborhood of a bottleneck are the main sources for local disturbances whose subsequent growth in metastable synchronized flow leads to spontaneous S→J transitions (see also simulations of [26] associated with the effect of lane changing on spontaneous S→J transitions upstream of an off-ramp bottleneck).

It should be noted that in metastable synchronized flow there can also be an *induced* S→J transition. In contrast with a spontaneous S→J transition, induced wide moving jam emergence in synchronized flow is caused by the upstream propagation of a nucleus required for an S→J transition, which has initially occurred within a *different* road of a traffic network connected with the road under consideration.

- Thus rather than the nature of wide moving jam emergence, the terms *spontaneous* and *induced* wide moving jam emergence distinguish different sources of the occurrence of a nucleus required for the initiating of an S→J transition.

5.3 Dual Role of Lane Changing in Synchronized Flow: Maintenance of Synchronized Flow or Wide Moving Jam Emergence

Lane changing can lead either to a nucleus required for traffic breakdown or, in contrast, to the maintenance of free flow (Sect. 3.4). Similar opposite effects can occur through vehicle lane changing in synchronized flow: Lane changing can either maintain synchronized flow or lead to wide moving jam emergence in synchronized flow. This dual role of lane changing in synchronized flow is as follows:

- Lane changing can lead *either* to the occurrence of a nucleus for the emergence of a wide moving jam in a metastable state of synchronized flow *or* to the dissolution of a moving jam that propagates in synchronized flow.

Empirical examples of these effects are shown in Fig. 5.5 [27]. In this case, moving jams emerge and dissolve in synchronized flow that affects two closely located adjacent downstream off-ramp and upstream on-ramp bottlenecks (Fig. 5.5 (a)). In Fig. 5.5 (b), a moving jam emerges in synchronized flow due to lane changing of vehicles (labeled by fat black trajectories 1 and 2) from the lane 5 to lane 4. As a result, a moving jam is forming in the lane 4. This effect is similar to empirical moving jam emergence in synchronized flow due to lane changing found in [24].

5.3 Dual Role of Lane Changing in Synchronized Flow

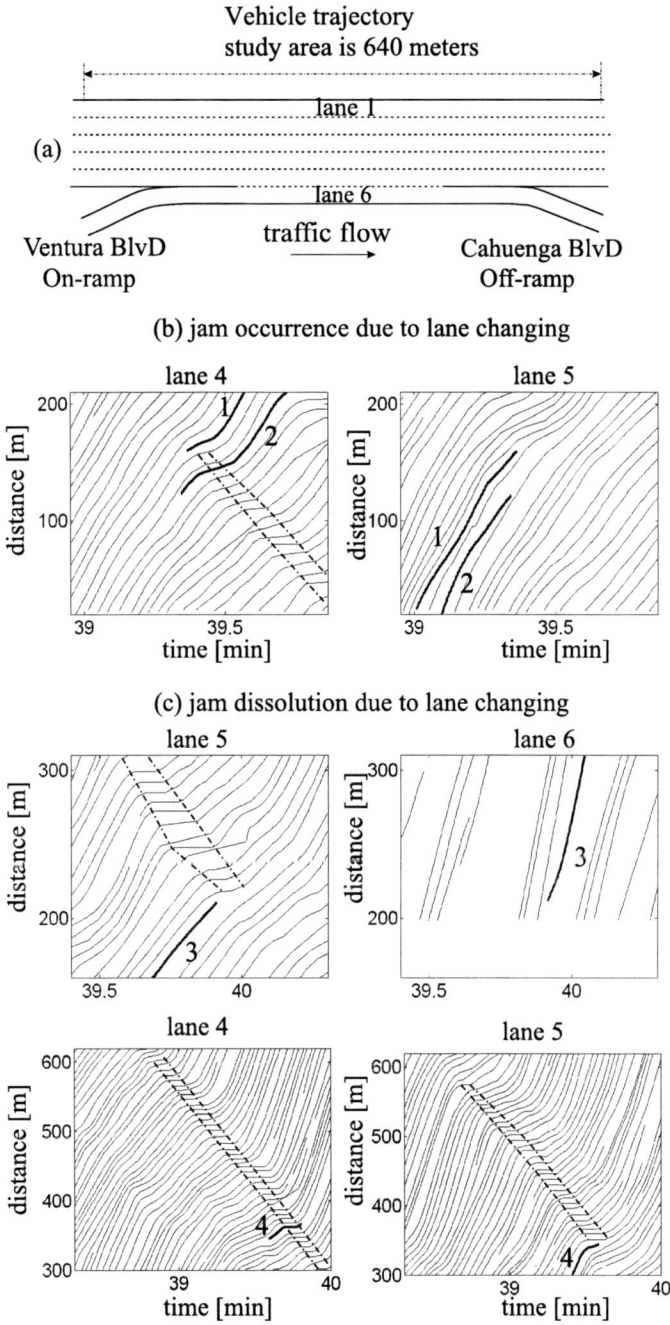

Fig. 5.5 Empirical study of dual role of lane changing in synchronized flow: (a) Simplified scheme of a section of the road US 101 on which data has been measured. (b, c) Vehicle trajectories showing the emergence of a moving jam (b) and the dissolution of moving jams (c) due to lane changing. Moving jams are marked-off by dashed curves. NGSIM-single vehicle data from June 15, 2005 (without motorcycle data) [28]. In figures time $t = 0$ is related to 7:50 am in raw NGSIM-data. Taken from [27]

In contrast, in Fig. 5.5 (c) we show two different examples of the jam dissolution due to lane changing: (i) a moving jam propagating in the lane 5 dissolves after a vehicle labeled by number 3 changes from the lane 5 to lane 6; (ii) a moving jam propagating in the lane 5 dissolves after a vehicle labeled by number 4 changes from the lane 5 to lane 4.

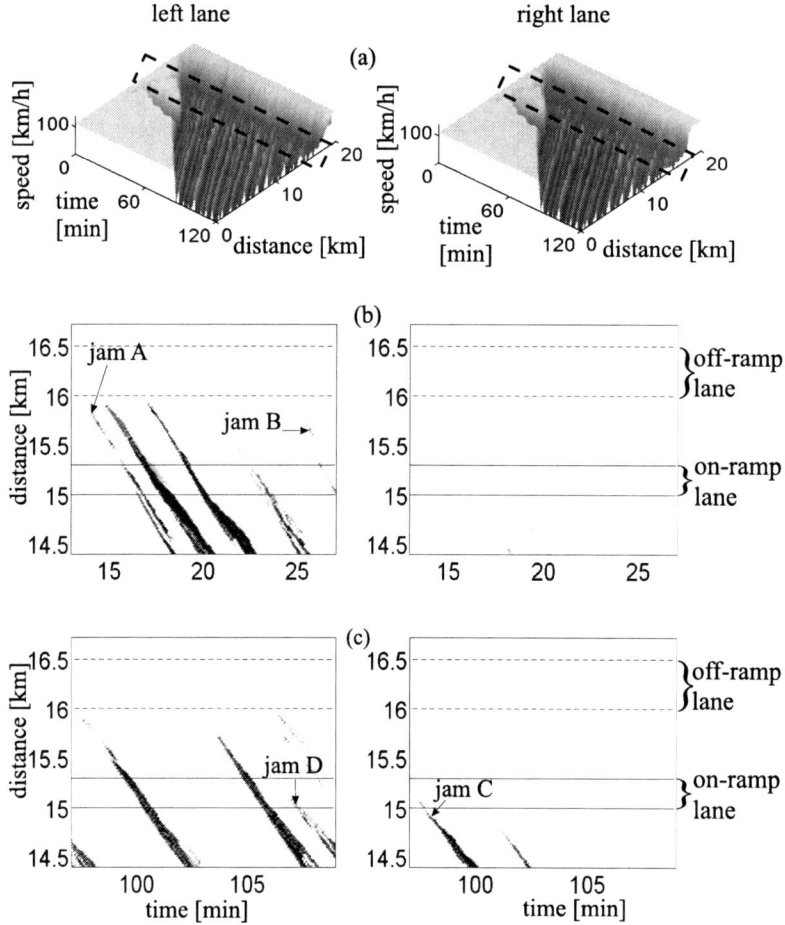

Fig. 5.6 Simulations of a congested pattern at sequence of five on-ramp and five off-ramp bottlenecks in the left (figures left) and right road lanes (right): (a) 1-min average data for speed in time and space. (b, c) Single-vehicle speed data for fragments within the pattern in (a) presented by regions with variable darkness (the lower the speed, the darker the region; in white regions the speed is higher than 3 km/h, in black regions the speed is zero). Taken from [27]

This dual role of lane changing for moving jam emergence and dissolution is also found in simulations with a three-phase traffic flow model [27]. To discuss simulation results that are compatible with empirical data shown in Fig. 5.5, we consider emergence and dissolution of moving jams in synchronized flow that affects two closely located adjacent downstream off-ramp and upstream on-ramp bottlenecks; as in empirical examples shown in Fig. 5.5, the distance between effective locations of these on- and off-ramp bottlenecks is about 500 m. These two bottlenecks are the farthest downstream on- and off-ramp bottlenecks in a sequence of closely located on-ramp and off-ramp bottlenecks on a two-lane road used in simulations (Fig. 5.6). In other words, we discuss here only a part of a complex congested pattern labeled by dashed region in Fig. 5.6 (a) between the farthest downstream on- and off-ramp bottlenecks of the sequence of several on- and off-ramp bottlenecks. Fragments of this pattern part that are studied here are shown in Fig. 5.6 (b, c)[5].

To prove that lane changing can lead to the dissolution of moving jams, specifically to the maintenance of synchronized flow, we consider two of the emergent moving jams in the left lane (labeled by "jam A" and "jam B" in Figs. 5.6 (b) and 5.7, left). These jams dissolve during their upstream propagation resulting in the maintenance of synchronized flow. To see that this maintenance of synchronized flow is indeed caused by lane changing, we consider vehicle trajectories. In Fig. 5.7 (a), a vehicle approaching "jam A" changes from the left lane to the right lane (vehicle trajectory labeled by number 1). This lane changing increases the space gap between vehicles at the upstream jam front; as a result, "jam A" dissolves resulting in synchronized flow labeled by "synchronized flow" in Fig. 5.7 (a), left. A qualitatively similar effect of jam dissolution resulting in synchronized flow occurs when a vehicle approaching "jam B" in the left lane changes from the left lane to the right lane (vehicle trajectory 2 in Fig. 5.7 (b)).

The emergence of moving jams in synchronized flow is shown in Fig. 5.8. In the first example, a vehicle merges from on-ramp onto the right lane on the main road (vehicle trajectory is labeled by number 3 in Fig. 5.8 (a)). This vehicle merging causes deceleration of the following vehicles in the right lane. As a result, a growing narrow moving jam occurs in the right lane (jam labeled by "jam C" in Fig. 5.8 (a), right). A qualitatively similar effect of jam emergence occurs in another example when two following each other vehicles change from the right lane to the left lane in which synchronized flow has been before (trajectories 4 and 5 in Fig. 5.8 (b)). Due to lane changing space gaps between following vehicles in the left lane and, therefore, the vehicle speed decrease considerably. This leads to the emergence of a growing narrow moving jam labeled by "jam D" in Fig. 5.8 (b), left.

5.4 Comparison of F→S and S→J Transitions

A spontaneous F→S transition (Sect. 3.3) and a spontaneous S→J transition occur in metastable states of free flow and synchronized flow, respectively. In both cases, nuclei required for the associated phase transitions should appear whose subsequent growth lead to the related phase transition. However, there is a qualitative difference

---

[5] The complete complex congested pattern shown in Fig. 5.6 (a) will be discussed in Sect. 7.3.2.

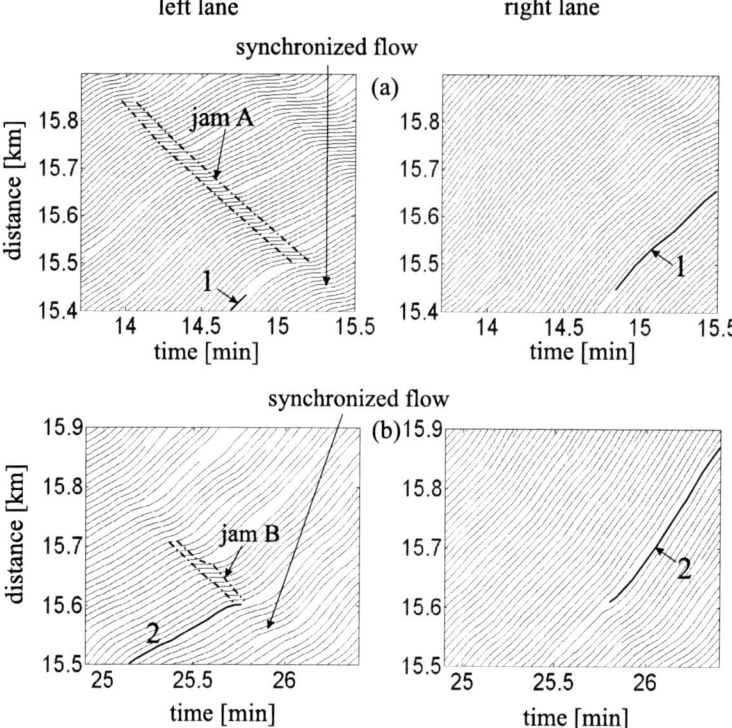

Fig. 5.7 Simulations of moving jam dissolution resulting in the maintenance of synchronized flow due to lane changing. Vehicle trajectories for two congested pattern fragments shown in Fig. 5.6 (b). Figures left and right are related to the left and right lanes, respectively. Taken from [27]

between these two phase transitions. This difference is associated with the kinetics of the growth of the nuclei required for the associated phase transitions.

When a nucleus for traffic breakdown (F→S transition) appears in free flow at a bottleneck, the growth of this nucleus and resulting traffic breakdown occurs usually also at the bottleneck. This is explained by a permanent (deterministic) disturbance, which is on average motionless and localized in free flow at the bottleneck. This disturbance increases probability of the nucleus occurrence within the disturbance and, therefore, traffic breakdown probability at the bottleneck considerably.

In contrast, a growing narrow moving jam, which is a nucleus required for wide moving jam emergence, propagates upstream in metastable synchronized flow. It takes some time delay before the growth of the jam leads to a wide moving jam, i.e., an S→J transition occurs. Thus we can make the following conclusions:

- As a result of the upstream propagation of the growing narrow moving jam, the S→J transition occurs usually upstream of the road location at which the nucleus for the S→J transition, i.e., the growing narrow moving jam has initially appeared.

5.5 Empirical Double Z-Characteristic for Phase Transitions in Traffic Flow

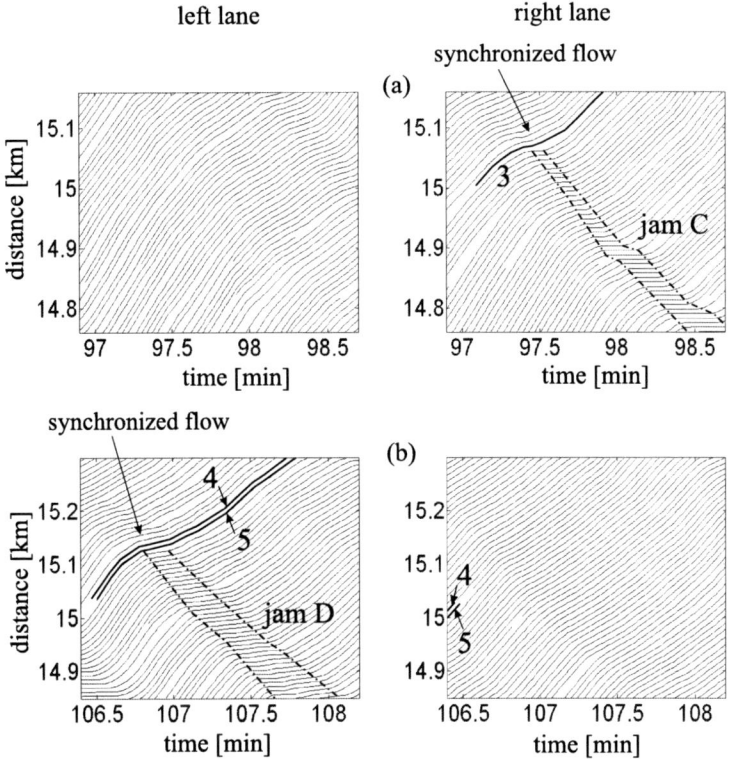

Fig. 5.8 Simulations of moving jam emergence in synchronized flow due to vehicle merging (a) and lane changing (b). Vehicle trajectories for two congested pattern fragments shown in Fig. 5.6 (c). Figures left and right are related to the left and right lanes, respectively. Taken from [27]

- Even if a nucleus for an S→J transition has appeared in metastable synchronized flow at a bottleneck, an S→J transition resulting from the growth of the nucleus occurs usually upstream of this bottleneck[6].

5.5 Empirical Double Z-Characteristic for Phase Transitions in Traffic Flow

Both F→S and S→J transitions can be illustrated by a double Z-characteristic for the F→S→J transitions (Fig. 5.9). The double Z-characteristic consists of an Z-characteristic for an F→S transition (between states of free flow and synchronized flow) and an Z-characteristic for an S→J transition (between states for synchronized

---

[6] The exclusion can be a very heavy bottleneck, which limits the average flow rate in congested traffic considerably. A discussion of traffic congestion at heavy bottlenecks appears in Sect. 7.2.5.

flow and low speed states within wide moving jams) as well as the states associated with the critical speeds required for the phase transitions (dashed curves in Fig. 5.9).

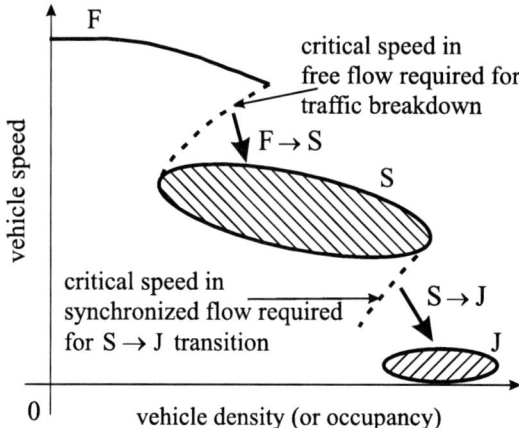

Fig. 5.9 Qualitative double Z-characteristic for F→S→J transitions that includes states of free flow (F), synchronized flow (S) and low speed states associated with moving blanks within wide moving jams (J) [3]

An empirical double Z-characteristic measured during GP formation at an on-ramp bottleneck (Fig. 5.10 (a)) is presented in Fig. 5.10 (b). Firstly, an F→S transition occurs at the bottleneck (arrows labeled by F→S in Figs. 5.10 (a, b)). Synchronized flow propagates upstream. In the pinch region of the synchronized flow, growing narrow moving jams emerge propagating upstream. Some of the jams transform into wide moving jams, i.e., S→J transitions occur (arrows S→J in Fig. 5.10 (a, b)). The S→J transitions occur later and upstream of the road location of the F→S transition. Within wide moving jams low speed states are observed that are probably associated with moving blanks (Sect. 2.6.4).

Later the GP begins to dissolve. In particular, some of wide moving jams associated with the GP dissolve and either synchronized flow or free flow occur (J→S or J→F transitions). An example of an J→F transition is shown in Fig. 5.10 (a). Finally, an S→F transition occurs at the bottleneck and the GP disappears (arrows labelled S→F in Fig. 5.10).

We see that there are many transitions between three traffic phases. The phase transitions exhibit hysteresis effects on the double Z-characteristic (Fig. 5.10 (b)). However, the phase transitions occur at different road locations, therefore, they cannot be distinguished from each other without knowledge of the whole *spatiotemporal dynamics* of the GP (Fig. 5.10 (a))[7]:

---

[7] Because the speed and flow rate within the GP are measured at spatial separated road detectors, the locations of phase transitions shown in Fig. 5.10 (a) are known only with the accuracy of about 1 km.

5.5 Empirical Double Z-Characteristic for Phase Transitions in Traffic Flow

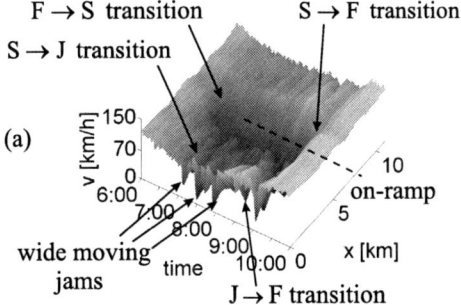
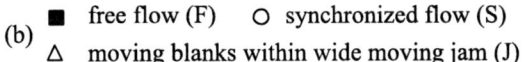
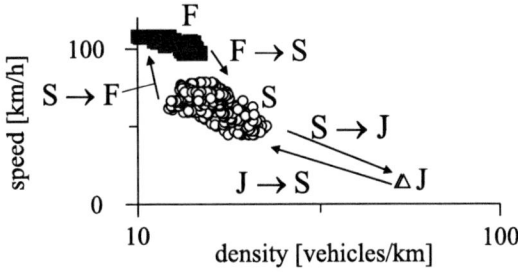

Fig. 5.10 Average speed in empirical GP at on-ramp bottleneck (a) and related empirical double Z-characteristic for phase transitions (b). Low speed states within wide moving jam are labeled as "moving blanks" in (b)

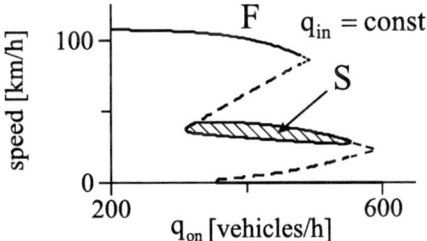

Fig. 5.11 Simulated double Z-characteristic for F→S→J transitions at an on-ramp bottleneck. Taken from [3]

- In general, a solely analysis of different congested traffic states and phase transitions in the flow–density plane (as well as in speed–density, space-gap–density planes, etc.) is not adequate with empirical features of congested traffic. This is because most of these empirical spatiotemporal features are *lost* in this congested traffic analysis. Thus an initial study of the whole spatiotemporal dynamics of the congested pattern is needed.

Features of the empirical double Z-characteristic for the F→S→J transition have been confirmed in numerical simulations of a three-phase traffic flow model (Fig. 5.11). For simplicity, it has been assumed in the simulations that the speed within wide moving jams is equal to zero, i.e., states J shown in Fig. 5.9 have been neglected.

References

1. B.S. Kerner, in *Proceedings of the 3rd Symposium on Highway Capacity and Level of Service*, ed. by R. Rysgaard. Vol 2 (Road Directorate, Ministry of Transport – Denmark, 1998), pp. 621–642
2. B.S. Kerner, Phys. Rev. Lett. **81**, 3797–3400 (1998)
3. B.S. Kerner, *The Physics of Traffic*, (Springer, Berlin, New York, 2004)
4. B.S. Kerner, in *Encyclopedia of Complexity and System Science*, ed. by R.A. Meyers. (Springer, Berlin, 2009), pp. 9302–9355
5. B.S. Kerner, in *Traffic and Granular Flow' 97*, ed. by M. Schreckenberg, D.E. Wolf. (Springer, Singapore, 1998), pp. 239–267
6. B.S. Kerner, Trans. Res. Rec. **1678**, 160–167 (1999)
7. B.S. Kerner, in *Transportation and Traffic Theory*, ed. by A. Ceder. (Elsevier Science, Amsterdam 1999), pp. 147–171
8. B.S. Kerner, Physics World **12**, 25–30 (August 1999)
9. B.S. Kerner, P. Konhäuser, Phys. Rev. E **50**, 54–83 (1994)
10. B.S. Kerner, H. Rehborn, Phys. Rev. E **53**, R1297–R1300; R4275–R4278 (1996); Phys. Rev. Lett. **79**, 4030–4033 (1997)
11. B.S. Kerner, S.L. Klenov, Phys. Rev. E **68**, 036130 (2003)
12. B.S. Kerner, S.L. Klenov, D.E. Wolf, J. Phys. A: Math. Gen. **35**, 9971–10013 (2002)
13. D.C. Gazis, R. Herman, R.W. Rothery, Oper. Res. **9**, 545–567 (1961)
14. R. Herman, E.W. Montroll, R.B. Potts, R.W. Rothery, Oper. Res. **7**, 86–106 (1959)
15. E. Kometani, T. Sasaki, J. Oper. Res. Soc. Jap. **2**, 11 (1958)
16. E. Kometani, T. Sasaki, Oper. Res. **7**, 704–720 (1959)
17. N.H. Gartner, C.J. Messer, A. Rathi (editors) *Traffic Flow Theory: A State-of-the-Art Report*, (Transportation Research Board, Washington DC, 2001)
18. D.C. Gazis, *Traffic Theory*, (Springer, Berlin, 2002)
19. D. Chowdhury, L. Santen, A. Schadschneider, Physics Reports **329**, 199 (2000)
20. D. Helbing, Rev. Mod. Phys. **73**, 1067–1141 (2001)
21. T. Nagatani, Rep. Prog. Phys. **65**, 1331–1386 (2002)
22. A.D. May, *Traffic Flow Fundamentals*, (Prentice-Hall, Inc., New Jersey, 1990)
23. B.S. Kerner, S.L. Klenov, J. Phys. A: Math. Gen. **35**, L31–L43 (2002)
24. S. Ahn, M.J. Cassidy, in *Transportation and Traffic Theory*, ed. by R.E. Allsop, M.G.H. Bell, B.G. Hydecker. (Elsevier, Amsterdam, 2007), pp. 691–710
25. B.S. Kerner, S.L. Klenov, J. Phys. A: Math. Gen. **39**, 1775–1809 (2006)
26. B.S. Kerner, Transp. Res. Rec. **1999**, 30–39 (2007)
27. B.S. Kerner, S.L. Klenov, unpublished
28. Next Generation Simulation Programs, http://ngsim.camsys.com/

Chapter 6
Origin of Hypotheses and Terms of Three-Phase Traffic Theory

In Chaps. 3–5, hypotheses of three-phase traffic theory have been discussed and applied to explain traffic breakdown and moving jam emergence in vehicular traffic observed in real measured traffic data. In this chapter, we show that the origin of some of these hypotheses is the traffic phase definitions, i.e., the empirical criteria for traffic phases in congested traffic [S] and [J] considered in Sect. 2.4.1. In addition, we will try to explain why terms of natural science used in three-phase traffic theory are needed for transportation engineering.

6.1 Hypotheses of Three-Phase Traffic Theory as The Result of Empirical Criteria for Traffic Phases

Reasons why hypotheses of three-phase traffic theory [1] result from the traffic phase definitions [S] and [J] are as follows:

1. The line J in the flow–density plane results from the definition of wide moving jam [J] (Sect. 2.5.2). Any point on the line J can be a final steady traffic state for vehicles accelerating at the downstream front of a wide moving jam. If the speed in this state is lower than the minimum speed that is possible in free flow, the state is a synchronized flow steady state that lies on the line J. Thus there should be infinite synchronized flow states lying on the line J (Fig. 3.3 (a)).
2. The definition of synchronized flow [S] means that downstream fronts of synchronized flow regions do *not* exhibit the jam characteristic feature [J]. For this reason, in addition to synchronized flow steady states lying on the line J (item 1) there should be synchronized flow steady states that are outside of the line J in the flow–density plane. Indeed, only in this case the downstream front between two different states of synchronized flow does not exhibit the characteristic jam velocity given by the slope of the line J. Thus there should be also steady states of synchronized flow outside of the line J, i.e., that there is a 2D-region of the steady states in the flow–density plane. This explains the fundamental hypothesis

of three-phase traffic theory about the 2D-region of the steady states of synchronized flow (Fig. 3.3 (a)).

3   The definition [S] is associated with empirical results for moving jam behavior within synchronized flow [2]: depending on synchronized flow characteristics upstream of a moving jam, the jam can grow or dissolve propagating in the synchronized flow over time. We have noted (item 1) that there are the infinite number of steady states of synchronized flow that lie on the line J. If these states were to build the upper boundary of the 2D-region of the synchronized flow steady states, i.e., there were no steady states above the line J, then, as explained in Sect. 5.2.1, a wide moving jam should dissolve propagating in the states of synchronized flow. This contradicts a possibility of the jam growth. In contrast, if the states on the line J were to build the low boundary of the 2D-region of the synchronized flow steady states, i.e., there were no steady states below the line J, then no wide moving jam can dissolve propagating in the states of synchronized flow. This also contradicts mentioned empirical results about possible jam dissolution. Thus we should assume that there are synchronized flow steady states above *and* below the line J (Fig. 5.2).

4   A hypothesis that synchronized flow states, which are below the line J, are stable with respect to wide moving jam emergence, whereas synchronized flow states, which are on and above the line J, are metastable states with respect to wide moving jam emergence (Fig. 5.2) has already been proven in Sect. 5.2.1.

5   To explain a hypothesis that the outflow rate q_{out} from a wide moving jam is smaller than the maximum flow rate that is possible in free flow (Fig. 5.2), we assume that this hypothesis is not satisfied. In this case, the jam outflow rate q_{out} would be always greater than the jam inflow rate while the jam propagates through free flow in which the flow rate is smaller than the maximum flow rate in free flow; as a result, any wide moving jam propagating in free flow would dissolve over time. This contradicts the feature [J] of the jam propagation through different states of free flow. This explains this hypothesis.

6.2 Are Terms of Natural Science used in Three-Phase Traffic Theory Needed for Transportation Engineering?

Explaining empirical traffic data in Chaps. 2–5, we have used many terms of three-phase traffic theory like *nucleation* of traffic breakdown, a *nucleus* required for traffic breakdown, *metastable* states of free flow with respect to an F→S transition, etc. These terms of three-phase traffic theory, which appear from natural science, are not usually used in transportation engineering.

Thus a question arises:

- are these terms of natural science used in three-phase traffic theory needed for transportation science and transportation engineering in general?

6.2 Are Terms of Three-Phase Traffic Theory Needed for Transportation Engineering?

To answer this question, we mention that a term for a traffic phenomenon should reflect features of the phenomenon found in empirical observations, i.e., in real measured traffic data. In addition, the "best" term should also contain information about the phenomenon that could be used in transportation engineering.

Examples of such well-known terms are *free flow*, *traffic pattern*, *traffic congestion*, *congested traffic pattern*, *traffic breakdown*, *discharge flow rate*, and many other terms of transportation engineering (see references in [3–12]) that are also used in three-phase traffic theory. In particular, the term *traffic breakdown* reflects an empirical feature of the onset of traffic congestion in an initial free flow that during the emergence of traffic congestion, the vehicle speed drops abruptly.

However, there are also many other wide-accepted and well-known terms in transportation science that we do not use. The main reason for this is that these terms do not reflect important empirical features of traffic phenomena found in real measured traffic data. Examples of such terms are *stop-and-go traffic*, *oscillations in congested traffic*, *shock waves in traffic flow*, and *highway capacity* considered as a *particular (fixed or stochastic) value*[1].

Here, we try to explain some terms of natural science used in three-phase traffic theory. The necessity of these terms for transportation science and transportation engineering can be explained as follows:

- Earlier traffic flow theories failed to explain empirical macroscopic spatiotemporal traffic phenomena (see explanations of this critical statement in Chap. 10). In contrast, three-phase traffic theory explains all up to now known real measured traffic data.

6.2.1 Traffic Phase or Traffic State?

The term a *phase* has been introduced in non-equilibrium distributed systems natural science like hydrodynamics, solid state physics, non-linear optics, semiconductors and semiconductor devises, gas plasma, chemical reactions, biological systems, etc. [13–18] as a generic term for a multitude of various *system states* that exhibit spatiotemporal features that are unique only for the same system phase. The most general and important of these unique features of the *system phase* are general properties of *phase transitions* between different phases of the system (see e.g., [14]).

Thus if system phases can be distinguished in a spatiotemporal system, then each of the phases can consist of many system states satisfying the spatiotemporal features that are unique only for all various system states of the same traffic phase. Thus the term *system state* is a species, i.e., infraordinate term in comparison with the generic term *system phase*. The term *system phase* has a sense, if in a system there can be distinguished at least two different system phases. For example, there are two following very different traffic states associated with the synchronized flow phase:

---

[1] Discussions of two last terms appear in Sects. 10.2.3 and 10.6.

- a narrow moving jam (Sect. 2.6.3)
- a synchronized flow state between narrow moving jams within the pinch region of synchronized flow (Sect. 5.1).

The generic term *traffic phase* is also the reason that and why we use the terms *synchronized flow pattern (SP)* and *general pattern (GP)* as the generic terms in the classification of congested patterns: the definitions of the SP and the GP are based on which traffic phase(s) a congested traffic pattern consists of (Sect. 2.4.6). For this reason, all different traffic congested patterns are referred to different types of the SPs and GPs; respectively, different congested states are referred to different states within the SPs and GPs.

6.2.2 Traffic Phase Nucleation and Metastable System State

In accordance with natural science (see, e.g., [14]), the terms *metastable traffic state*, *first-order phase transition* in traffic flow, and *metastable traffic flow* are defined as follows[2].

- A metastable traffic state is a traffic state in which all small enough local disturbances decay over time. However, if a great enough local disturbance appears within the traffic state, then the disturbance grows leading to the emergence of a new traffic phase, i.e., to a phase transition.
- A first-order phase transition in traffic is a phase transition occurring in a metastable state of an initial traffic phase.
- Metastable traffic flow is traffic flow that is in a metastable traffic state with respect to a phase transition to another traffic phase.

It is shown in a general theory of metastable systems of natural science (see, e.g., [14]), i.e., the systems that are in a metastable system state that if a system phase is a metastable one with respect to a phase transition, then a new phase occurs due to *nucleus occurrence* in the metastable state of the system. For example, if a local disturbance occurs in an initial free flow within which the speed is equal to or lower than a critical speed (density is equal to or greater than a critical density), then traffic breakdown occurs, i.e., the synchronized flow phase emerges within the initial free flow phase; such a growing local disturbance is called a *nucleus* required for traffic breakdown (Sect. 3.3). Thus the terms *traffic phase nucleation* and *metastable system* are associated with the same nature of free flow to exhibit a *first-order phase transition* from the free flow phase to the synchronized flow phase that occurs in the metastable state of free flow due to the synchronized flow nucleation, i.e., due to the occurrence of a nucleus required for traffic breakdown.

This discussion of metastable traffic flow explains the terms *nucleus required for a phase transition* in traffic flow and *phase transition nucleation* in traffic flow used in three-phase traffic theory as follows:

---

[2] In application to traffic breakdown and moving jam emergence in synchronized flow, these definitions have already been discussed in Chaps. 3 and 5, respectively.

- A nucleus required for a phase transition is a local disturbance within which the speed is equal to or lower than a critical speed for the phase transition (density is equal to or greater than a critical density). The nucleus grows leading to the phase transition in traffic flow.
- Phase transition nucleation is the occurrence and subsequent growth of a nucleus required for a phase transition.

6.2.3 Spontaneous or Induced Phase Transition?

In accordance with a general terminology of natural science (see, e.g., [14]), the terms a *spontaneous* phase transition and an *induced* phase transition distinguish the *source* of the occurrence a nucleus whose subsequent growth leads to a phase transition.

- If a nucleus required for a phase transition occurs due to random disturbances within an initial traffic phase, then the associated phase transition resulting from the growth of this nucleus is a spontaneous phase transition.

This is independent of whether the random nucleus appears at a system non-homogeneity (for traffic flow, the role of a road non-homogeneity is played by a road bottleneck) or outside of the non-homogeneity. It is only important that the nucleus appears within the same system phase without influence of either other traffic phases or an external source for nucleus occurrence.

Thus traffic breakdown (F→S transition) occurring at a bottleneck under condition that there are initially free flows both at the bottleneck and downstream and upstream of the bottleneck is an example of a spontaneous traffic breakdown. This is independent of the fact that a random nucleus for traffic breakdown appears with the greatest probability at the bottleneck because the bottleneck introduces a permanent non-homogeneity in free flow.

- If a nucleus required for a phase transition occurs due to the propagation of another traffic phase, then the phase transition is an *induced phase transition*.

For example, an induced traffic breakdown occurs, if a nucleus required for traffic breakdown at a bottleneck is associated with congested pattern propagation that reaches the bottleneck. Then in accordance with the terminology of natural science, in three-phase traffic theory this traffic breakdown is called the induced traffic breakdown at the bottleneck.

6.2.4 Nucleation or Triggering of Phase Transition?

The term *nucleation* of a phase transition in traffic flow cannot be replaced by the term *triggering*, which is often used in the transportation research literature.

To explain this, firstly we note that the term *nucleation* is related to different traffic phases between which phase transitions are possible. The term *triggering* is related to a device called in engineering a *trigger* that has two distinct different device states "0" and "1" between them transitions are possible.

A trigger must ensure a clear transition between two states "0" and "1" of the trigger through the application of an *external signal or disturbance* called *triggering*. This transition should not be influenced by *internal* disturbances (fluctuations) of the trigger. Otherwise, if internal disturbances (fluctuations) in the trigger lead to transitions between trigger states "0" and "1" *without* application of some external signal (or disturbance), then such a trigger cannot be used for applications.

To understand this statement, we consider one of the most known fields of trigger application in a computer memory. If internal disturbances (fluctuations) of the trigger lead to transitions between trigger states "0" and "1" *without* application of some external signal, then the triggering between states "0" and "1" cannot be solely determined by external signals. Therefore, such a trigger cannot be used for the computer memory: states of triggers in the memory would be changed randomly over time, i.e., no memory feature would be possible to achieve with this trigger.

Thus the term *triggering* is related to a transition between two system states, which must not be influenced by internal system disturbances (fluctuations) at all. The most important feature of the triggering is that the transition must *not* depend on whether system disturbances of small or large amplitude occur randomly in the system.

In contrast, the term *nucleation* is related to a phase transition between two system phases, which can occur *either* through external *or* through internal, in particular random disturbances. In contrast with the triggering, the most important feature of the term *nucleation* is that the phase transition must depend on whether internal system disturbances of small or large amplitude occur in the system.

In application to traffic flow, we consider lane changing that can lead to a random local speed (density) disturbance in an initial traffic flow. A possible growth of the disturbance caused by lane changing with the subsequent phase transition in the traffic flow cannot be considered the triggering of the phase transition. There are two reasons for this statement:

1) There can be driving situations in which lane changing causes disturbances of very small or even negligible speed decrease (density increase); for example, this is the case in free flow of small density. As a result, in these cases no phase transitions occur through lane changing at all. This is an important feature of the phase transition *nucleation*. In contrast, the term *triggering* means that the transition must occur, when a signal (disturbance) is applied to a trigger.

2) As a result of the disturbance occurrence caused by lane changing, a phase transition occurs only, if the speed within the disturbance is equal to or lower than the critical speed (density is equal to or greater than the critical density) required for the phase transition. Otherwise, if lane changing leads to a disturbance of a higher speed than the critical one (smaller density than the critical density), no phase transition can occur. This is another important feature of the phase transition nucleation that contradicts the features of a trigger explained above.

Thus rather than the triggering, lane changing can lead to the occurrence of a nucleus required for a phase transition in traffic flow, i.e., to the phase transition nucleation.

6.2.5 Discussion of Terms "Stop-and-Go Traffic" and "Oscillations in Congested Traffic"

In literature, a sequence of moving jams in congested traffic are often called *stop-and-go traffic* or *oscillations in congested traffic*. In three-phase traffic theory, these well-known terms are not used. This is because a sequence of wide moving jams propagating through synchronized flow *and* a sequence of narrow moving jams propagating in synchronized flow are associated with stop-and-go traffic or with oscillations in congested traffic.

However, as we have shown above based on empirical spatiotemporal analysis of measured data in Chap. 2, wide moving jams belong to the wide moving jam phase, while the sequence of narrow moving jams propagating in synchronized flow belongs to the synchronized flow phase. As shown in Chaps. 2–5 and 7, the wide moving jam phase exhibits qualitatively different spatiotemporal features in comparison with the synchronized flow phase.

Rather than the use for the designation of two phenomena with qualitatively different empirical spatiotemporal features the same terms *stop-and-go traffic* or *oscillations in congested traffic*, to understand real traffic and make efficient engineering traffic applications one should use those terms, which reflect and emphasize these qualitatively different empirical spatiotemporal features of traffic. This explains why rather than the terms *stop-and-go traffic* or *oscillations in congested traffic*, in three-phase traffic theory the terms *wide moving jam* and *narrow moving jam* are used[3].

References

1. B.S. Kerner, in *Proceedings of the 3rd Symposium on Highway Capacity and Level of Service*, ed. by R. Rysgaard. Vol 2 (Road Directorate, Ministry of Transport – Denmark, 1998), pp. 621–642; Phys. Rev. Lett. **81**, 3797–3400 (1998); in *Traffic and Granular Flow' 97*, ed. by M. Schreckenberg, D.E. Wolf. (Springer, Singapore, 1998), pp. 239–267; Trans. Res. Rec. **1678**, 160–167 (1999); in *Transportation and Traffic Theory*, ed. by A. Ceder. (Elsevier Science, Amsterdam 1999), pp. 147–171; Physics World **12**, 25–30 (August 1999); J. Phys. A: Math. Gen. **33**, L221-L228 (2000); Networks and Spatial Economics, **1**, 35 (2001); in *Progress in Industrial Mathematics at ECMI 2000*, (Springer, Berlin, 2001), pp. 286–292; Mathematical and Computer Modelling, **35**, 481–508 (2002)
2. B.S. Kerner, *The Physics of Traffic*, (Springer, Berlin, New York, 2004)
3. A.D. May, *Traffic Flow Fundamentals*, (Prentice-Hall, Inc., New Jersey, 1990)

---

[3] A criticism of applications of queuing theories to freeway traffic congestion has been discussed in Sect. 3.3.4 of the book [2]. Therefore we do not also use the term *vehicle queue* for traffic congestion occurring at bottlenecks on unsignalized freeways.

4. *Highway Capacity Manual 2000* (National Research Council, Transportation Research Board, Washington, D.C., 2000)
5. I. Prigogine, R. Herman, *Kinetic Theory of Vehicular Traffic*, (American Elsevier, New York, 1971)
6. W. Leutzbach, *Introduction to the Theory of Traffic Flow*, (Springer, Berlin, 1988)
7. N.H. Gartner, C.J. Messer, A. Rathi (editors), *Traffic Flow Theory: A State-of-the-Art Report*, (Transportation Research Board, Washington DC, 2001)
8. G.B. Whitham, *Linear and Nonlinear Waves*, (Wiley, New York, 1974)
9. G.F. Newell, *Applications of Queuing Theory*, (Chapman Hall, London, 1982)
10. C.F. Daganzo, *Fundamentals of Transportation and Traffic Operations*, (Elsevier Science Inc., New York, 1997)
11. D.C. Gazis, *Traffic Theory*, (Springer, Berlin, 2002)
12. D. Helbing, Rev. Mod. Phys. **73**, 1067–1141 (2001)
13. S. Chandrasekhar, *Hydrodynamic and Hydromagnetic Stability*, (Oxford University Press, Oxford 1961)
14. C.W. Gardiner, *Handbook of Stochastic Methods*, (Springer, Berlin, 1994)
15. H. Haken, *Synergetics*, (Springer, Berlin 1977); *Advanced Synergetics*, (Springer, Berlin 1983); *Information and Self-Organization*, (Springer, Berlin 1988)
16. V.A. Vasil'ev, Yu.M. Romanovskii, D.S. Chernavskii, V.G. Yakhno, *Autowave Processes in Kinetic Systems*, (Springer, Berlin 1990)
17. A.S. Mikhailov, *Foundations of Synergetics Vol. I*, (Springer, Berlin 1994), 2nd ed.; A.S. Mikhailov, A.Yu. Loskutov, *Foundation of Synergetics II. Complex patterns*, (Springer, Berlin 1991)
18. B.S. Kerner, V.V. Osipov, *Autosolitons: A New Approach to Problems of Self-Organization and Turbulence*, (Kluwer, Dordrecht, Boston, London 1994); Sov. Phys. Usp. **32**, 101–138 (1989); Sov. Phys. Usp. **33**, 679–719 (1990)

Chapter 7
Spatiotemporal Traffic Congested Patterns

There is a great variety of complex spatiotemporal congested patterns that appear due to traffic breakdown at a bottleneck. The nucleation nature of traffic breakdown and moving jam emergence in synchronized flow discussed in Chaps. 3 and 5 is the fundament for the understanding of pattern features. In this chapter, we present a brief review of some congested pattern features found in real measured traffic data[1].

7.1 Simplified Diagram of Congested Patterns at Isolated Bottleneck

Although an isolated effectual bottleneck (isolated bottleneck for short) is obviously a rough idealization of a real bottleneck[2], on real highways there can be road sections with adjacent effectual bottlenecks that are located far enough from each other. Then before a congested pattern, which has emerged at a downstream bottleneck, reaches an upstream bottleneck, the pattern can be considered a pattern at an isolated bottleneck. Such a rough simplification of the reality allows us to understand some important common spatiotemporal features of congested traffic patterns.

Characteristics of a congested pattern resulting from traffic breakdown at an isolated effectual bottleneck depend on a bottleneck strength:

- The bottleneck strength characterizes the influence of the bottleneck on traffic breakdown and resulting traffic congestion: for the same congested pattern type, the greater the bottleneck strength, the smaller the *average* flow rate on the main

---

[1] A more detailed analysis of congested traffic pattern features can be found in [1–4].

[2] An isolated effectual bottleneck is an effectual bottleneck (Sect. 2.4.2.1), which is located on a road far enough from other effectual adjacent bottlenecks. One or several road non-homogeneities can be responsible for the existence of an isolated effectual bottleneck. Thus an isolated effectual bottleneck can consist of a combination of several on- and off-ramps and other road non-homogeneities closely located, if *no spatially separated* traffic breakdowns are observed at each of the non-homogeneities separately.

B.S. Kerner, *Introduction to Modern Traffic Flow Theory and Control*,
DOI 10.1007/978-3-642-02605-8_7, © Springer-Verlag Berlin Heidelberg 2009

road within the congested pattern upstream of the bottleneck. We denote this average flow rate by $q^{(\mathrm{cong})}$.

The averaging time interval T_{av} for the average flow rate $q^{(\mathrm{cong})}$ is assumed to be considerably longer than time distances between any moving jams within a congested pattern[3]. At given other bottleneck characteristics, for an on-ramp bottleneck the bottleneck strength increases with an increase in the flow rate q_{on} to the on-ramp[4]; for an off-ramp bottleneck the bottleneck strength increases with the increase in the percentage of vehicles leaving the main road to the off-ramp.

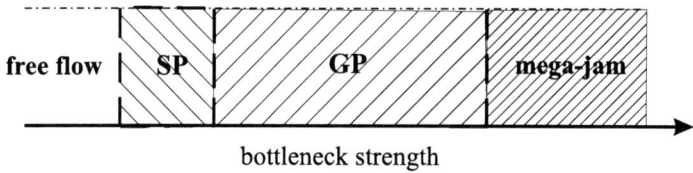

Fig. 7.1 Simplified qualitative diagram of congested patterns at an isolated effectual bottleneck that represents regions of the spontaneous emergence of synchronized flow patterns (SP), general patterns (GP), and mega-jam at the bottleneck as a function of the bottleneck strength at a given large enough flow rate in free flow on the main road upstream of these congested patterns [1, 2].

If the flow rate in free flow q_{in} on the main road upstream of a bottleneck is a given value that is a great enough, then under a continuous gradual increase in the bottleneck strength, firstly a synchronized flow pattern (SP) results from traffic breakdown at the bottleneck, secondly, the SP transforms into a general pattern (GP)[5], and finally, the GP transforms into a mega-wide moving jam (mega-jam) (Fig. 7.1) [1, 2].

On real freeways there are many effectual adjacent bottlenecks. When an SP or an GP has occurred at a bottleneck, the pattern can propagate further reaching an adjacent effectual bottleneck. At this bottleneck another congested pattern can either already exist or be induced by the propagation of the SP or GP. Thus in real traffic the SP or GP can usually be approximately considered congested patterns at the isolated bottleneck only for a finite time interval, before the SP or GP reaches the adjacent bottleneck. Complex spatiotemporal dynamics of congested traffic occurring at adjacent effectual bottlenecks will be considered in Sect. 7.3.

---

[3] For example, $T_{\mathrm{av}} \gtrsim 30$ min.

[4] However, this is only valid as long as no congested pattern occurs in the on-ramp lane; otherwise, with the increase in q_{on} the flow rate $q^{(\mathrm{cong})}$ reaches a limit (minimum) value [1].

[5] Full diagrams of the occurrence and evolution of a variety of different SPs and GPs at on- and off-ramp bottlenecks as functions of q_{in} and the bottleneck strength can be found in [1].

7.2 Variety of Congested Patterns at Isolated Bottleneck

7.2.1 Synchronized Flow Patterns

As a result of traffic breakdown at an isolated bottleneck, various synchronized flow patterns (SPs) can occur at the bottleneck. There are three main types of SPs [1]:

(1) Localized SP (LSP for short).
(2) Widening SP (WSP for short).
(3) Moving SP (MSP for short).

The LSP, WSP, and MSP are distinguished through qualitatively different spatiotemporal behavior of their fronts, which separate free flow outside of the SP and synchronized flow within the SP. At the downstream front of the SP vehicles accelerate from synchronized flow upstream of the front to free flow downstream of the front. At the upstream front of the SP vehicles decelerate from free flow upstream of the front to synchronized flow within the SP.

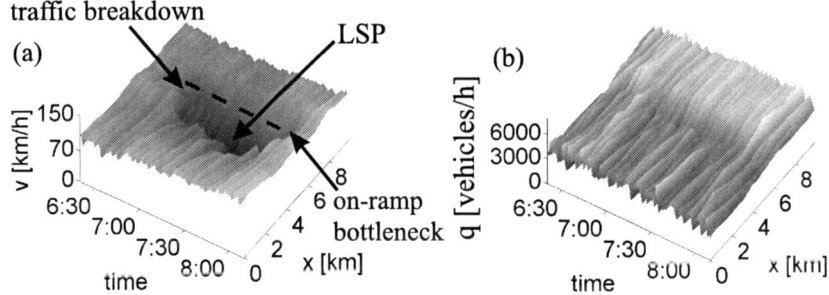

Fig. 7.2 Empirical LSP at on-ramp bottleneck. (a) Averaged vehicle speed in the LSP in space and time. (b) Total flow rate across road in space and time. The dashed line in (a) shows the location of the bottleneck. Taken from [1]

The definition of an LSP is as follows.

- The downstream front of the LSP is fixed at the bottleneck and the upstream front of the LSP does not continuously propagate upstream over time: this front is localized at some distance upstream of the bottleneck (Figs. 7.2 and 7.3).

The location of the upstream front of synchronized flow in the LSP and, therefore, the LSP width $L_{\rm LSP}$ (in the longitudinal direction) can exhibit great changes over time (even at time-independent flow rates $q_{\rm in}$ and $q_{\rm on}$, $L_{\rm LSP}^{\rm (mean)}(t)$ varies between 0.5 and 4 km over time in Fig. 7.3 (d)).

The definition of an WSP is as follows.

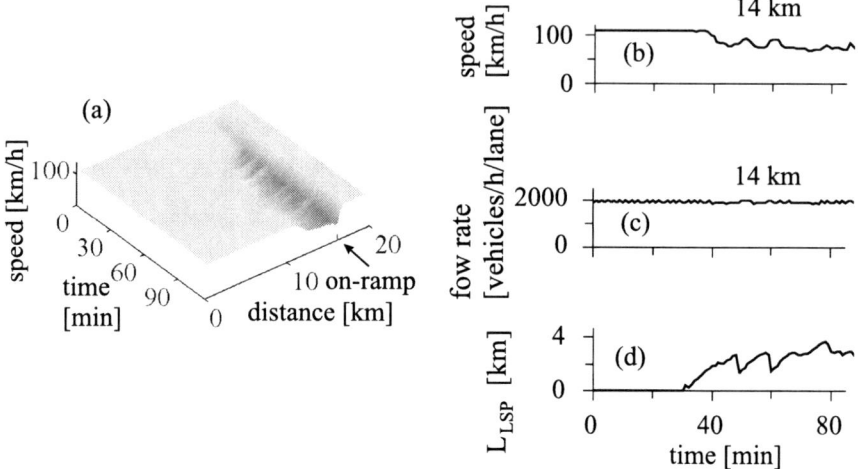

Fig. 7.3 Simulated LSP at on-ramp bottleneck: (a) Averaged vehicle speed in the LSP in space and time. (b, c) Speed (b) and flow rate (c) at road location $x = 14$ km as time-functions. (d) LSP width $L_{\rm LSP}$ as a time-function. Taken from [1]

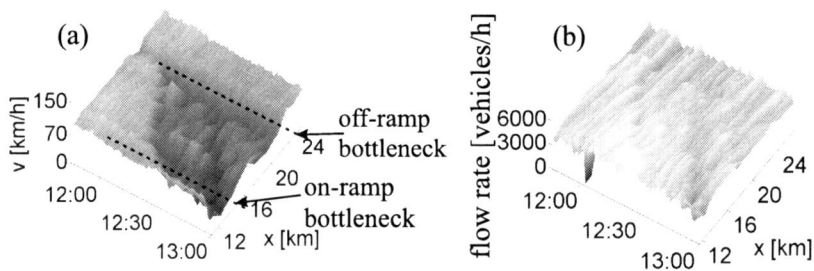

Fig. 7.4 Empirical example of an WSP: (a, b) Average speed (a) and flow rate (b) in space and time. Taken from [1]

- As in the LSP, the downstream front of an WSP is fixed at the bottleneck. In contrast to the LSP, the upstream front of the WSP propagates upstream continuously over time (Figs. 7.4 and 7.5).

It must be noted that due to synchronized flow widening, the upstream front of synchronized flow of an WSP should eventually reach an upstream adjacent effectual bottleneck. At this bottleneck another congested pattern either can already exist or be induced. Thus in real traffic the WSP can usually exist only for a finite time, before the WSP reaches the upstream effectual bottleneck.

The definition of an MSP is as follows.

7.2 Variety of Congested Patterns at Isolated Bottleneck

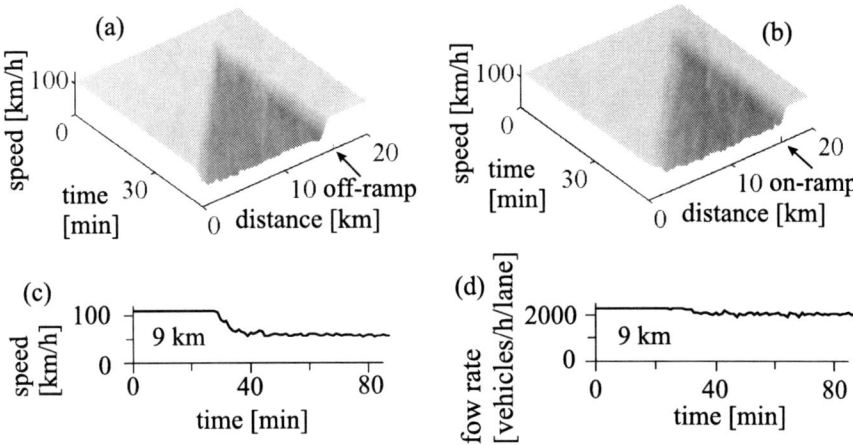

Fig. 7.5 Simulations of WSP at off-ramp (a) and on-ramp (b–d) bottlenecks: (a, b) Averaged vehicle speed in the WSP in space and time. (c, d) Speed (c) and flow rate (d) as time-functions at road location $x = 9$ km within the WSP in (b). Taken from [1]

- In contrast to the LSP and WSP, both the downstream and upstream fronts of the MSP propagate on the road as long as the MSP does not reach an adjacent effectual bottleneck.

Depending on flow rate distribution within an MSP and in free flows outside of the MSP, each of the MSP fronts can propagate either upstream or downstream. This means that there can be three different types of MSPs:

(i) An MSP whose downstream front propagates downstream but the upstream front propagates upstream. The width (in the longitudinal direction) of such an MSP increases continuously over time.
(ii) An MSP that propagates downstream as a whole localized structure; in this case, both MSP fronts propagate downstream. Such an MSP occurs if the average flow rate within the MSP is greater than the flow rates in free flows upstream and downstream of the MSP.
(iii) An MSP that propagates upstream as a whole localized structure; in this case, both MSP fronts propagate upstream (Figs. 7.6–7.8).

Independent of the MSP type, one of the MSP fronts can reach an adjacent effectual bottleneck at which free flow is in a metastable state with respect to traffic breakdown (Sect. 3.3). In this case, the MSP is caught at the bottleneck, i.e., the catch effect occurs (Sect. 3.1.2) resulting in traffic breakdown at the bottleneck. Thus the MSP can exist only for a finite time, before the MSP reaches a nearest adjacent effectual bottleneck at which traffic breakdown can be induced.

Considering MSPs below, we limit attention to MSPs of type (iii), i.e., to MSPs that propagate upstream. Because MSPs emerge often at off-ramp bottlenecks, we discuss this effect in more detail (Figs. 7.6 and 7.7 (a, b)). Traffic breakdown (F→S

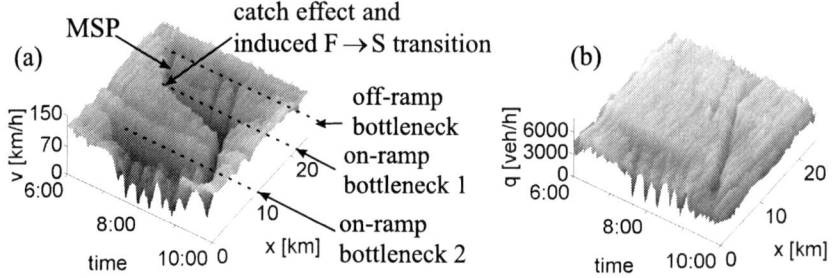

Fig. 7.6 Empirical example of MSP with subsequent catch effect and induced F→S transition at on-ramp bottleneck 1: (a, b) Average speed (a) and flow rate (b) in space and time. Bottlenecks are the same as those in Fig. 3.1. Taken from [1]

transition) at the off-ramp bottleneck, which leads to the emergence of MSPs shown in Fig. 7.7 (a, b), is explained by lane changing [1] (see Sect. 3.3.1.2): Vehicles, which move initially in the passing lane(s) and going to the off-ramp, change to the right lane upstream of the off-ramp increasing the density in the right lane and fluctuations upstream of the off-ramp. For this reason, when an increase in traffic demand upstream of the off-ramp bottleneck occurs, traffic breakdown is observed upstream of the off-ramp. Consequently, the downstream front of the resulting synchronized flow is also fixed upstream of the off-ramp while the upstream front of synchronized flow propagates upstream. However, a return phase transition from synchronized flow to free flow (S→F transition) can occur relatively easily at the off-ramp bottleneck [1]. For example, the increase in traffic demand upstream of the off-ramp bottleneck, which has caused traffic breakdown and resulting synchronized flow emergence at the bottleneck, can last a short time interval only. Then the resulting S→F transition can lead to the restoration of free flow only in a neighborhood of the off-ramp whereas there is still a region of synchronized flow upstream of the effective location of the off-ramp bottleneck. This causes the upstream propagation of the downstream front of the synchronized flow region, i.e., MSP formation.

Between MSPs shown in Fig. 7.7 (a, b) there are free flows during long time intervals and therefore it is not clearly visible whether these MSPs emerge at a bottleneck or without any bottlenecks. To prove that the MSPs emerge at the off-ramp bottleneck, we present empirical Fig. 7.7 (c) in which congested pattern formation occurs on the same road section as that in Fig. 7.7 (a, b). Then the fact that the MSPs shown in Fig. 7.7 (a, b) indeed emerge at the off-ramp bottleneck is clearly seen in Fig. 7.7 (c): The effective location of the off-ramp bottleneck at which the downstream front of synchronized flow is fixed during a long time interval is exactly *the same one* as that of MSP emergence shown in Fig. 7.7 (a, b).

Empirical MSPs and WSPs discussed above exhibit a *common empirical feature* that is as follows:

- After an F→S transition has occurred at a bottleneck, *firstly* the downstream front of resulting synchronized flow is *fixed* at an effective location of the bottleneck.

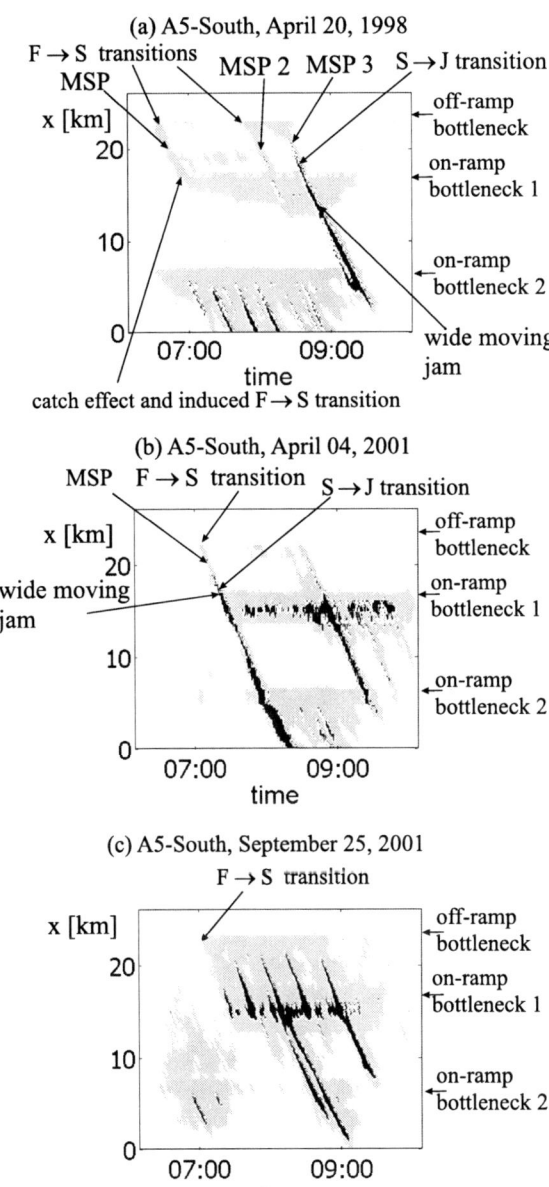

Fig. 7.7 Empirical examples illustrating traffic breakdown at off-ramp bottleneck as well as qualitative difference between MSPs and wide moving jams (see Sect. 7.2.3): Approximate distributions of free flow (white), synchronized flow (gray), and moving jams (black) in space and time for three different days on the same road section. Empirical example shown in Fig. (a) is the same as that in Fig. 7.6. Bottlenecks are the same as those in Figs. 3.1 and 7.6. Taken from [1, 5]

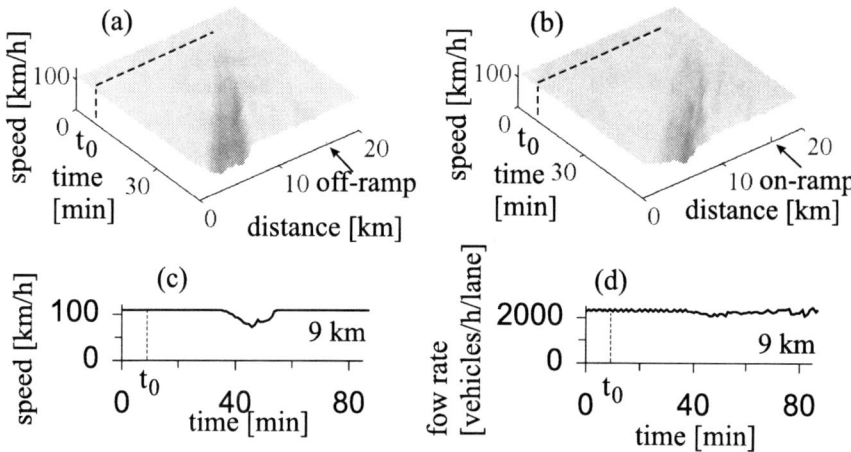

Fig. 7.8 Simulations of MSPs at off-ramp (a) and on-ramp (b–d) bottlenecks in the framework of three-phase traffic theory: (a, b) Averaged vehicle speed in the MSPs in space and time. (c, d) Speed (c) and flow rate (d) as time-functions at road location $x = 9$ km within the MSP in (b). Taken from [1]

In real traffic, a time interval denoted by $T_{\rm syn}^{(B)}$ within which the downstream front of synchronized flow is fixed at the bottleneck is *limited*: After this time interval, a return S→F transition occurs at the bottleneck. In some real measured data, this S→F transition can lead to free flow only in a neighborhood of the bottleneck whereas there is still a region of synchronized flow upstream of the effective location of the bottleneck; as a result, the downstream front of this synchronized flow region begins to propagate in the upstream direction.

- Therefore, an MSP and an WSP occurring at the bottleneck are distinguished from each other *only* through considerably different values of $T_{\rm syn}^{(B)}$: If $T_{\rm syn}^{(B)}$ is appreciably shorter than the time it takes for the upstream front of an SP to reach an upstream adjacent effectual bottleneck, the SP should be classified as an MSP (Fig. 7.7 (a, b)); otherwise, when the downstream front of an SP is fixed at the bottleneck until the upstream front of the SP reaches the upstream bottleneck, the SP should be classified as an WSP.

Finally, we should note that within SPs the speed is lower than in free flow, however, the flow rate can be approximately the same as that in free flow (Figs. 7.2–7.6 and 7.8). The latter feature of SPs can be used in traffic control (Sect. 9.2).

7.2.2 Explanations to Terms "Moving Synchronized Flow Pattern" and "Narrow Moving Jam"

Both an MSP and a narrow moving jam are associated with the synchronized flow phase. Moreover, from the definition of a narrow moving jam made in Sect. 2.6.3 we can conclude that if a narrow moving jam is surrounded both upstream and downstream by free flows, i.e., the jam propagates in free flow, then the jam is an MSP of type (iii).

To explain why rather than an MSP of type (iii) we use a specific term a *narrow moving jam*, we should note that there are two types of narrow moving jams that exhibit qualitatively different features:

(i) A narrow moving jam, which is surrounded by free flows, i.e., the jam propagates in free flow. In this case, the jam can be caught at the bottleneck resulting in traffic breakdown at the bottleneck.
(ii) A narrow moving jam, which is surrounded by synchronized flows, i.e., the jam propagates in synchronized flow. This is usually the case when narrow moving jams emerge spontaneously within a pinch region of synchronized flow (Sect. 5.1).

In contrast with an MSP, a narrow moving jam, which propagates in synchronized flow, cannot lead to traffic breakdown when it reaches a bottleneck. This is because the propagation of a narrow moving jam in synchronized flow means that before the jam has reached the bottleneck, synchronized flow has already been at the bottleneck. Thus rather than an MSP, the narrow moving jam, which is surrounded both upstream and downstream by other states of synchronized flows, is a special state of synchronized flow.

Even if a narrow moving jam propagates in free flow, the jam is only a *particular case* of the MSPs of type (iii). This is because in a general case the width (in the longitudinal direction) of an MSP can change over time considerably. In particular, the MSP width can be considerably greater than the widths of the MSP fronts. In contrast, a narrow moving jam consists of the jam fronts.

7.2.3 Qualitative Difference between Moving Synchronized Flow Pattern and Wide Moving Jam

In real measured data, an MSP that has initially occurred at a bottleneck, i.e., due to an F→S transition at the bottleneck can transform into a wide moving jam over time, i.e., an S→J transition occurs within the MSP. This is because the average speed within synchronized flow of the MSP can decrease considerably while the MSP propagates on the road and, therefore, the pinch effect with the subsequent wide moving jam emergence can occur spontaneously. The width (in the longitudinal direction) of the MSP can be as great as the width of the resulting wide moving jam.

Thus a question arises: what is a qualitative difference between an MSP, which propagates upstream on the road (MSP of type (iii) of Sect. 7.2.1), and a wide moving jam that propagates also upstream on the road? Answers on this question are as follows:

- Considering *averaged empirical data*, the qualitative difference between the MSP and wide moving jam is that the MSP is caught at an upstream adjacent bottleneck, when the MSP reaches the bottleneck at which free flow is in a metastable state with respect to traffic breakdown (Fig. 3.2). In contrast, the wide moving jam propagates through this bottleneck while maintaining the mean velocity of the downstream jam front (Fig. 3.1).
- Considering *single vehicle empirical data*, the qualitative difference between the MSP and wide moving jam is that there is no flow interruption interval within the MSP, i.e., condition (2.23) is not satisfied. In contrast, there is at least one flow interruption interval within the wide moving jam, i.e., condition (2.23) is satisfied.

To illustrate the qualitative difference between the MSP and wide moving jam, we consider results of empirical observations of MSP emergence and the subsequent MSP evolution shown in Fig. 7.7 (a). There are three different MSPs (labeled by MSP, MSP 2, and MSP 3 in Fig. 7.7 (a)) that emerge at an off-ramp bottleneck (see explanations of MSP emergence in Sect. 7.2.1). These MSPs propagate upstream. Two of them (MSP and MSP 2) are caught at an upstream on-ramp bottleneck (labeled by "on-ramp bottleneck 1"). However, the MSP 3 transforms into a wide moving jam *before* the MSP 3 reaches the on-ramp bottleneck 1. This wide moving jam emergence is labeled by "S→J transition" in Fig. 7.7 (a). The wide moving jam propagates subsequently through the on-ramp bottleneck 1 as well as through an upstream on-ramp bottleneck (labeled by "on-ramp bottleneck 2") while maintaining the mean downstream jam front velocity. The same sequence of F→S→J transitions is also observed on this road section on another day (Fig. 7.7 (b)): Firstly, an MSP is formed. Later, during the subsequent upstream propagation of the MSP, it transforms into a wide moving jam (S→J transition) before the MSP reaches the on-ramp bottleneck 1.

It must be noted that there is a time delay of an S→J transition. This time includes a random time delay of spontaneous nucleation of a narrow moving jam in synchronized flow and a random time of the jam growth until the jam transforms into a wide moving jam. The greater the density and the lower the speed in synchronized flow, the shorter the mean time delay of an S→J transition denoted by $T_{SJ}^{(\mathrm{mean})}$. For this reason, in empirical observations $T_{SJ}^{(\mathrm{mean})}$ can change between values that are considerably shorter than 1 min (Fig. 5.5 (b)) to values that are about 10 min (see Fig. 7.9). In turn, the mean duration of traffic breakdown is usually about 1 min or shorter (see, e.g., Fig. 2.5 (a)).

Thus when the density of synchronized flow within an MSP is great enough (respectively, the speed is very low) and, as a result, $T_{SJ}^{(\mathrm{mean})} \lesssim 1$ min, then an initial F→S transition leading to MSP emergence and an S→J transition leading to the subsequent transformation of the MSP into a wide moving jam cannot be distinguished in 1-min averaged traffic data, in particular when the data is measured at spatially separated road locations. This is because the time interval between these F→S and S→J transitions can be too short to be found in the averaged data. Therefore, the formation of this wide moving jam, which occurs due to a sequence of the F→S→J transitions, can be *misinterpreted* as a phase transition from free flow to a wide moving jam (F→J transition).

7.2.4 General Congested Patterns

We compare two examples of empirical GPs. In the first example, an WSP that has initially emerged at an off-ramp bottleneck tranforms into an GP over time (Fig. 7.9). This WSP transformation occurs downstream of an upstream on-ramp bottleneck.

Another empirical example of an GP that emerges at an on-ramp bottleneck is shown in Fig. 7.10. In the GP the frequency of wide moving jam emergence in synchronized flow is considerably greater than the one for the GP shown in Fig. 7.9. This difference in GP formation is explained by a stronger self-compression of synchronized flow in the pinch region of the GP at the on-ramp bottleneck (Fig. 7.10) in comparison with the GP at the off-ramp bottleneck (Fig. 7.9). These and other empirical spatiotemporal features of GP formation have been reproduced in simulations of three-phase traffic flow models (Fig. 7.11) [1].

7.2.5 Congested Patterns at Heavy Bottlenecks

In general, the average flow rate $q^{(\mathrm{cong})}$ within a congested traffic pattern upstream of a bottleneck is the smaller, the greater the bottleneck strength. Empirical data show that the flow rate $q^{(\mathrm{cong})}$ within GPs occurring at usual bottlenecks like on- and off-ramp bottlenecks discussed above is approximately within a range

$$1100 \lesssim q^{(\mathrm{cong})} \lesssim 1700 \text{ vehicles/h/lane}. \tag{7.1}$$

In contrast with the usual bottlenecks, due to for example bad weather conditions or accidents *heavy* bottlenecks can occur, which exhibit a much greater influence on traffic (greater bottleneck strength) that limits $q^{(\mathrm{cong})}$ to small values, sometimes as low as zero, i.e., condition (7.1) is not satisfied. We define a heavy bottleneck as follows.

- A heavy bottleneck is a highway bottleneck that limits the average flow rate $q^{(\mathrm{cong})}$ within a congested traffic pattern upstream of the bottleneck to such small values at which no sequence of wide moving jams can be found in 1-minute (and longer) averaged empirical data.

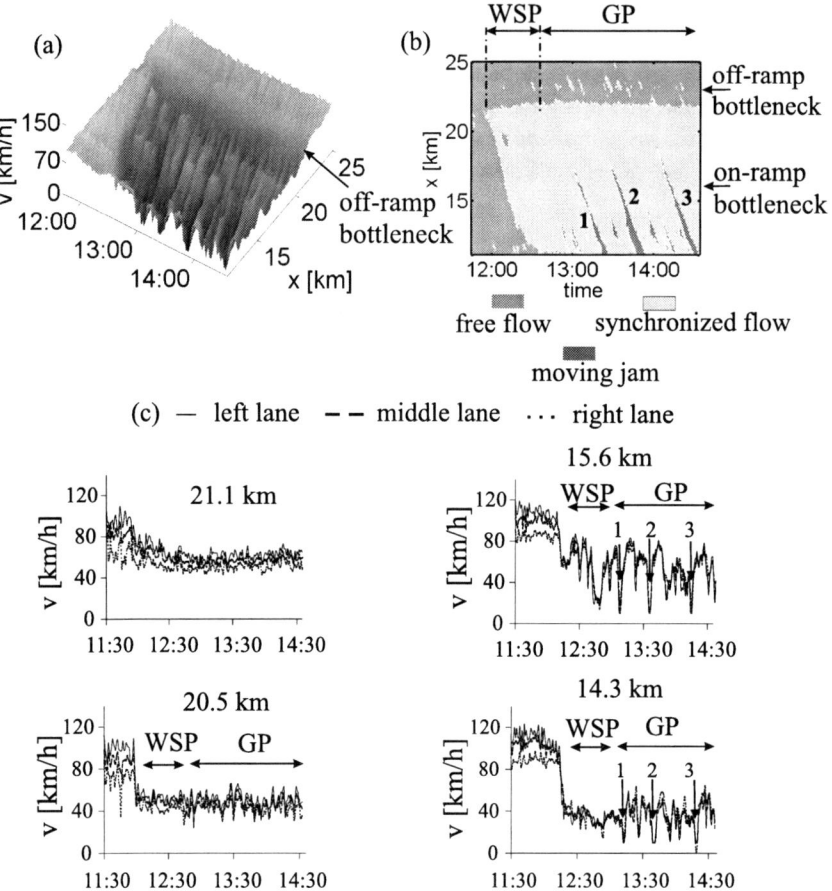

Fig. 7.9 Empirical example of WSP transformation into an GP over time at off-ramp bottleneck: (a) Average speed in space and time. (b) Graph of (a) with overview of free flow, synchronized flow, and wide moving jams labeled by numbers 1, 2, and 3. (c) Time dependence of speed at some fixed road locations. Taken from [1]

An empirical example of such a congested traffic pattern is shown in Fig. 7.12. Rather than the regular structure of traffic congestion within an GP consisting of the pinch region of synchronized flow and a sequence of wide moving jam upstream (Fig. 7.10), in empirical data associated with snow and ice, i.e., bad weather conditions a non-regular spatiotemporal structure of congestion is observed in which *no* sequence of wide moving jams can be distinguished (Fig. 7.12). The average speeds measured in different road lanes (Fig. 7.12 (b)) show that the structure of traffic congestion is very non-homogeneous in space and time. Downstream of the bottleneck speed is high ($x = 4.07$ km in Fig. 7.12 (a)), however, the flow rate $q^{(\mathrm{cong})} \approx 513$ vehicles/h/lane averaged during $T_{\mathrm{av}} = 40$ min is considerably smaller

7.2 Variety of Congested Patterns at Isolated Bottleneck 119

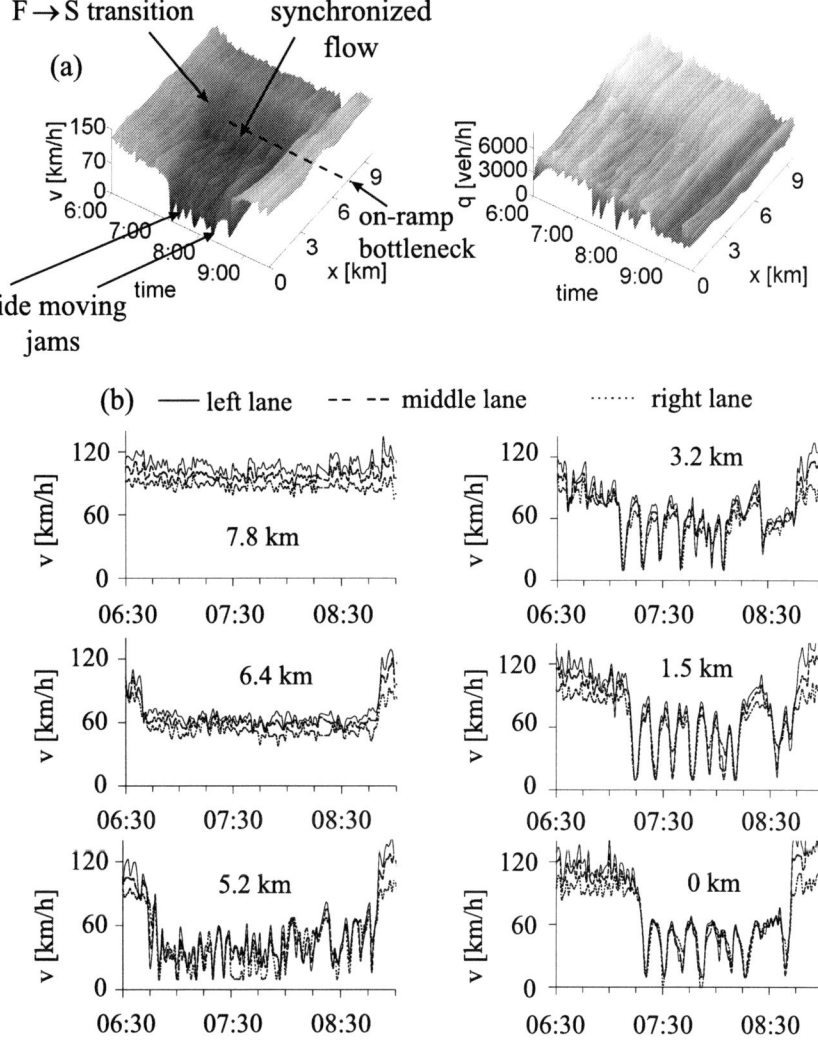

Fig. 7.10 Empirical example of GP at on-ramp bottleneck. (a) Average vehicle speed (left) and total flow rate (right) across road in space and time. (b) Speed in the three road lanes at different locations (1 min data). Taken from [1]

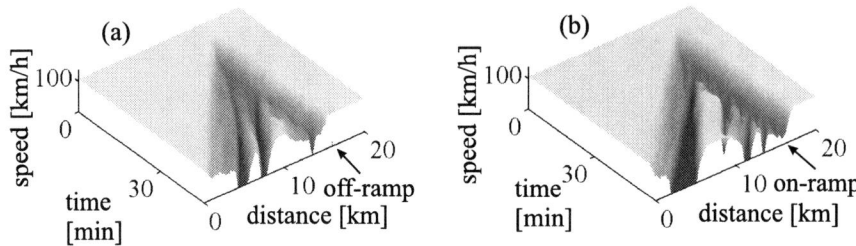

Fig. 7.11 Simulations of GPs at off-ramp (a) and on-ramp bottlenecks (b): Average vehicle speed in space and time. Taken from [1]

than the minimum flow rate in the pinch region of GPs observed in measured traffic data at usual bottlenecks (7.1).

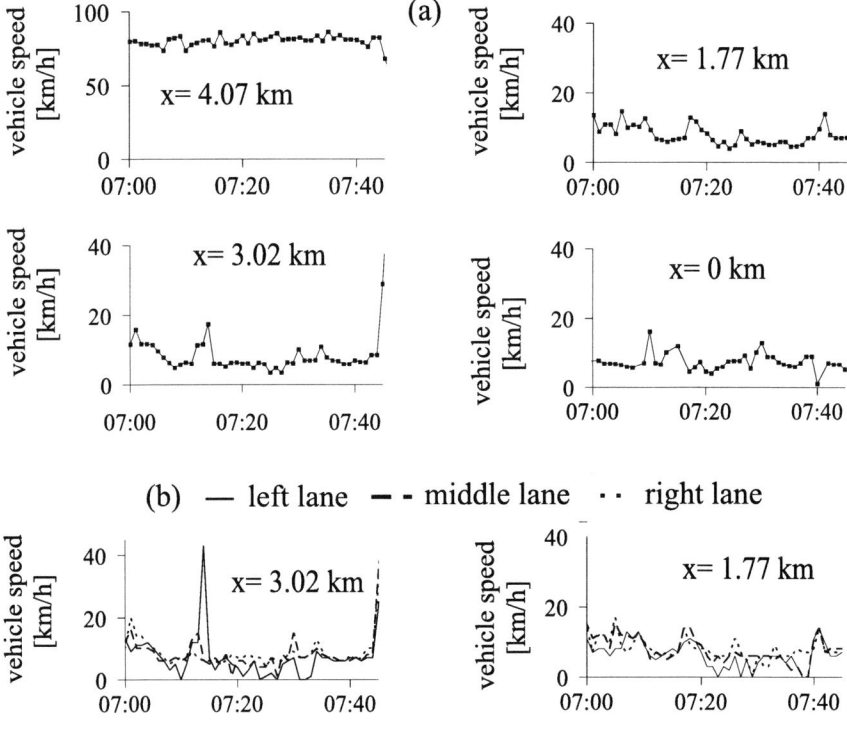

Fig. 7.12 Empirical structure of congestion caused by snow and ice: (a) Average speed across road at different locations. (b) Average speed in different road lanes at two locations. 1-min average data. Taken from [2]

To understand how and why spatiotemporal features of traffic congestion change qualitatively under a continuous decrease in the flow rate $q^{(cong)}$ to small enough values through a continuous increase in the bottleneck strength, in the remaining of this section we discuss briefly results of a theory of traffic congestion at heavy bottlenecks developed in [2].

As discussed in Sect. 7.2.4, under condition (7.1) when an GP is formed at the bottleneck (Fig. 7.13 (a, b)), the GP consists of the pinch region of synchronized flow and a sequence of wide moving jams upstream. In the example of the GP, the pinch region width $L^{(pinch)}(t)$ changes over time between about 1 and 2 km (Fig. 7.13 (c)). $L^{(pinch)}$ is defined as the distance between the downstream front of synchronized flow fixed at the bottleneck and the road location upstream at which a wide moving jam has just occurred.

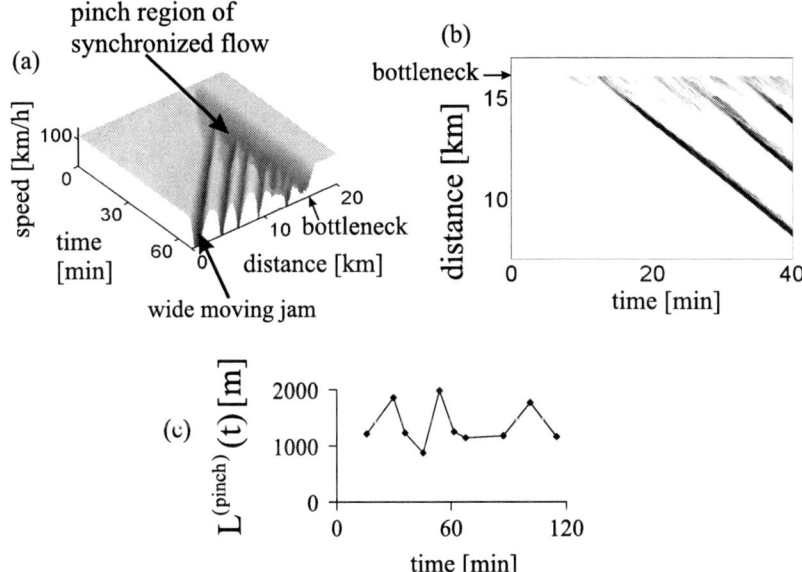

Fig. 7.13 Simulated GP characteristics: (a) Average speed in space and time. (b) Single-vehicle speed data presented by regions with variable darkness (the lower the speed, the darker the region; in white regions the speed is higher than 30 km/h). (d) $L^{(pinch)}(t)$. The flow rate $q^{(cong)} = 1546$ vehicles/h/lane. Taken from [2]

When the flow rate $q^{(cong)}$ decreases due to an increase in the bottleneck strength, the mean pinch region width $L_{mean}^{(pinch)}$ of the GP decreases (Fig. 7.14). At some critical flow rate $q^{(cong)}$ associated with a critical strength of the bottleneck, there are random time intervals when the pinch region disappears, i.e., $L^{(pinch)} = 0$ (Fig. 7.15). This means that there are time instants at which there is no pinch region and wide moving jams emerge directly at the bottleneck, whereas for other time intervals the pinch region appears again; this means that $L^{(pinch)}(t)$ becomes a non-regular time-

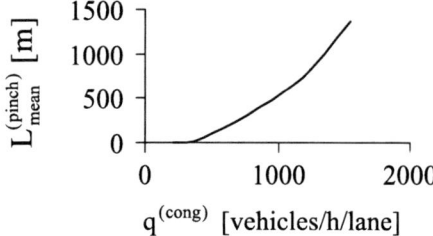

Fig. 7.14 Simulated dependence of the mean length of the pinch region of GPs as function of the flow rate $q^{(\text{cong})}$. Taken from [2]

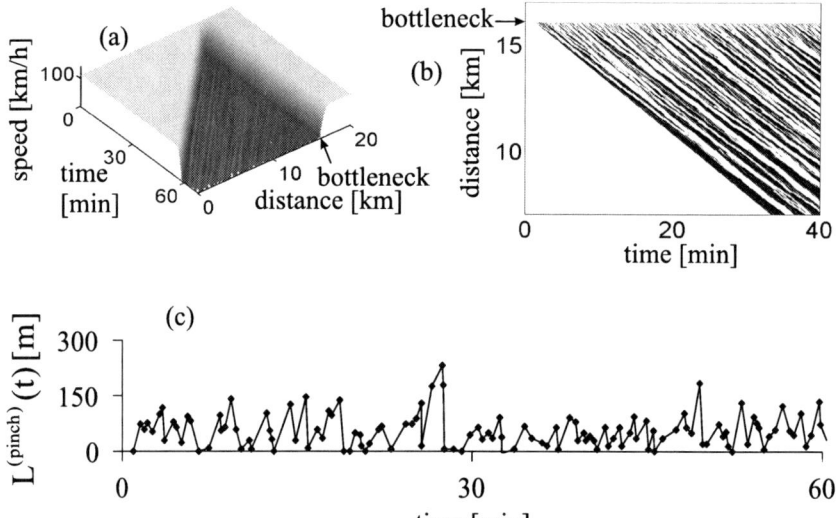

Fig. 7.15 Simulated characteristics of GP with non-regular pinch region: (a) Average speed in space and time. (b) Single-vehicle speed data presented by regions with variable darkness (the lower the speed, the darker the region; in white regions the speed is higher than 5.4 km/h). (c) $L^{(\text{pinch})}(t)$. The flow rate $q^{(\text{cong})} = 440$ vehicles/h/lane. Taken from [2]

function (Fig. 7.15 (c)). Such an GP at the bottleneck can be considered the *GP with a non-regular pinch region*.

When the flow rate $q^{(\text{cong})}$ decreases further due to a greater bottleneck strength, then $L_{\text{mean}}^{(\text{pinch})}$ decreases continuously (Fig. 7.14). $L_{\text{mean}}^{(\text{pinch})}$ reaches zero at a threshold flow rate $q^{(\text{cong})}$ associated with a threshold bottleneck strength at which the pinch region of GPs does not exist.

When an GP with a non-regular pinch region is realized, wide moving jams of the GP can exhibit complex and non-regular spatiotemporal dynamic behavior (Fig. 7.16). This non-regular jam dynamics is associated with the following effects:

7.2 Variety of Congested Patterns at Isolated Bottleneck

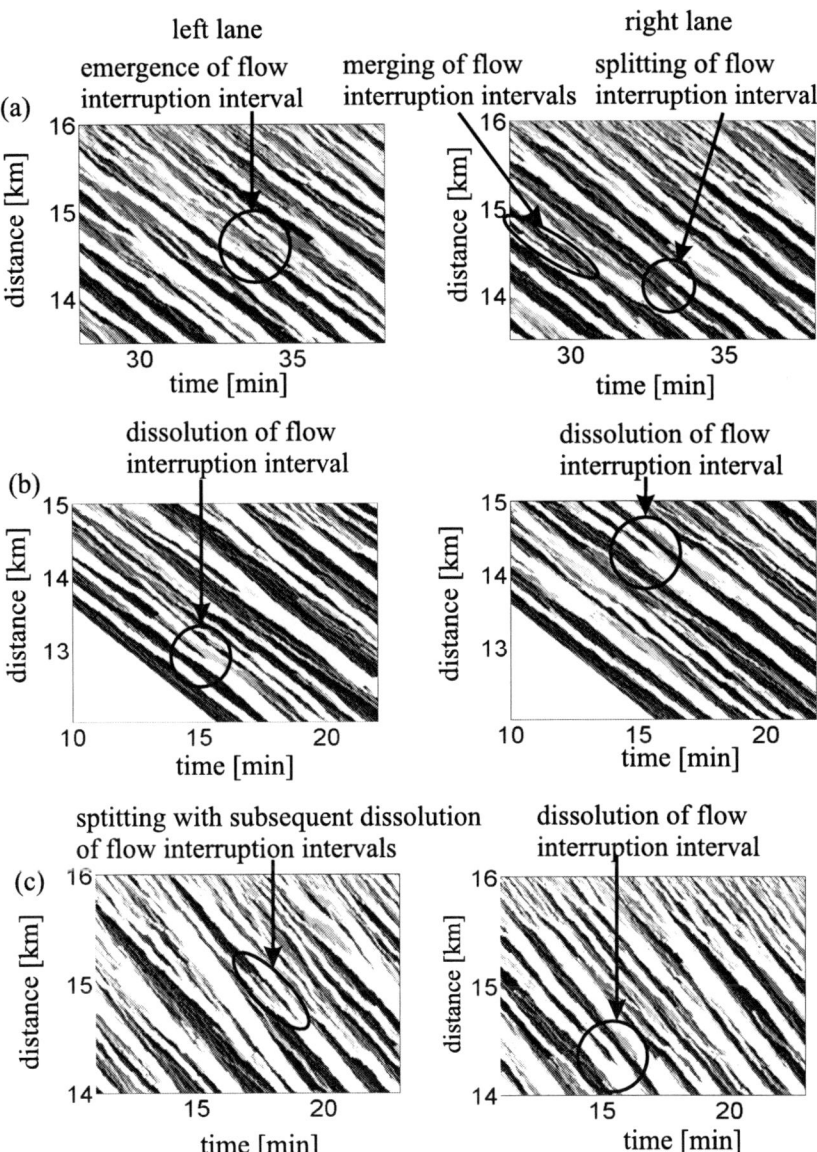

Fig. 7.16 Fragments of Fig. 7.15 (b) in larger scales in time and space in the left (figures left) and right lanes (right). White regions (single-vehicle speeds are equal to or higher than 5.4 km/h) are related to synchronized flows and moving blanks. Black regions (single-vehicle speeds are equal to zero) are related to flow interruption intervals. Taken from [2]

(1) The effect of the *splitting* of a flow interruption interval into two (or more) flow interruption intervals (Fig. 7.16 (a)).
(2) The effect of the *emergence* of a new flow interruption interval (Fig. 7.16 (a)).
(3) The effect of the *merging* of two (or more) traffic flow interruption intervals (Fig. 7.16 (a)).
(4) The effect of the *dissolution* of a flow interruption interval (Fig. 7.16 (b, c)).

The dynamic effects (1)–(4) occur spontaneously during upstream propagation of wide moving jams. In different lanes, these effects occur often independently of each other and at different road locations. A spatiotemporal competition of these effects determines a non-regular dynamic behavior of wide moving jams.

This *fine* non-regular spatiotemporal jam behavior can also explain the result that we cannot distinguish wide moving jams in 1-min average measured data (Fig. 7.12): the microscopic non-regular jam dynamics is averaged in this 1-min average measured traffic data. This averaging of non-regular microscopic spatiotemporal moving jam dynamics leads to non-regular and non-homogeneous speed distributions in which this real fine spatiotemporal jam dynamics cannot be found.

When due to the subsequent increase in the bottleneck strength the flow rate $q^{(\mathrm{cong})}$ decreases further and becomes equal to or smaller than another critical value, wide moving jams *merge* into a mega-wide moving jam (mega-jam for short). In this case, traffic congestion upstream of an isolated heavy bottleneck cannot be considered as an GP any more. Simulations show that

- as the microscopic structure of a wide moving jam, the microscopic spatiotemporal structure of a mega-jam consists of an alternation of flow interruption intervals and moving blanks (Fig. 7.17 (a–d)).

This explains the term *mega-wide moving jam*:

- A mega-jam is a wide moving jam with an extremely great width growing continuously over time.

As for GPs with a non-regular pinch region (Fig. 7.15), within the mega-jam we found a very complex and non-regular spatiotemporal dynamics of flow interruption intervals and moving blanks (Fig. 7.17 (e–h)). The non-regular spatiotemporal mega-jam dynamics is associated with the same dynamic effects (1)–(4) listed above. When the bottleneck strength increases further, the microscopic spatiotemporal structure of traffic congestion does not qualitatively change any more. However, time intervals within which moving blanks occur become shorter.

Thus the complexity of traffic congestion at a heavy bottleneck caused for example by bad weather conditions or accidents is associated with the following traffic phenomena [2]:

- A non-regular dynamic behavior of wide moving jams as well as random disappearance and appearance of the pinch region of synchronized flow within an GP upstream of the bottleneck (Figs. 7.15 and 7.16).
- The merger of wide moving jams of the GP into a mega-jam (Fig. 7.17).

7.2 Variety of Congested Patterns at Isolated Bottleneck 125

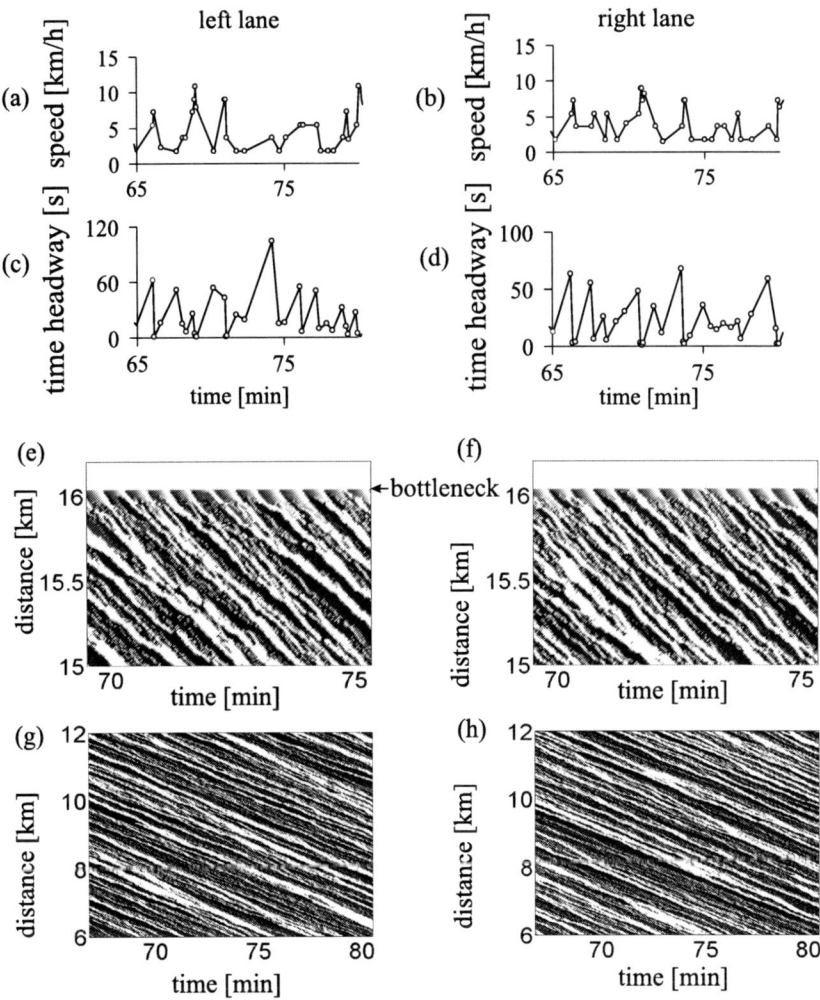

Fig. 7.17 Simulated mega-jam: (a–d) Single-vehicle speed and time headways at location 5 km. (e–h) Single-vehicle speed presented by regions with variable darkness (the lower the speed, the darker the region; in white regions the speed is higher than 1.8 km/h). Taken from [2]

These phenomena determine the evolution of a spatiotemporal structure of traffic congestion at the heavy bottleneck, which occurs when the average flow rate $q^{(\mathrm{cong})}$ within congested traffic decreases due to an increase in the bottleneck strength:

- There is some small average flow rate within traffic congestion at which the mean length of the pinch region of the GP decreases up to zero (Fig. 7.14); at such a heavy bottleneck, the pinch region does not exist any more.
- When this flow rate decreases further due to a greater bottleneck strength, then, beginning at some very small flow rate $q^{(\mathrm{cong})}$, traffic congestion upstream of the

bottleneck consists of the mega-jam only (Fig. 7.17). In this case, synchronized flow remains only within its downstream front fixed at the bottleneck. This synchronized flow front separates free flow downstream and the mega-jam upstream of the bottleneck.

7.3 Complex Congested Patterns at Adjacent Bottlenecks

When there are two or more adjacent bottlenecks, there can be many traffic phenomena associated with complex spatiotemporal dynamics of congested patterns.

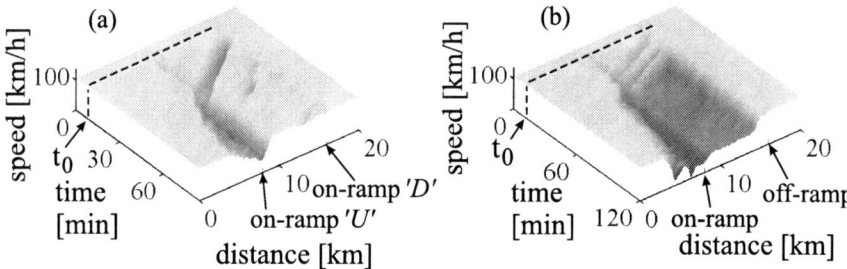

Fig. 7.18 Simulations of induced traffic breakdowns at upstream bottlenecks with the catch effect due to the upstream propagation of SPs. Taken from [1]

In particular, some of these effects discussed above and in [1] are as follows:

(i) *Induced traffic breakdown* caused by upstream congested pattern propagation. A congested pattern, which occurs at the downstream bottleneck, can induce another congested pattern at the upstream bottleneck at which free flow is realized before (Figs. 3.1 and 3.2).

(ii) The *catch effect* that is one of the possible results of an induced traffic breakdown: Rather than to propagate through an effectual bottleneck, an SP is caught at the bottleneck with the subsequent formation of a new congested pattern at this bottleneck (Figs. 3.2, 7.4, 7.6, and 7.18).

(iii) The propagation of *foreign wide moving jam* through a congested pattern at the upstream bottleneck. These foreign wide moving jams occur downstream of the congested pattern within an GP at a downstream bottleneck. A foreign wide moving jam is defined as follows.

- A foreign wide moving jam is a wide moving jam, which has initially occurred downstream of an effectual bottleneck and propagates through a congested pattern upstream of this bottleneck.

7.3 Complex Congested Patterns at Adjacent Bottlenecks 127

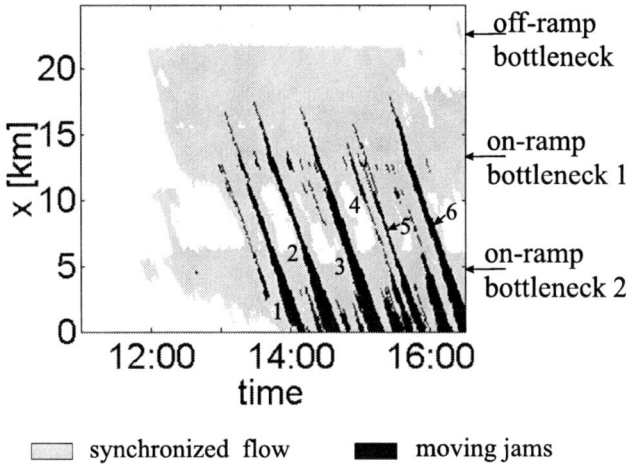

Fig. 7.19 Empirical example of an expanded congested pattern (EP). Free flow (white), synchronized flow (gray), moving jams (black). Taken from [1]

Wide moving jams labeled by numbers 1, 2, ..., 6 in Fig. 7.19 that initially emerged downstream of the on-ramp bottleneck 1 are foreign wide moving jams when they propagate through synchronized flows upstream of the on-ramp bottlenecks 1 and 2. Simulations of foreign wide moving jam propagation is shown in Fig. 7.20.

(iv) An intensification of downstream congestion due to the onset of upstream congestion (see Sect. 19.4 of the book [1]).

In this section, we limit a discussion of so-called expanded congested patterns (EP). We show that due to lane changing and vehicle merging at adjacent bottlenecks a very complex non-regular moving jam dynamics within synchronized flow of EPs on multi-lane roads can be observed.

7.3.1 Expanded Congested Patterns (EP)

An expanded traffic congested pattern (EP) is defined as follows.

- An EP is a congested traffic pattern whose synchronized flow affects at least two effectual adjacent bottlenecks.

If congested traffic within an EP consists only of the synchronized flow phase, the EP is an example of an SP that affects at least two bottlenecks.

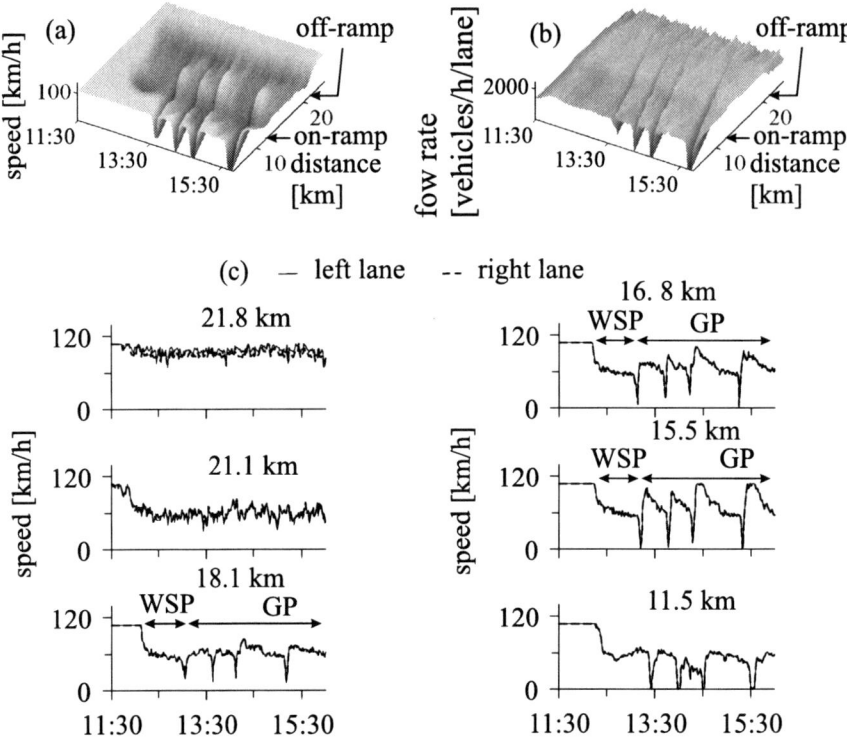

Fig. 7.20 Simulations of a part of empirical EP shown in Figs. 7.9 and 7.19: (a, b) Speed (a) and flow rate (b) in space and time. (c) Time-functions of speed at different locations. Taken from [3]

If congested traffic within an EP consists of both traffic phases of congested traffic, synchronized flow and wide moving jam, then the EP is an example of an GP whose synchronized flow affects at least two bottlenecks:

- The term *expanded congested pattern* emphasizes only that rather than the SP or the GP occurs at an isolated effectual bottleneck, the synchronized flow of the SP or the GP affects two or more effectual adjacent highway bottlenecks.

To emphasize this point, we use for these different EPs two terms *ESP* and *EGP* as follows:

- An expanded synchronized flow traffic pattern (ESP for short) is an SP whose synchronized flow affects two or more effectual adjacent highway bottlenecks.
- An expanded general congested traffic pattern (EGP for short) is an GP whose synchronized flow affects two or more effectual adjacent highway bottlenecks.

A typical empirical example of EP formation is shown in Fig. 7.4: An WSP that has initially occurred at the downstream off-ramp bottleneck exists only for a relatively short time, which is determined by the propagation of the upstream front of the WSP to the upstream on-ramp bottleneck. In this case, the lifetime of the WSP is about 20 min. After synchronized flow of the WSP is caught at the upstream bottleneck, a new congested pattern is formed. Synchronized flow in this pattern affects both bottlenecks. Thus this EP is an ESP.

It should be noted that the ESP formation can also be considered the transformation of the initial WSP into the ESP. Congested pattern transformation, i.e., the transformation of one type of a congested pattern into another type of a congested pattern is observed very often in real measured traffic data (see Chaps. 13 and 14 of the book [1]).

If the above empirical congested pattern (Fig. 7.4) is considered during a considerably longer time interval and on the whole road section on which measurements are available, we find a typical complex EP with several wide moving jams propagating through the EP and two upstream on-ramp bottlenecks while maintaining the mean velocity for the downstream jam front that are the same for these different jams (jams labeled by numbers 1,..., 6 in Fig. 7.19). Thus in this case, the ESP transforms into an EGP whose synchronized flow affects three effectual bottlenecks.

7.3.2 Non-Regular Spatiotemporal Traffic Dynamics on Multi-Lane Roads

If at least two adjacent bottlenecks are located very closely to each other, as this is the case on the road US 101 (Fig. 5.5 (a)), and an EP occurs whose synchronized flow affects these bottlenecks, a complex moving jam dynamics is observed within synchronized flow of the EP [5, 7]:

- Moving jams can emerge and dissolve in synchronized flow randomly at different road locations in different road lanes (Fig. 7.21).

This empirical complexity of the spatiotemporal dynamics of moving jams can be explained by numerical simulations of EGPs occurring on a two-lane road section with several closely located adjacent on-ramp and off-ramp bottlenecks (Figs. 7.22, 7.23, and 7.24). Some results of this theoretical study [5, 7] are as follows:

1. At a given flow rate upstream of the bottlenecks and different sets of on-ramp inflow rates and percentages of vehicles going to different off-ramps we find qualitatively different EGPs:

   - At the same sequence of adjacent bottlenecks, the *infinite number* of various EPs with the complex dynamics of congested traffic can be expected at different sets of on-ramp inflow rates and percentage of vehicles going to off-ramps (only three examples are shown in Figs. 7.22, 7.23, and 7.24).

2. The complex dynamics of congested traffic within the EPs that is often a non-regular one can be considered resulting from spatiotemporal combinations and interactions of the diverse dynamics of synchronized flow and wide moving jams.

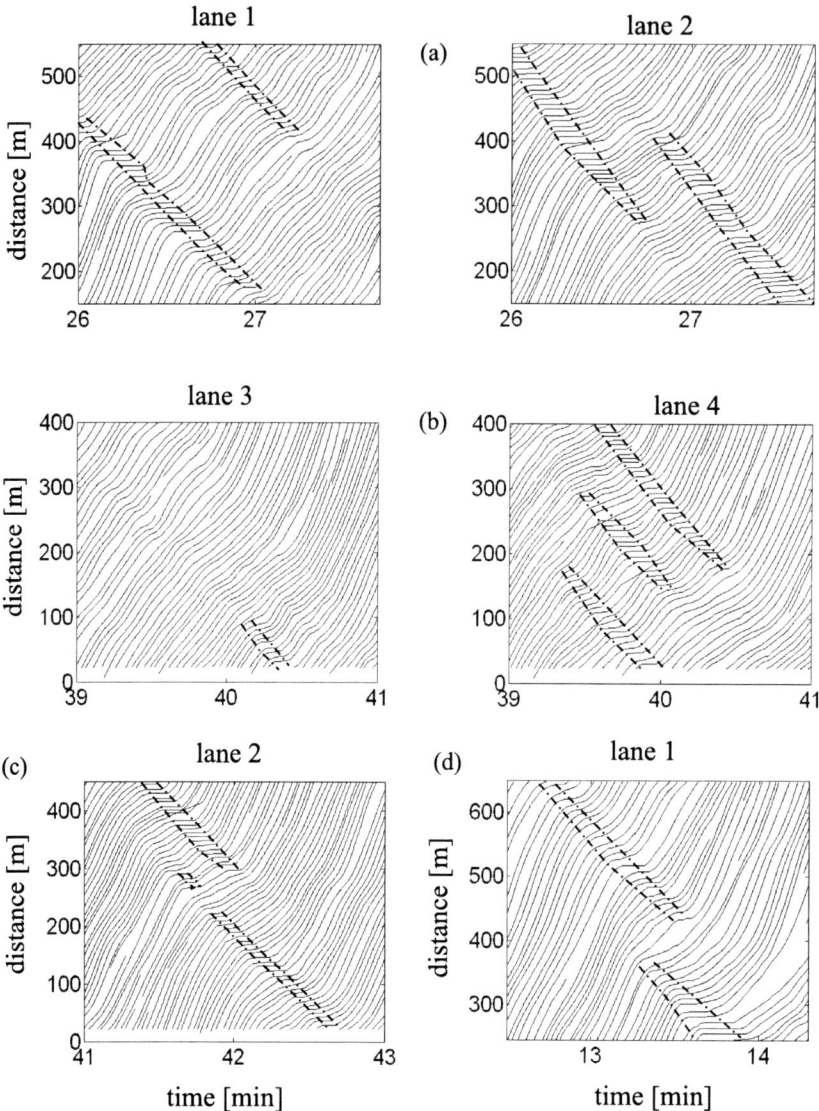

Fig. 7.21 Empirical vehicle trajectories showing complex non-regular spatiotemporal dynamics of moving jams within synchronized flow that affects the upstream on-ramp and downstream off-ramp bottlenecks on the road US 101 shown in Fig. 5.5 (a). Moving jams are marked-off by dashed curves. NGSIM-single vehicle data measured on June 15, 2005 [6]. In figures time $t = 0$ is related to 7:50 am in raw NGSIM-data. Taken from [5]

7.3 Complex Congested Patterns at Adjacent Bottlenecks

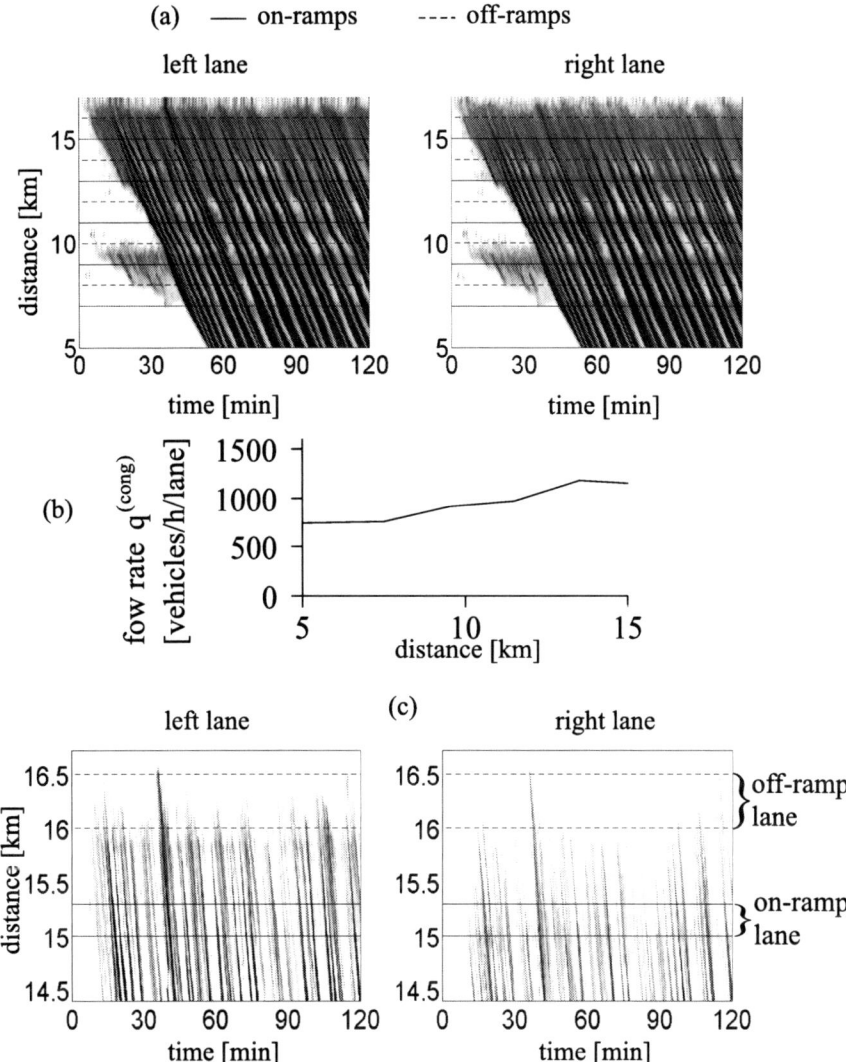

Fig. 7.22 Analysis of EGP shown in Fig. 5.6 (a): (a) Single-vehicle speed data presented in space and time by regions with variable darkness (the lower the speed, the darker the region; in white regions, speeds are higher than 80 km/h; in black regions, speeds are equal to zero). Beginnings of merging regions of the on-ramps and off-ramps are at locations $x_{\text{on},i} = 7, 9, 11, 13, 15$ km and $x_{\text{off},j} = 8, 10, 12, 14, 16$ km, respectively. (b) Flow rate $q^{(\text{cong})}$ within the EGP as a function of road location. (c) Jam dynamics in the neighborhood of the farthest on- and off-ramp bottlenecks; in white regions speeds are higher than 20 km/h. Taken from [5, 7]

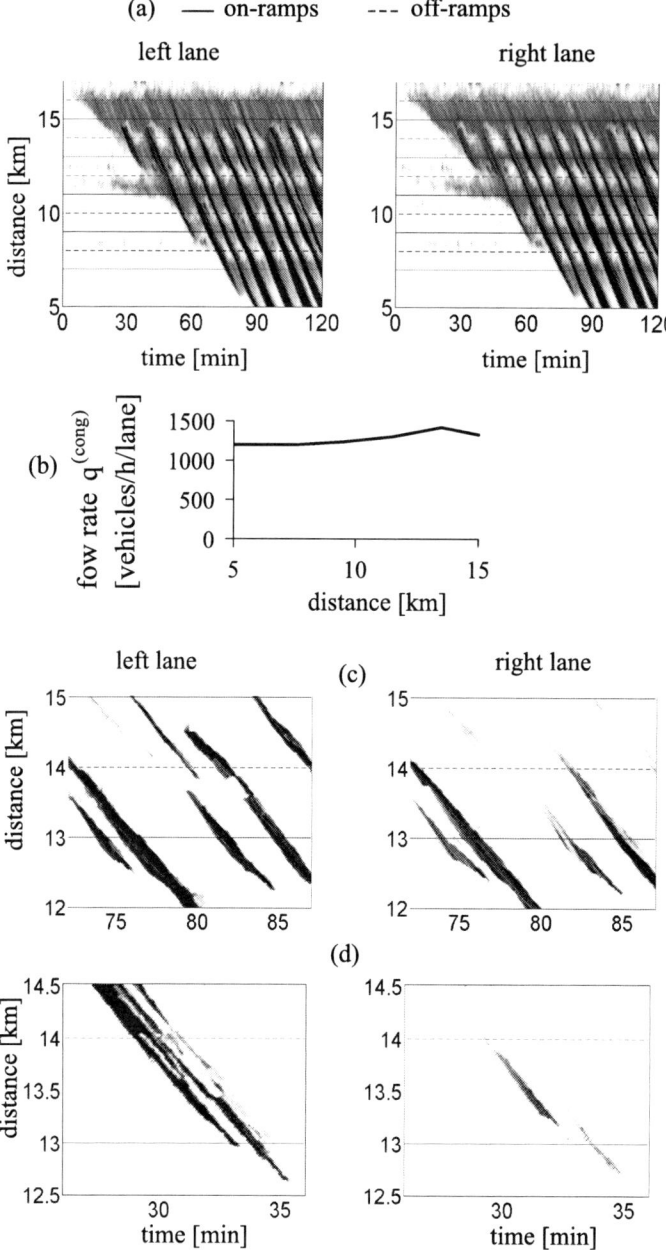

Fig. 7.23 Simulations of EGP: (a, b) Single-vehicle speed data (a) (in white regions, speeds are higher than 80 km/h; in black regions, speeds are equal to zero) and flow rate $q^{(\mathrm{cong})}$ (b) associated with the same adjacent bottlenecks as those in Fig. 7.22, however, at another set of flow rates to on-ramps and percentages of vehicles going to off-ramps. (c, d) Two fragments of (a) in greater scales in space and time (in white regions, speeds are higher than 7.2 km/h; in black regions, speeds are equal to zero). Taken from [5, 7]

7.3 Complex Congested Patterns at Adjacent Bottlenecks   133

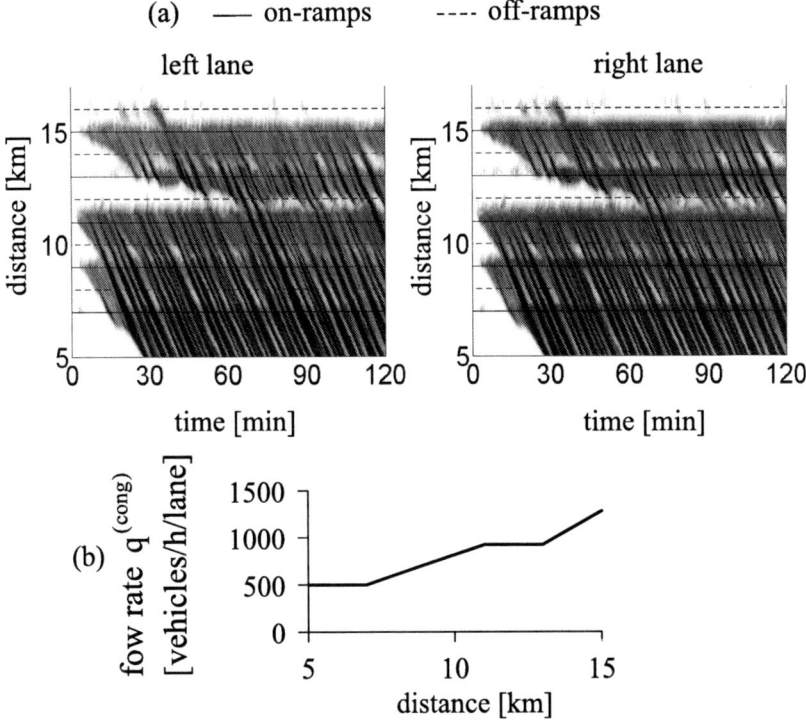

Fig. 7.24 Simulations of EGP: The same adjacent bottlenecks as those in Figs. 7.22 and 7.23, however, at another set of flow rates to on-ramps and percentages of vehicles going to off-ramps. Taken from [5, 7]

3. In general, the non-regular dynamics of wide moving jams within the EGPs is associated with the same effects (splitting, merging, emergence, and dissolution of flow interruption intervals) as those found in congested traffic at an isolated heavy bottleneck discussed in Sect. 7.2.5. Due to closely located adjacent bottlenecks these effects can occur even if within an EGP the flow rate $q^{(\mathrm{cong})}$ is not very small, i.e., it satisfies condition (7.1).

4. In addition with these effects, we find a variety of *nucleation-interruption effects* in synchronized flow of the EGPs [1]:

- A nucleation-interruption effect is a sequence of the emergence and dissolution of a narrow moving jam in synchronized flow:
  - firstly, a growing narrow moving jam occurs in synchronized flow;
  - later, this jam dissolves, i.e., the narrow moving jam persists only during a time interval, which is too short for the transformation of the narrow moving jam into a wide moving jam.

A variety of the nucleation-interruption effects found in simulations (Fig. 7.25) can explain empirical complex dynamics of narrow moving jams within syn-

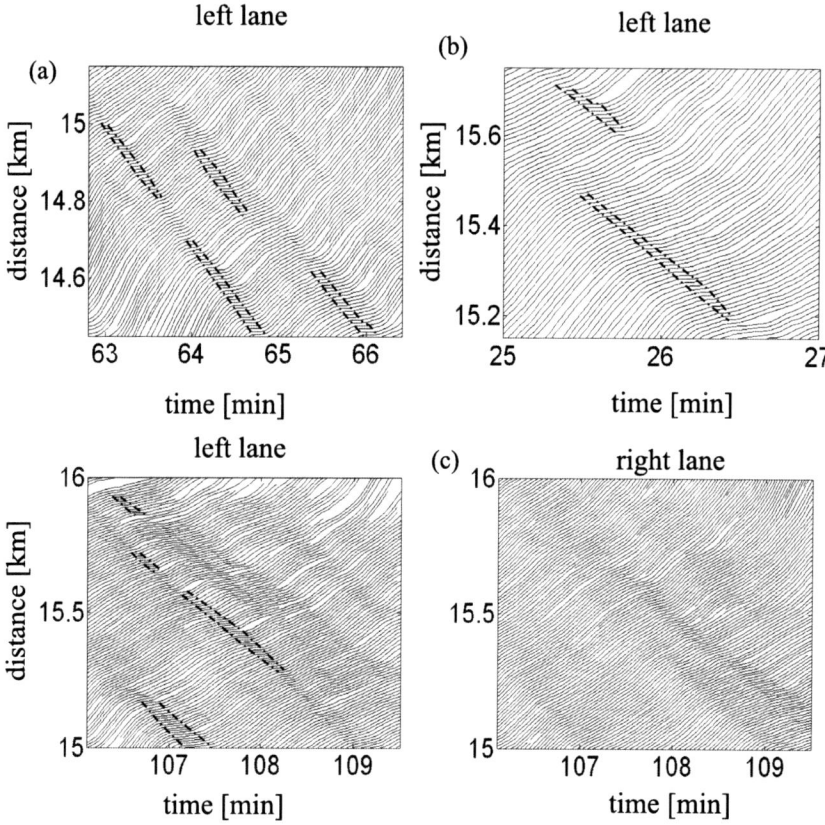

Fig. 7.25 Simulations of nucleation-interruption effects resulting in complex non-regular spatiotemporal dynamics of narrow moving jams within an EGP in a neighborhood of the farthest on- and off-ramp bottlenecks shown in Fig. 7.22 (c): Vehicle trajectories in space and time in the left lane only (a, b) and in both lanes (c) for three pattern fragments; narrow moving jams are marked-off by dashed curves. Taken from [5]

chronized flow in which the jams occur and dissolve at different road locations randomly and often independent of each other in the different lanes (Fig. 7.21).

5. These diverse nucleation-interruption effects found in empirical data (Fig. 7.21) and simulations (Figs. 7.23 (c, d) and 7.25) can be explained by the following traffic phenomena:

- the dual role of lane changing in synchronized flow (Sect. 5.3):
  - lane changing can lead to the occurrence of a nucleus for an S→J transition, i.e., to the occurrence of a growing narrow moving jam;
  - in contrast, lane changing can lead to the dissolution of a narrow moving jam, i.e., to the maintenance of synchronized flow,

- the increase in the frequency of lane changing in neighborhoods of on- and off-ramp bottlenecks within the EGPs.

6. In addition to lane changing between lanes on the main road, there are the following effects at the on- and off-ramp bottlenecks that can cause nucleation-interruption effects:

- vehicle merging from an on-ramp onto the main road can lead to the deceleration of the following vehicles resulting in the occurrence of a growing narrow moving jam in synchronized flow on the main road;
- vehicles leaving the main road to an off-ramp can lead to the increase in space gaps between vehicles moving on the main road resulting in the dissolution of a growing narrow moving jam in synchronized flow on the main road.

7. At different sets of the on-ramp inflow rates and percentages of vehicles going to off-ramps we find that

- either the average flow rate in congested traffic $q^{(\mathrm{cong})}$ satisfies condition (7.1) for usual bottlenecks or $q^{(\mathrm{cong})}$ is very small as that for a heavy bottleneck;
- EGPs can occur for which on some road section condition (7.1) for usual bottlenecks is satisfied (road locations $13 < x < 15$ km in Fig. 7.24), whereas at other road locations within the EGPs congested traffic should be referred to traffic congestion at heavy bottlenecks ($x < 10$ km in Fig. 7.24).

References

1. B.S. Kerner, *The Physics of Traffic*, (Springer, Berlin, New York, 2004)
2. B.S. Kerner, J. Phys. A: Math. Theor. **41**, 215101 (2008); 369801 (2008)
3. B.S. Kerner, S.L. Klenov, A. Hiller, J. Phys. A: Math. Gen. **39**, 2001–2020 (2006)
4. B.S. Kerner, S.L. Klenov, A. Hiller, H. Rehborn, Phys. Rev. E **73**, 046107 (2006)
5. B.S. Kerner, S.L. Klenov, unpublished
6. Next Generation Simulation Programs, http://ngsim.camsys.com/
7. B.S. Kerner, S.L. Klenov, Trans. Rec. Rec. (2009) (in press)

Part II
Impact of Three-Phase Traffic Theory on Transportation Engineering

Chapter 8
Introduction to Part II: Compendium of Three-Phase Traffic Theory

Empirical spatiotemporal features of traffic patterns determine all traffic flow characteristics, which are the basis for reliable methods of traffic control and dynamic management. As shown in Part I, three-phase traffic theory explains the pattern features. In this chapter, we summarize the main definitions and hypotheses of this theory [1–3]:

(1) There are two traffic phases in congested traffic: synchronized flow and wide moving jam. Therefore, there are three traffic phases:

  1. Free flow (F).
  2. Synchronized flow (S).
  3. Wide moving jam (J).

(2) The wide moving jam and synchronized flow phases in congested traffic are defined through the empirical criteria [J] and [S]:

  [J] A wide moving jam is a moving jam that maintains the mean velocity of the downstream front of the jam v_g while propagating through highway bottlenecks.
  [S] In contrast to the wide moving jam, the downstream front of synchronized flow is usually fixed at a bottleneck.

  The characteristic feature of wide moving jam propagation [J] can be presented by a line J in the flow–density plane whose slope is equal to the velocity v_g (Fig. 8.1 (a)).

(3) Steady states of synchronized flow cover a two-dimensional (2D) region in the flow–density plane (dashed region in Fig. 8.1 (a)): at a given speed in synchronized flow, a driver accepts different space gaps within a limited gap range $g_{safe} \leq g \leq G$ (G and g_{safe} are the synchronization and safe space gaps, respectively, see Fig. 8.1 (b)) and does not control a fixed space gap to the preceding vehicle.

(4) Traffic breakdown at a bottleneck is a local phase transition from free flow to synchronized flow (F→S transition). Traffic breakdown is explained by a com-

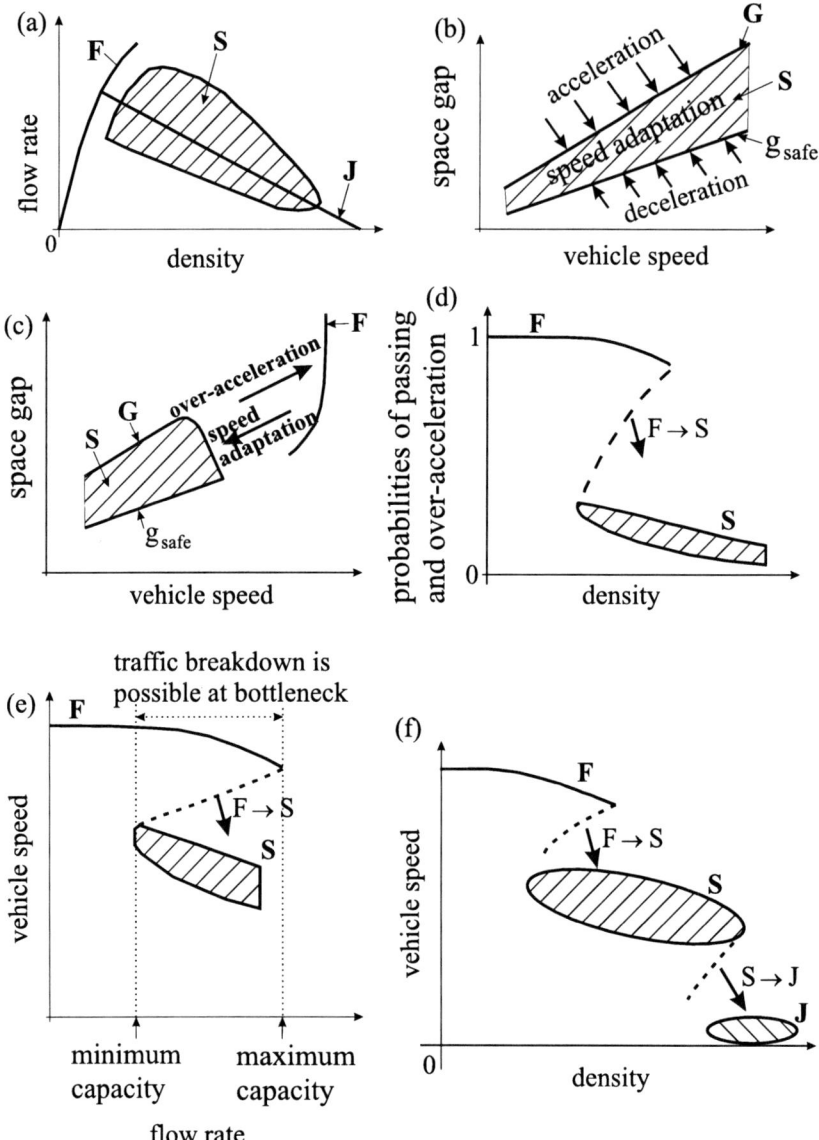

Fig. 8.1 Hypotheses of three-phase traffic theory about phase transitions in traffic flow [1, 2]. (a) Qualitative presentation of free flow states F, a 2D-region of steady states of synchronized flow S, and the line J in the flow–density plane. (b) Qualitative explanation of the speed adaptation effect. (c) Qualitative explanation of a competition between speed adaptation and over-acceleration in the space-gap–speed plane. (d) Qualitative Z-shaped function of probabilities of passing and over-acceleration on density. (e) Qualitative shape of the Z-characteristic for traffic breakdown in the speed–flow plane. (f) Qualitative shape of the double Z-characteristic for F→S→J transitions in the speed–density plane. In (b–f), dashed regions S are parts of the 2D-region for synchronized flow in (a)

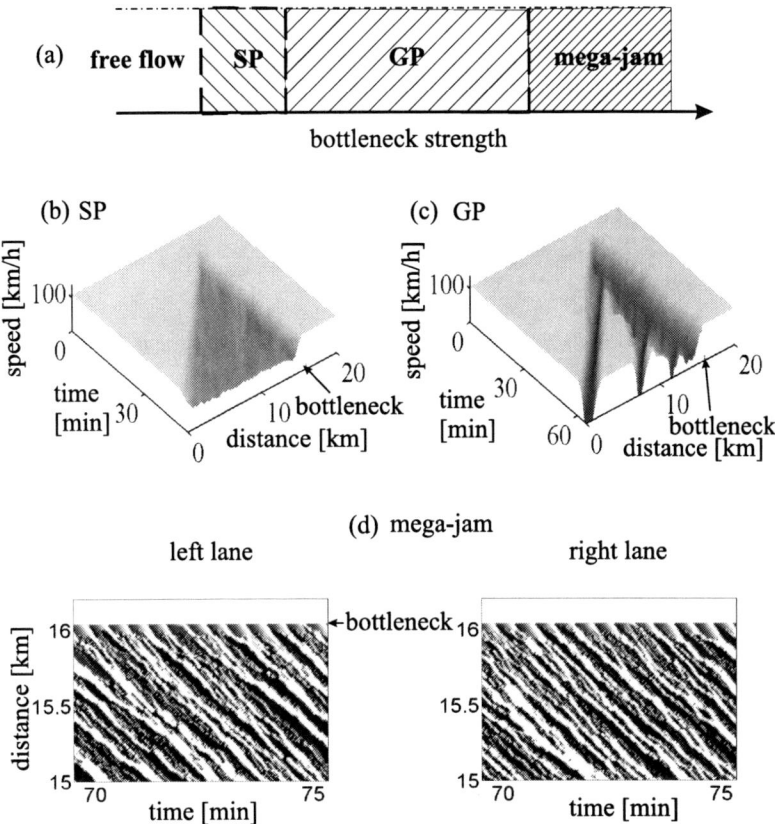

Fig. 8.2 Hypotheses of three-phase traffic theory about congested patterns at a bottleneck [1, 3]. (a) Qualitative simplified diagram of congested patterns at the bottleneck that represents regions of the spontaneous emergence of synchronized flow patterns (SP), general patterns (GP), and mega-jam at the bottleneck as a function of the bottleneck strength at a given great enough flow rate in free flow on the main road upstream of these congested patterns. (b, c) Speed in space and time in one of the possible SPs (b) and GPs (c). (d) Flow interruption intervals (black regions) and moving blanks, i.e., moving upstream regions without vehicles (white regions) within a mega-jam in the left (figure left) and right (right) road lanes. In (b–d), results of numerical simulations are shown

petition between two opposing tendencies occurring within a local disturbance in free flow at the bottleneck in which the speed is lower and vehicle density is greater than in an initial free flow:

(i) a tendency towards synchronized flow due to vehicle deceleration associated with a speed adaptation effect (arrow towards synchronized flow in Fig. 8.1 (c));

(ii) a tendency towards the initial free flow due to vehicle acceleration associated with an over-acceleration effect (arrow towards free flow in Fig. 8.1 (c)).

The speed adaptation effect occurring in car-following is the adaptation of the vehicle speed to the speed of the preceding vehicle to any space gap within the 2D-region of synchronized flow states of item (3) (Fig. 8.1 (b)). The over-acceleration effect is driver maneuver leading to a higher speed from an initial car-following at a lower speed. In particular, the over-acceleration effect is associated with vehicle acceleration for passing from car-following, i.e., lane changing to a faster lane.

(5) Probabilities of passing and over-acceleration are Z-shaped density functions (Fig. 8.1 (d)).

(6) There is the infinite number of highway capacities of free flow at the bottleneck. The capacities are within a capacities range between the minimum capacity and maximum capacity (Fig. 8.1 (e)).

(7) The line J (Fig. 8.1 (a)) divides all steady states of synchronized flow into (i) states, which are below the line J, in which no wide moving jams emerge and (ii) states, which are above the line J, in which a wide moving jam can emerge spontaneously (S→J transition).

(8) In free flow, wide moving jams emerge due to a sequence of F→S→J transitions that can be illustrated by a double Z-characteristic (Fig. 8.1 (f)). The first Z-shaped relationship, which includes free flow F and synchronized flow S, is associated with an F→S transition, i.e., traffic breakdown (labeled by arrows F→S in Figs. 8.1 (d–f)). The second Z-shaped relationship, which includes synchronized flow S and low speed states within wide moving jams J, is associated with an S→J transition (labeled by arrow S→J in Fig. 8.1 (f)).

(9) If the bottleneck strength, which characterizes the influence of a bottleneck on traffic breakdown and resulting traffic congestion, increases gradually, firstly a synchronized flow pattern (SP) emerges upstream of the bottleneck (Fig. 8.2 (a, b)). Congested traffic within an SP consists of synchronized flow only. At a greater bottleneck strength, the SP transforms into a general congested pattern (GP) (Fig. 8.2 (a, c)). Congested traffic within an GP consists of synchronized flow and wide moving jams that emerge in synchronized flow. If the bottleneck strength increases further, the GP transforms into a mega-wide moving jam (mega-jam) (Fig. 8.2 (a, d)) while synchronized flow remains only within the downstream front of synchronized flow fixed at the bottleneck; the front separates free flow downstream and the mega-jam upstream.

References

1. B.S. Kerner, in *Proceedings of the 3rd Symposium on Highway Capacity and Level of Service*, ed. by R. Rysgaard. Vol 2, (Road Directorate, Ministry of Transport – Denmark, 1998), pp. 621–642; Phys. Rev. Lett. **81**, 3797–3400 (1998); Trans. Res. Rec. **1678**, 160–167 (1999); in *Transportation and Traffic Theory*, ed. by A. Ceder. (Elsevier Science, Amsterdam 1999), pp. 147–171; Physics World **12**, 25–30 (August 1999); Phys. Rev. E **65**, 046138 (2002)
2. B.S. Kerner, *The Physics of Traffic*, (Springer, Berlin, New York, 2004)
3. B.S. Kerner, J. Phys. A: Math. Theor. **41**, 215101 (2008); 369801 (2008)

Chapter 9
Freeway Traffic Control based on Three-Phase Traffic Theory

9.1 Reconstruction and Tracking of Congested Patterns

For most traffic control methods, a reliable traffic pattern reconstruction and prediction is of a great importance. For this reason, before we start with a consideration of traffic control methods, we briefly[1] discuss methods for reconstruction, tracking, and prediction of spatiotemporal congested traffic patterns.

9.1.1 FOTO and ASDA Models

The FOTO and ASDA models perform automatic recognition and tracking of spatiotemporal congested traffic patterns. These models have been introduced by the author in 1996–1998 and further developed and implemented in on-line applications together with Rehborn, Aleksić, and Haug [2, 4–7]. Recently, Rehborn and Palmer have made a further development implementation of the FOTO and ASDA models for on-line applications and various road installations [8–11].

The FOTO and ASDA models are based on a spatiotemporal traffic phase classification made in three-phase traffic theory. A main feature of the FOTO and ASDA models is that they perform reliably *without any validation of model parameters* in different environmental and traffic conditions.

The FOTO (**F**orecasting of **T**raffic **O**bjects) model identifies traffic phases and tracks synchronized flow. The ASDA model (**A**utomatische **S**tau**d**ynamik**a**nalyse: Automatic Tracking of Moving Jams) is devoted to the tracking of moving jam propagation.

In the FOTO model, firstly based on local measurements of traffic the local identification of the traffic phases is performed (Fig. 9.1). Recall that in averaged, i.e.,

---

[1] A more detailed consideration of reconstruction, tracking, and prediction of congested patterns and their on-line applications in real installations can be found in Chaps. 21 and 22 of the book [1] as well as in [2, 3].

B.S. Kerner, *Introduction to Modern Traffic Flow Theory and Control*,
DOI 10.1007/978-3-642-02605-8_9, © Springer-Verlag Berlin Heidelberg 2009

macroscopic measured data an accurate identification of the synchronized flow and wide moving jam phases can be made based on the criteria [S] and [J]. In order to do this, the evolution of the data in space and time has to be analyzed (see Sect. 2.4). For this reason, to recognize and track traffic congested patterns in *on-line applications*, rather than the use of the criteria [S] and [J], the FOTO model performs an *approximate* traffic phase identification based on the following algorithm. At each time step, FOTO analyses averaged speeds and flow rates measured by detectors. Traffic phases are identified based either on a set of fuzzy rules of [1, 2] or with the following more simple rules: (i) FOTO detects synchronized flow, when

$$v \leq v_{\text{syn}}; \tag{9.1}$$

(ii) FOTO detects a wide moving jam, when

$$v < v_{\text{jam}} \text{ and } q < q_{\text{jam}}; \tag{9.2}$$

(iii) FOTO detects free flow, when

$$v > v_{\text{syn}}, \tag{9.3}$$

where speeds v_{syn}, v_{jam}, and the flow rate q_{jam} are model parameters.

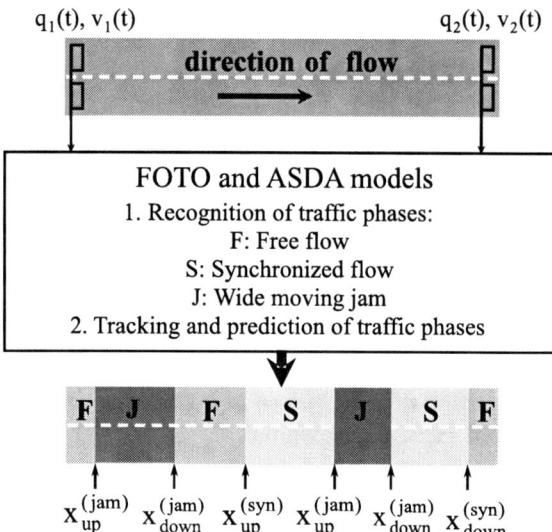

Fig. 9.1 Illustration of the FOTO and ASDA model approach. Taken from [1]

Secondly, fronts of wide moving jams ($x_{\text{up}}^{(\text{jam})}$, $x_{\text{down}}^{(\text{jam})}$) and of synchronized flow ($x_{\text{up}}^{(\text{syn})}$, $x_{\text{down}}^{(\text{syn})}$) are determined. Finally, based on the FOTO and ASDA models the tracking and forecasting of these fronts in time and space is calculated, i.e., the posi-

9.1 Reconstruction and Tracking of Congested Patterns

tions of all fronts $x_{\text{up}}^{(\text{jam})}$, $x_{\text{down}}^{(\text{jam})}$, $x_{\text{up}}^{(\text{syn})}$, and $x_{\text{down}}^{(\text{syn})}$ as functions of time are found. Note that for traffic forecasting, historical time series at least for flow rates are necessary.

This tracking is related to the *on-line* description of the upstream and downstream fronts of synchronized flow and wide moving jams as well as the average speed within these traffic phases. This tracking is also carried out *between* road detectors (or other road locations at which the phases have been identified), i.e., when synchronized flow and wide moving jams cannot be measured at all. Note that the FOTO and ASDA models enable us to predict a merging and/or a dissolution of two or more initially different synchronized flow regions and of two or more initially different wide moving jams that occur between detector locations.

Using the average vehicle speed and density (or occupancy) within the traffic phases that are known from empirical studies and current traffic measurements one can calculate other traffic characteristics, e.g., trip travel times and/or vehicle trajectories.

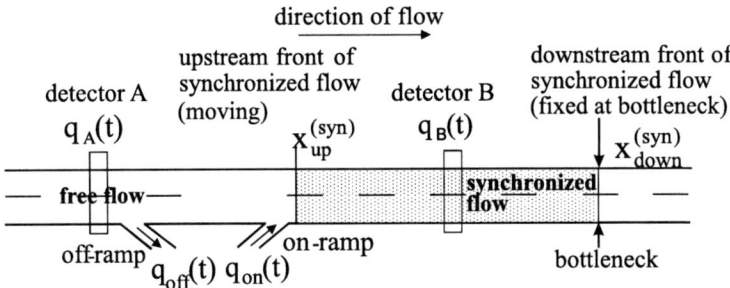

Fig. 9.2 Sketch of synchronized flow tracking with the FOTO model. Taken from [1]

If synchronized flow has been identified by the FOTO model, the upstream front location of this object is tracked by the FOTO model via a cumulative flow rate approach (Fig. 9.2). We assume that due to the upstream propagation of synchronized flow whose downstream front is fixed at a bottleneck, the synchronized flow has been identified at road detector B at time $t = t_{\text{syn, B}}$. At $t > t_{\text{syn, B}}$ a time-dependence of the location $x_{\text{up}}^{(\text{syn})}(t)$ of the upstream front of synchronized flow between any two detectors A and B (flow rates measured at these detectors are denoted by q_A and q_B, respectively) can be calculated through the formula:

$$x_{\text{up}}^{(\text{syn})}(t) = \mu \Delta M(t), \qquad (9.4)$$

where

$$\Delta M(t) = \Delta M_{\text{total}}(t)/n, \qquad (9.5)$$

n is the number of road lanes, μ is a model parameter, $\Delta M_{\text{total}}(t)$ is the net vehicle number leaving the road section between detectors A and B that is equal to (Fig. 9.2)

$$\Delta M_{\text{total}}(t) = \int_{t_{\text{syn, B}}}^{t} q_{\text{B}}(t)dt - \int_{t_{\text{syn, B}}}^{t} q_{\text{A}}^{*}(t)dt \quad \text{at} \quad t > t_{\text{syn, B}}, \tag{9.6}$$

$$q_{\text{A}}^{*}(t) = q_{\text{A}}(t) + q_{\text{on}} - q_{\text{off}}. \tag{9.7}$$

If a wide moving jam has been identified, the front locations of this jam are tracked with the ASDA model. For example, the locations of the upstream and downstream fronts of the wide moving jam $x_{\text{up}}^{(\text{jam})}(t)$ and $x_{\text{down}}^{(\text{jam})}(t)$, respectively, are found through the use of Stokes's shock wave formula[2]:

$$x_{\text{up}}^{(\text{jam})}(t) = -\int_{t_0}^{t} \frac{q_1(t) - q_{\min}(t)}{\rho_{\max}(t) - (q_1(t)/v_1(t))} dt, \quad t \geq t_0, \tag{9.8}$$

$$x_{\text{down}}^{(\text{jam})}(t) = -\int_{t_1}^{t} \frac{q_2(t) - q_{\min}(t)}{\rho_{\max}(t) - (q_2(t)/v_2(t))} dt, \quad t \geq t_1. \tag{9.9}$$

In (9.8) and (9.9), $q_1(t)$ and $v_1(t)$, $q_2(t)$ and $v_2(t)$ are the measured flow rate and average speed upstream and downstream of the jam, respectively (Fig. 9.1); t_0 is the time instant when the moving jam is detected at the downstream detector, the time t_1 determines the appearance of the downstream jam front at this detector; $q_{\min}(t)$ is the measured flow rate within the jam that is usually taken as zero; $\rho_{\max}(t)$ is the vehicle density within the jam.

- Both the cumulative flow approach and Stokes's formula are well-known in traffic science. The feature of the FOTO and ASDA models is that the models identify firstly two different traffic phases in congested traffic, synchronized flow and wide moving jam, and then synchronized flow is tracked with the cumulative flow approach, while in contrast wide moving jams are tracked based on Stokes's formula.

A typical congested pattern reconstructed and tracked by the FOTO and ASDA models in California (USA) is illustrated in Fig. 9.3. On the interstate freeway I405-South in Orange County located south from Los Angeles several detectors measure traffic data on the mainly four lane road. The space-time diagram from March 4, 2003 shows large regions of synchronized flow and several moving jams propagating up to 10 km on the road.

The FOTO and ASDA models have been installed over the entire freeway network of the German Federal State of Hessen with about 2500 detectors and 1200 km road distance. The models have to cope with different detector infrastructures (distances between detectors up to 10 km), different measurement intervals (1 and

---

[2] It must be noted that the flow rates and densities in (9.8) and (9.9) are found independent of each other based on measured data, i.e., *no* fundamental diagram (no flow–density relationship) is used in these formulae: Rather the formula for the shock-wave velocity of the Lighthill-Whitham-Richards (LWR) theory, formulae (9.8) and (9.9) follow from Stokes's shock wave formula derived in 1848 [12]. Explanations of the fundamental difference between the LWR and Stokes's shock wave formulae appear in Sect. 10.2.

9.1 Reconstruction and Tracking of Congested Patterns 147

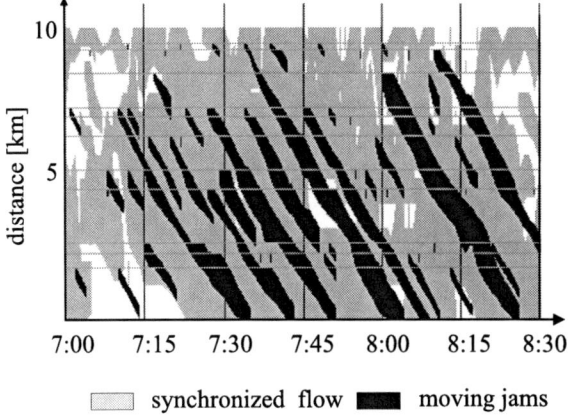

Fig. 9.3 Congested traffic pattern recognition and tracking through the use of FOTO and ASDA on the I405-South freeway in Orange County in the USA. Taken from [1]

5 minutes) and with different kinds and mixtures of local, regional, long-distance, and leisure traffic.

It has been found in the on-line application that the FOTO and ASDA models are reliable without any validation of model parameters in all situations [2, 7–11]. A possible dissolution of wide moving jams and synchronized flow, which occurs between detectors, can be predicted by the FOTO and ASDA models, when the traffic phases cannot be measured. A choice of optimal detector locations with regard to locations of effectual bottlenecks yields significant savings in roadside infrastructure investment. In a four month field trial we found that with only 23% of all given measurements on the freeway A5 approximately 60% of relevant information can be reconstructed. An example of the traffic phase recognition and tracking presented in Figs. 9.4 and 9.5 (a) shows some typical results from the on-line operation of FOTO and ASDA at the Traffic Control Centre of the German Federal State of Hessen.

From a comparison of spatiotemporal congested patterns measured on freeways in the USA (Fig. 9.3), Germany (Fig. 9.5 (a)), and the UK (Fig. 9.5 (b)) we can conclude that

- the FOTO and ASDA models can recognize and track very different congested patterns without changes in the model parameters;
- although quantitative characteristics of congested patterns in these three different countries are very different, qualitatively the same spatiotemporal structure of congested traffic patterns is observed; this *general* structure consists of synchronized flow whose downstream front is fixed at a bottleneck and wide moving jams that emerge in the synchronized flow propagating subsequently upstream while maintaining the mean velocity of their downstream fronts. This explains also the term *general congested pattern (GP)* used in three-phase traffic theory for these different traffic patterns.

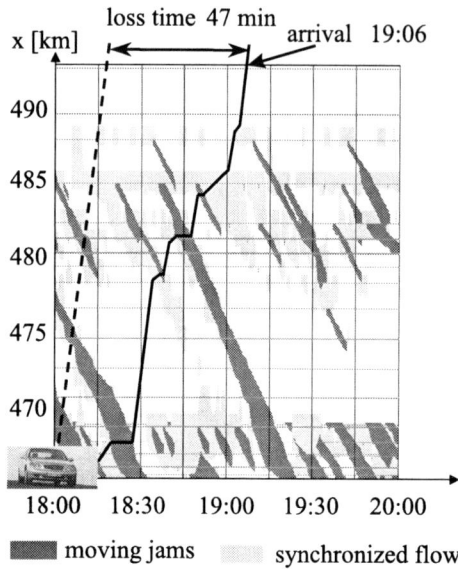

Fig. 9.4 Congested traffic pattern recognition and tracking through the use of FOTO and ASDA at the Traffic Control Centre of the German Federal State of Hessen. Reconstruction and tracking of synchronized flow and moving jams with FOTO and ASDA including vehicle trip time trajectory (solid line). Taken from [1]

Traffic pattern recognition with the FOTO and ASDA models can further be used for prediction in a traffic network as discussed in [1, 3]. For this objective in addition with current traffic pattern recognition, a historical database about effectual bottlenecks and resulting traffic congested pattern in the network is required.

9.1.2 Vehicle Onboard Traffic Prediction

In above consideration of traffic pattern recognition, the FOTO and ASDA models have used traffic data measured by road detectors. Current time dependence of vehicle speed measured by a vehicle can also be used for an approximate recognition of traffic phases.

The recognition of a *new* traffic phase in a vehicle can approximately be made based on conditions that during a time interval T that is longer than a given one the vehicle speed v is lower or higher than a given value [13]: A vehicle detects synchronized flow, when

$$v_{\text{jam}} < v \leq v_{\text{syn}} \text{ during } T > T_{\text{syn}}. \qquad (9.10)$$

A vehicle detects a wide moving jam, when

9.1 Reconstruction and Tracking of Congested Patterns

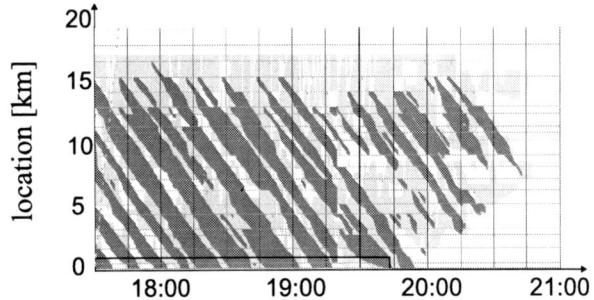

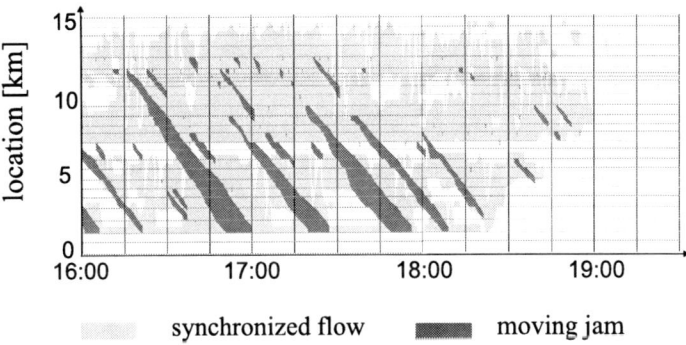

Fig. 9.5 Comparison of congested traffic pattern recognition and tracking through the use of FOTO and ASDA on the A5 freeway in Germany (a) and on the M42 freeway in the UK (b). Taken from [8–10]

$$v \leq v_{\text{jam}} \text{ during } T > T_{\text{jam}} \quad (9.11)$$

A vehicle propagating initially within synchronized flow or a wide moving jam detects free flow, when

$$v > v_{\text{free}} \text{ during } T > T_{\text{free}}, \quad (9.12)$$

where T_{free}, T_{syn}, T_{jam}, v_{free}, v_{syn}, and v_{jam} are model parameters.

When a historical database about effectual bottlenecks and congested patterns in a traffic network is stored in the vehicle, this current congested pattern recognition in the vehicle can be applied for a prognosis of future congested traffic patterns that should occur downstream of the vehicle (Fig. 9.6). This traffic prognosis can be made onboard autonomous, i.e., even if no data from a traffic control center is available.

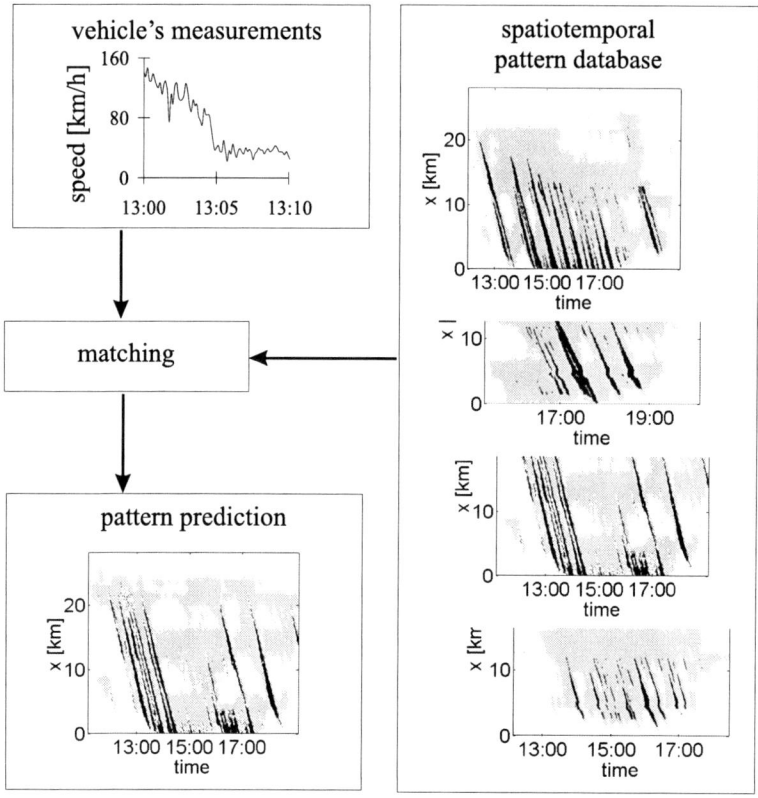

Fig. 9.6 Simplified qualitative scheme of a vehicle onboad autonomous traffic prediction. Taken from [1]

Onboard traffic prediction can be improved considerably through the use of traffic information from a traffic control center as well as from vehicle-vehicle communication building a vehicle ad-hoc network that discussion appears in Sect. 9.5.

9.2 Feedback On-Ramp Metering

One of the most used traffic control methods is on-ramp metering at an on-ramp bottleneck [14–18]. Various models and methods based on the "free flow control approach" to on-ramp metering have been developed. The basic idea of this approach is to maintain free flow on the main road at the bottleneck (e.g., [14, 16, 17, 19–23]). Free flow at the bottleneck should be maintained at the maximum possible throughput in free flow downstream of the bottleneck. In other words, traffic breakdown and resulting congested patterns at the bottleneck should be prevented. An example of this approach is the ALINEA method of Papageorgiou *et al.* [17, 19–24] used in many real installations.

However, due to the probabilistic nature of traffic breakdown at bottlenecks (Chap. 3) ALINEA exhibits an application limitation (see Sect. 10.6.5): in a wide range of the flow rate in free flow at a bottleneck traffic breakdown can randomly occur. This leads to congested pattern emergence with subsequent congestion propagation upstream of the bottleneck. This critical conclusion is also related to many other free flow control methods [25–29].

For this reason, the author introduced a different "congested pattern control approach" to feedback on-ramp metering called ANCONA [1, 30, 31]. In ANCONA, congestion at the bottleneck is allowed to set in. However, average vehicle speed within a congested pattern that emerges at the bottleneck should be maintained through the use of ANCONA at relatively high level (higher than about 60 km/h). Moreover, ANCONA ensures that congestion does not propagate upstream; rather than propagating upstream the congested pattern should be localized on the main road in the neighborhood of the bottleneck.

9.2.1 Theoretical Background of ANCONA

To explain ANCONA, let us firstly recall that due to an F→S transition at the bottleneck, firstly an SP appears. If the synchronized flow speed within the SP is high enough, then probability of wide moving jam emergence in this synchronized flow is negligible, i.e., an GP is not formed *and* the flow rate within the SP can be as great as in an initial free flow (Sect. 7.2.1).

This fact is used in ANCONA. ANCONA allows SP emergence at the bottleneck, but it suppresses GP formation. To prevent wide moving jam emergence within the SP, ANCONA ensures a high enough speed (higher than about 60 km/h) within the emergent SP. In addition, ANCONA rules ensure that the emergent SP is localized on the main road in the neighborhood upstream of the bottleneck, i.e., the SP should not propagate continuously upstream. To register SP formation at the bottleneck, in ANCONA feedback detector location is upstream of the effective bottleneck location (Fig. 9.7)[3].

In comparison with other on-ramp metering strategies that try to maintain free flow at the bottleneck, benefits of the ANCONA approach are considerably greater throughput and considerably shorter waiting times in the on-ramp lane(s). We will examine these ANCONA benefits in a comparison of ANCONA with ALINEA and UP-ALINEA in Sect. 10.7.

---

[3] In [28] has been found that the closer the feedback detector location to the effective bottleneck location, the more efficient is ANCONA application. However, in any case the feedback detector location in ANCONA must be *upstream* of the effective bottleneck location. For simplicity, we study here only a case [1, 31] in which a feedback detector location in ANCONA is upstream of the bottleneck as shown in Fig. 9.7.

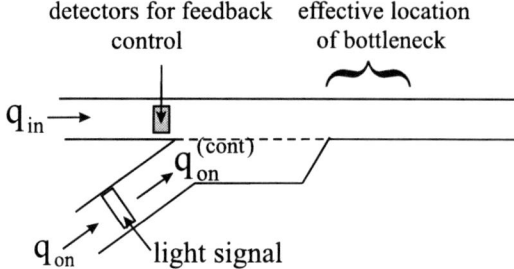

Fig. 9.7 Qualitative scheme of on-ramp bottleneck with upstream location of detector for on-ramp feedback metering control.

In contrast with standard on-ramp feedback metering methods, in which free flow should be maintained in a small neighborhood of a target (optimal) occupancy (or target flow rate, speed, or else density) (Sect. 10.6.5), ANCONA does not restrict on-ramp inflow as long as free flow is measured at the bottleneck. This ANCONA feature is associated with the *probabilistic nature* of traffic breakdown at the bottleneck as well as with a *discontinues character* of traffic breakdown: there is a wide range of flow rates downstream of the bottleneck within which the probability of spontaneous traffic breakdown at the bottleneck is greater than zero (Chap. 3). For this reason, either traffic breakdown occurs regardless of feedback on-ramp metering or a control method has to decrease on-ramp inflow drastically to suppress the breakdown.

- Thus at great enough flow rates in a traffic network when congestion should occur somewhere in the network, rather than to control free flow at the bottleneck a much more efficient use of the network could be achieved by a control of congested patterns resulting from spontaneous traffic breakdown at the bottleneck.

Only after synchronized flow is set in at the bottleneck, ANCONA begins to control this synchronized flow through the feedback decrease in on-ramp inflow. This feedback congested pattern control should ensure that either the SP dissolves over time and free flow returns at the bottleneck or the SP should spatially be localized at the bottleneck.

Thus in ANCONA there is no on-ramp control as long as free flow is measured at the bottleneck. This means that the flow rate of vehicles denoted by $q_{\mathrm{on}}^{(\mathrm{cont})}$, which can merge from the on-ramp onto the main road, is equal to

$$q_{\mathrm{on}}^{(\mathrm{cont})} = q_{\mathrm{on}}, \qquad (9.13)$$

where q_{on} is the flow rate to the on-ramp (Fig. 9.7). When a spontaneous traffic breakdown occurs, synchronized flow appears at the bottleneck. In this case, the average speed denoted by $v^{(\mathrm{det})}$ measured at the feedback control detector upstream of the bottleneck (Fig. 9.7) decreases. This speed decrease is explained by lower speed within synchronized flow and by upstream propagation of the upstream syn-

9.2 Feedback On-Ramp Metering

chronized flow front. If the average speed $v^{(\text{det})}$ is equal to or drops below a chosen "congestion speed" denoted by v_{cong}, i.e., when the average speed $v^{(\text{det})}$ satisfies the condition

$$v^{(\text{det})} \leq v_{\text{cong}}, \qquad (9.14)$$

feedback on-ramp metering via ANCONA is performed: via light signal operation ANCONA reduces the flow rate $q_{\text{on}}^{(\text{cont})}$ to a smaller flow rate denoted by $q_{\text{on 1}}$. As a result, instead of condition (9.13) the following condition is satisfied:

$$q_{\text{on}}^{(\text{cont})} = q_{\text{on 1}} < q_{\text{on}}. \qquad (9.15)$$

Thus ANCONA switches feedback on-ramp inflow control on only after spontaneous traffic breakdown has occurred at the bottleneck. Due to this decrease in the flow rate (9.15), ANCONA tries to achieve a return phase transition from synchronized flow to free flow (S→F transition) at the bottleneck. As a result of this S→F transition, the initial SP dissolves and the average speed at the detector $v^{(\text{det})}$ increases above v_{cong}:

$$v^{(\text{det})} > v_{\text{cong}}. \qquad (9.16)$$

When condition (9.16) is satisfied, via light signal operation a greater flow rate

$$q_{\text{on}}^{(\text{cont})} = q_{\text{on 2}}, \qquad (9.17)$$

is allowed, where the flow rate $q_{\text{on 2}} > q_{\text{on 1}}$ (flow rates $q_{\text{on 1}}$ and $q_{\text{on 2}}$ and speed v_{cong} are model parameters). If under condition (9.17) traffic breakdown occurs at the bottleneck once more, an incipient SP begins to propagate upstream of the bottleneck. As a result, the speed at the feedback control detector decreases. Therefore, condition (9.14) is satisfied once more. This leads to a new decrease in $q_{\text{on}}^{(\text{cont})}$, and so on.

- Thus to control traffic breakdown and resulting congested traffic patterns at the bottleneck, ANCONA performs a *discontinues* change in the flow rate $q_{\text{on}}^{(\text{cont})}(t)$ (Fig. 9.7). This ANCONA feature is associated with the discontinues character of traffic breakdown in real traffic.

If there are no traffic breakdowns at the bottleneck, i.e., condition (9.16) is satisfied during a relatively long time interval, ANCONA does not restrict the flow rate $q_{\text{on}}^{(\text{cont})}$ any more, i.e., condition (9.13) is satisfied. This is independent of how great occupancy and flow rate are.

ANCONA tries to achieve greater throughputs in the whole traffic network under very great traffic demand when congestion has to occur somewhere in the network. ANCONA should ensure a more homogeneous distribution of congestion among all bottlenecks in the network. Indeed, rather than the maintaining of free flow at the bottleneck at the expense of rapid growth of congestion in the on-ramp lane(s) and, therefore, in other links of the traffic networks, ANCONA tries to minimize influence of congestion both in the on-ramp lane and on the main road.

9.2.2 Numerical Simulations of ANCONA

The above explanations of ANCONA are illustrated in numerical simulations shown in Figs. 9.8 and 9.9. Without ANCONA application (Fig. 9.8), an GP emerges spontaneously at the bottleneck when $q_{\rm on}$ is an increasing time function (Fig. 9.8 (a)) and the flow rate in free flow upstream of the bottleneck $q_{\rm in}$ (Fig. 9.7) is constant. However, if ANCONA is applied (Fig. 9.9), then at the same flow rates $q_{\rm on}(t)$ and $q_{\rm in}$ as those in Fig. 9.8, GP emergence shown in Fig. 9.8 (b) is suppressed via ANCONA application; this can be seen from the averaged speed on the main road in space and time shown in Fig. 9.9 (a). To suppress this GP emergence, the flow rate of vehicles that can merge from the on-ramp onto the main road $q_{\rm on}^{(\rm cont)}$ is switched over time between two values $q_{\rm on\ 1}$ and $q_{\rm on\ 2}$ via ANCONA (Fig. 9.9 (b)).

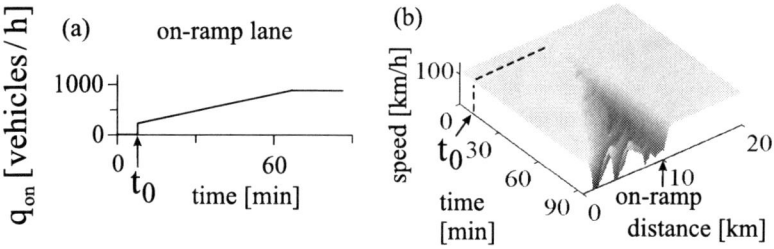

Fig. 9.8 Simulated GP at on-ramp bottleneck at which *no* on-ramp feedback metering control is applied. At $t = t_0$ on-ramp inflow is switched on. Taken from [1]

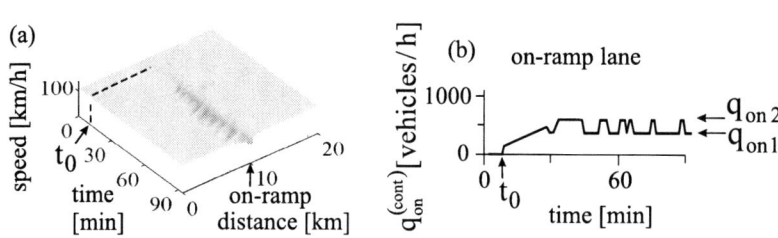

Fig. 9.9 ANCONA suppresses GP emergence under the same flow rates as those in Fig. 9.8: (a) Speed in time and space. (b) $q_{\rm on}^{(\rm cont)}$ as a time-function. Taken from [1]

We can see from Fig. 9.9 (a) that although traffic breakdown occurs spontaneously at the bottleneck, ANCONA application leads to a spatial localization of this congested pattern at the bottleneck, i.e., congestion does not propagate upstream continuously. Moreover, no wide moving jams emerge within this congested pattern, i.e., this pattern is an SP.

9.3 Speed Limit Control

In this section, based on simulations in the framework of three-phase traffic theory we show that speed limit control can result in traffic breakdown with the subsequent formation of wide moving jams or in contrast to wide moving jam dissolution. In other words, we find that the effect of speed limit control on traffic flow characteristics depends qualitatively on conditions at which it is applied.

9.3.1 Traffic Breakdown at Bottleneck through Application of Speed Limit Control

Here, we show that rather than improving traffic flow characteristics speed control in free flow can induce traffic breakdown at a bottleneck [32].

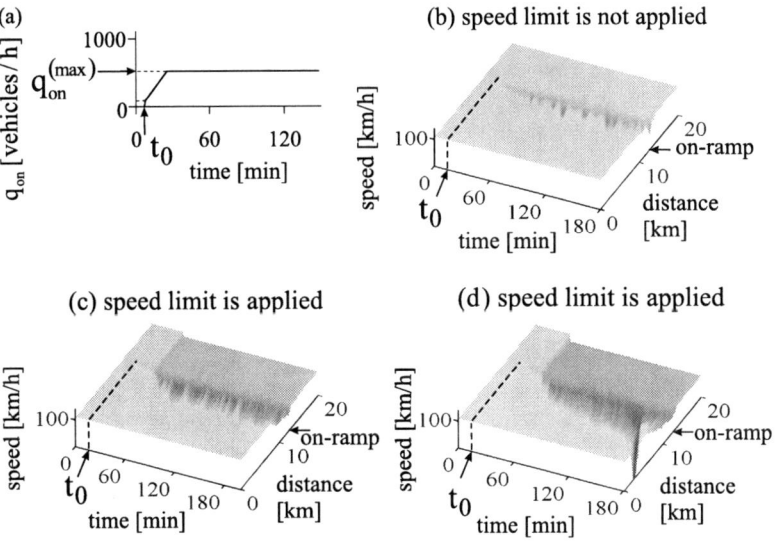

Fig. 9.10 Simulations of speed limit control in flow of identical vehicles: (a) Flow rate to the on-ramp as a time-function. (b–d) Speed without speed limit (b) and under speed limit (c, d). At $t = t_0$ on-ramp inflow is switched on. $q_{\text{on}}^{(\text{max})} = 540$ (b, d) and 420 (c) vehicles/h. Taken from [32]

We consider three scenarios on a two-lane road section with an on-ramp bottleneck. In the scenarios, the flow rate to on-ramp $q_{\text{on}}(t)$ increases over time from 90 vehicles/h to some $q_{\text{on}}^{(\text{max})}$ (Fig. 9.10 (a)); the flow rate in free flow on the main road upstream of the bottleneck q_{in} is time-independent.

In the first scenario, there is no speed limitation; vehicles move with the maximum free flow speed 108 km/h. Although on-ramp inflow increase to $q_{\text{on}}^{(\text{max})} = 540$ vehicles/h, free flow condition remains at the bottleneck (Fig. 9.10 (b)).

In the second scenario, we apply speed limitation 90 km/h downstream of the on-ramp bottleneck, i.e., all vehicles should move with speeds, which are equal or lower than 90 km/h. We find that if $q_{\text{on}}(t)$ increases to $q_{\text{on}}^{(\text{max})} = 420$ vehicles/h and the other conditions are as those in the first scenario, traffic breakdown is induced through speed limitation at the bottleneck. As a result, an LSP is formed upstream of the bottleneck (Fig. 9.10 (c)).

In the third scenario, as in the second one we also apply speed limitation 90 km/h downstream of the bottleneck. However, q_{on} is increased to a greater value $q_{\text{on}}^{(\text{max})} = 540$ vehicles/h. Firstly, as in Fig. 9.10 (c) traffic breakdown is induced and an LSP occurs. However, within the LSP speed becomes lower than the speed within the LSP in Fig. 9.10 (c); as a result, a wide moving jam emerges spontaneously within the LSP, i.e., the LSP transforms into an GP over time (Fig. 9.10 (d)). The wide moving jam propagates upstream continuously. Thus we see that rather than preventing congestion, speed limitation can induce SPs and GPs at the bottleneck.

SPs and GPs emergence through the use of speed limit control can be explained based on the double Z-characteristic for phase transitions (Fig. 5.9). There is a critical speed in free flow for traffic breakdown (dashed curve between states F and S in Fig. 5.9). When the speed within a local disturbance in the neighborhood of the bottleneck becomes lower than the critical speed, then this disturbance grows leading to traffic breakdown with the subsequent emergence of synchronized flow at the bottleneck. Speed limitation reduces speed at the bottleneck. Consequently, a smaller additional random speed decrease within a speed disturbance at the bottleneck can lead to traffic breakdown. Thus when speed limit is applied, traffic breakdown occurs with greater probability at the bottleneck at the same flow rates upstream of the bottleneck.

To explain GP emergence under speed control shown in Fig. 9.10 (d), we recall that there is a critical speed for the emergence of wide moving jams in synchronized flow (dashed curve between states S and J in Fig. 5.9). When the speed within a local disturbance in synchronized flow becomes lower than this critical speed, then this disturbance is a nucleus for an S→J transition whose growth leads to the emergence of a wide moving jam. The lower the speed, the greater the probability of the nucleus occurrence at the same other synchronized flow characteristics. The average speed within the LSP decreases when a greater flow rate $q_{\text{on}}^{(\text{max})} = 540$ vehicles/h is applied (Fig. 9.10 (d)). This explains why speed limitation induces firstly the LSP and later leads to wide moving jam occurrence upstream of the bottleneck.

9.3.2 Dissolution of Wide Moving Jams in Synchronized Flow under Speed Limit Control

Now we show that in some other cases, speed limitation can lead to moving jam dissolution in synchronized flow. We consider the same two-lane freeway section, however, with two different adjacent bottlenecks (Fig. 9.11 (a)): a downstream off-ramp bottleneck and an upstream on-ramp bottleneck. In addition, we simulate now heterogeneous flow consisting of three types of vehicles: fast, slow, and long vehicles. The maximum speed in free flow for fast vehicles (120 km/h) is assumed to be higher than the one for slow and long vehicles (90 km/h).

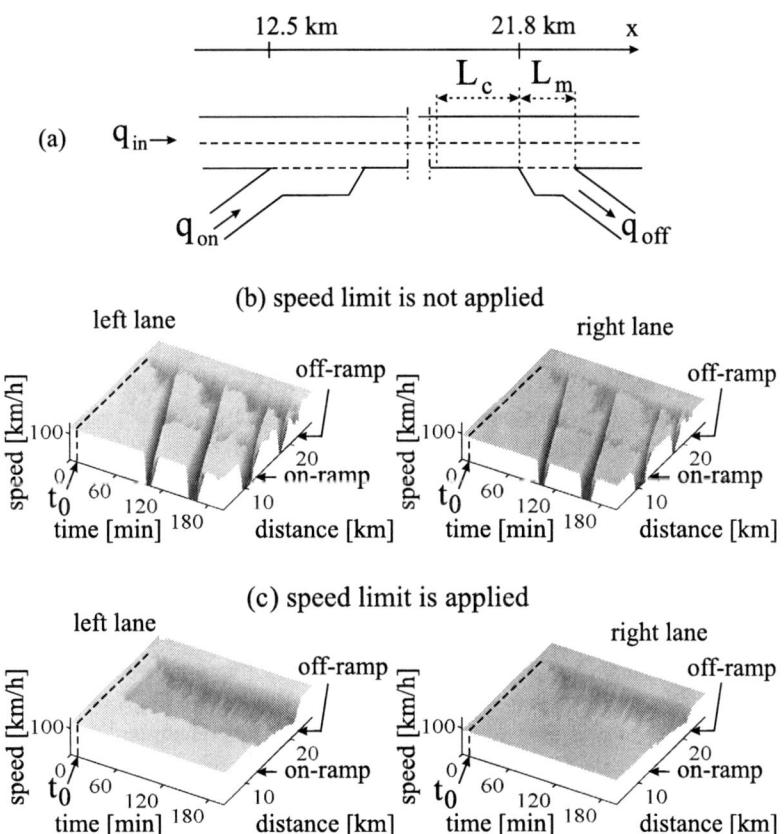

Fig. 9.11 Simulations of speed limit in heterogeneous flow: (a) Model of bottlenecks. (b, c) Speed in space and time. At $t = t_0$ on-ramp inflow is switched on. $q_{\text{on}}^{(\max)} = 420$ vehicles/h. The percentage of vehicles going to the off-ramp is 12%. Taken from [32]

Simulations show that free flow exists only at relatively small flow rates q_{on} and q_{in}. Slow and long vehicles move mostly in the right lane, whereas fast vehicles move mostly in the left (passing) lane. The average speed in free flow in the left lane is close to 120 km/h, whereas in the right lane the speed is close to 90 km/h.

When q_{in} is time-independent but the flow rate to on-ramp q_{on} increases over time from 60 to $q_{on}^{(max)} = 420$ vehicles/h (Fig. 9.10 (a)), traffic breakdown occurs at the off-ramp bottleneck with subsequent synchronized flow formation at the off-ramp bottleneck. Traffic breakdown occurs independent of whether there is speed limit (Fig. 9.11 (c)) or not (Fig. 9.11 (b)).

However, there is a fundamental difference in these two cases: whereas without speed limit, wide moving jams emerge spontaneously in synchronized flow (Fig. 9.11 (b)), there is no wide moving jam emergence in synchronized flow under speed limit 90 km/h (Fig. 9.11 (c)). The speed limit leads to wide moving jam dissolution; as a result, only an LSP remains upstream of the off-ramp bottleneck.

Before we explain this result, we recall the reason for traffic breakdown at the off-ramp bottleneck. Traffic breakdown occurs because vehicles in the left lane, which are going to the off-ramp, must change from the left (passing) to right lane. This lane changing occurs within a region on the main road (whose length is denoted by L_c in Fig. 9.11 (a)) upstream of the beginning of the off-ramp. The lane changing leads to speed disturbances in the right lane within the length L_c. The speed within some of these disturbances becomes lower than the critical speed (dashed curve between states F and S in Fig. 5.9). This explains traffic breakdown that occurs independent of whether there is a speed limit (Fig. 9.11 (c)) or not (Fig. 9.11 (b)).

Now we explain why wide moving jams emerge in synchronized flow when no speed limit is applied (Fig. 9.11 (b)), whereas no wide moving jams occur in synchronized flow under speed limit application (Fig. 9.11 (c)). When no speed limit is applied, upstream of the off-ramp bottleneck the speed of the fast vehicles in the left lane is considerably higher (120 km/h) than the speed in the right lane (90 km/h). When vehicles in the left lane can move with higher speeds than those in the right lane, we can assume that vehicles in the left lane going to the off-ramp try to stay in the left lane as long as possible: this theoretical assumption means that L_c should be relatively short. In Fig. 9.11 (b), we have chosen $L_c = 200$ m. Simulations show that at $L_c = 200$ m, many speed disturbances with relatively low speed appear in the right lane on the main road upstream of the off-ramp. This is because the shorter the length L_c of the lane changing region (Fig. 9.11 (a)), the more frequently fast vehicles going to the off-ramp must change from the left lane to the right lane within this region. These low speed disturbances caused by lane changing propagate upstream in synchronized flow leading to spontaneous wide moving jam emergence in synchronized flow (Fig. 9.11 (b)). Simulations show that the speed within some of the disturbances is lower than the critical speed for an S→J transition (dashed curve between states S and J in Fig. 5.9).

In contrast, when speed limit is applied downstream of the on-ramp bottleneck, both fast and slow vehicles move with a speed that is not higher than 90 km/h. We can make an assumption that fast vehicles in the left lane going to the off-ramp change to the right lane within a considerably longer section on the main road, i.e., under speed limit L_c increases: fast vehicles should not necessarily have a motivation any more to stay in the left lane as long as possible. This is because the speed in the left lane is now as low as in the right lane. For the scenario with speed limit, we have chosen $L_c = 800$ m. Then no wide moving jams occur between the bottlenecks (Fig. 9.11 (c)). Simulations show that at longer L_c the speed within speed disturbances associated with lane changing is higher than the critical speed for S→J transition (dashed curve between states *S* and *J* in Fig. 5.9). This explains why wide moving jams do not emerge in synchronized flow under speed limit (Fig. 9.11 (c)).

9.4 Cooperative Driving for Improving of Traffic Flow and Safety

Collective traffic control like on-ramp metering and speed limit discussed above are important methods for the improving of traffic flow characteristics. Nevertheless, it can be expected that a stronger effect on traffic flow characteristics and traffic safety could be achieved through the cooperative adaptation of individual driver behavior to the current driving situation based on driver assistance systems in vehicles in a *cooperation with* car-to-car communication. Such *cooperative driving* is one of the most important challenges for intelligent transportation systems in the future (e.g., [33–36]).

However, a question arises:

- What driver behavior should be changed to improve traffic flow and enhance traffic safety?

Answers to this question, which follow from three-phase traffic theory (Part I) [1, 37], are briefly discussed in this section. Some simulation results of cooperative driving for improving of traffic flow characteristics appear in Sect. 9.5.

There are the following main traffic phenomena, which lead to the deterioration of performance of traffic system, in particular to the reduction in traffic safety:

- Traffic breakdown, i.e., an F→S transition, which occurs mostly at a highway bottleneck.
- Wide moving jam emergence in synchronized flow resulting from traffic breakdown.

Therefore, these traffic phenomena should be prevented.

9.4.1 Driver Behavior for Prevention of Traffic Breakdown

As follows from three-phase traffic theory, traffic breakdown is associated with a competition between two opposing tendencies occurring within a local disturbance within which the speed is lower and vehicle density is greater than in an initial free flow – a tendency towards synchronized flow due to vehicle deceleration associated with the speed adaptation effect and a tendency towards free flow due to vehicle acceleration associated with the over-acceleration effect (Sect. 3.2). This competition leads to traffic breakdown, if a nucleus for traffic breakdown appears in free flow at the bottleneck. In turn, the nucleus is a local disturbance in free flow within which the speed is equal to or lower than a critical speed, respectively, the density is equal to or greater than a critical density required for traffic breakdown. Thus we see that changes in driver behavior for prevention of traffic breakdown should lead to

- the reduction in probability of the occurrence of a nucleus required for traffic breakdown through the decrease in speed fluctuations in free flow or/and
- the facilitation of the over-acceleration effect that increases the tendency towards free flow.

The decrease in speed fluctuations in free flow in a neighborhood of on- and off-ramp bottlenecks can be achieved through the use of cooperative driving in which

- vehicles moving on the main road in the neighbor lane to on- or off-ramps keep larger space gaps for those vehicles that merge from the on-ramp onto the main road or change lane to leave the main road to the off-ramp, respectively or/and
- vehicles decrease acceleration (and deceleration, if it is possible).

The facilitation of the over-acceleration effect can be achieved through the use of individual driver assistance systems (for example, adaptive cruise control (ACC) (Sect. 9.6) and lane changing assistance systems) that help drivers

- to increase lane changing probability to a faster lane in a neighborhood of the bottleneck or/and
- to increase space gaps (time headways) approaching a local disturbance in free flow at the bottleneck or/and
- to decrease a driver time delay in acceleration at the downstream front of the local disturbance.

It should be noted that when fluctuations in free flow at the bottleneck are great and, therefore, a nucleus required for traffic breakdown has occurred in free flow at the bottleneck, nevertheless, the abovementioned changes in driver behavior can lead to the interruption of the nucleus growth and, finally, to the dissolution of the nucleus.

However, we should mention that the increase in lane changing probability to a faster lane can lead to the opposite effect – the increase in a local disturbance and, consequently, to the occurrence of a nucleus required for traffic breakdown. This is associated with a dual role of lane changing in free flow discussed in Sect. 3.4. Indeed, lane changing can lead to the occurrence of a nucleus required for traffic breakdown, when due to lane changing a vehicle forces the following vehicle in the target lane to decelerate strongly.

The same opposite effect leading to the occurrence of a nucleus required for traffic breakdown can also occur through the decrease in a driver time delay in acceleration at the downstream front of a local disturbance in free flow. This is because if a driver accelerates with a too short time delay after the preceding vehicle has just begun to accelerate, then a possible subsequent (and unexpected) deceleration of the preceding vehicle can cause strong deceleration of the following drivers with the subsequent occurrence of the nucleus for traffic breakdown.

We see that very accurate and precise *cooperative* and well *coordinated* changes in individual driver behavior in different road lanes in a neighborhood of a highway bottleneck are required to reach the goal of the prevention of traffic breakdown.

Suggestions for changes in individual driver behavior can be made at each time instant and individual for every driver. The suggestions can include individual choices of a road lane, speed, space gap (or time headway), and acceleration for each of the vehicles in the neighborhood of the bottleneck. Such cooperative driving can be called *microscopic driver navigation*. In addition to the changes in individual driver behavior, the microscopic driver navigation can include individual suggestions for the route choice in a traffic network, if route change is possible.

9.4.2 Driver Behavior for Prevention of Moving Jam Emergence

If traffic demand in a traffic network is great enough, it can turn out that microscopic driver navigation in free flow, which can also be made in cooperation with collective traffic control, cannot prevent traffic breakdown at some of bottlenecks in the network. In this case, synchronized flow results from traffic breakdown.

To prevent wide moving jam emergence in the synchronized flow and, therefore, the subsequent propagation of wide moving jams, some additional changes in individual driver behavior can be required. This is because wide moving jam emergence is associated with a competition between two opposing tendencies occurring within a local disturbance in which the speed is lower and vehicle density is greater than in an initial synchronized flow – a tendency towards a wide moving jam due to the over-deceleration effect and a tendency towards synchronized flow due to the speed adaptation effect (Sect. 5.2). This competition leads to a wide moving jam, if a growing narrow moving jam, which is a nucleus required for wide moving jam emergence, appears in synchronized flow. Changes in driver behavior for the prevention of wide moving jam emergence in synchronized flow should lead to

- the reduction in probability of the occurrence of a growing narrow moving jam through the decrease in speed fluctuations in synchronized flow or/and
- the facilitation of the speed adaptation effect that increases the tendency towards synchronized flow.

The facilitation of the speed adaptation effect can be achieved through the use of individual driver assistance systems that help drivers

- to increase space gaps (time headways) required for speed adaptation in synchronized flow or/and
- to decrease a driver time delay in deceleration at the upstream front of a local disturbance in synchronized flow as well as within the disturbance or/and
- to decrease a driver time delay in acceleration at the downstream front of the local disturbance.

Additionally, if a growing narrow moving jam has occurred in one of the road lanes, vehicles can recognize the jam and send a message to upstream moving vehicles. Based on this message some of drivers in this lane can be suggested to change to other lanes with the goal to increase space gaps in the lane and, therefore, prevent the subsequent growth of the jam. Lane changing with the goal of the increase in space gaps for the following vehicles can be an important instrument for the interruption of the jam growth. This can destroy the jam that has earlier occurred in synchronized flow (see a more detailed explanation of the nucleation-interruption effect in Sect. 7.3.2).

However, we should mention that the increase in lane changing probability can lead to the opposite effect – the occurrence of a growing narrow moving jam in the target lane in which vehicles change. This is associated with a dual role of lane changing in synchronized flow discussed in Sect. 5.3. The nucleus in the target lane can occur, when lane changing forces the following vehicles in the target lane to decelerate strongly. Therefore, individual driver assistance systems should help drivers to avoid such badly adapted lane changing.

It should be mentioned that in many cases great disturbances in synchronized flow appear through the upstream propagation of speed (density) disturbances occurring initially in a neighborhood of a bottleneck at which synchronized flow has occurred. For this reason, microscopic driver navigation in the neighborhood of the bottleneck discussed in Sect. 9.4.1 can also be very important for the reduction in probability of the occurrence of nuclei required for wide moving jam emergence, i.e., growing narrow moving jams in synchronized flow upstream of the bottleneck.

9.5 Traffic Control based on Wireless Car Communication

Ad-hoc vehicle networks based on wireless car-to-car and car-to-infrastructure (C2X) communication is one of the important future fields of vehicle and traffic flow control. This is because there are many possible applications of ad-hoc vehicle networks, including various systems for danger warning, traffic adaptive assistance systems, traffic information and prediction in vehicles, improving of traffic flow characteristics through ACC-vehicles (Sect. 9.6), etc. [38–44].

However, the evaluation of ad-hoc vehicle networks requires many communicating vehicles moving in real traffic flow, i.e., field studies of ad-hoc vehicle networks are very complex and expensive. For this reason, to prove the performance of ad-hoc vehicle networks based on wireless vehicle communication, reliable simulations of ad-hoc vehicle networks are of importance and indispensable.

9.5.1 A Model of Ad-Hoc Vehicle Network

In general, for simulations of ad-hoc vehicle networks, both simulations of vehicle communication and traffic flow are needed [45–53]. Here we restrict a consideration of ad-hoc vehicle network simulations devoted to traffic flow control. As will be explained in Chap. 10, traffic flow models within the framework of the fundamental diagram approach cannot show the fundamental empirical features of traffic breakdown (Chap. 3). For this reason, we consider ad-hoc vehicle network simulations in the context of three-phase traffic flow theory [54].

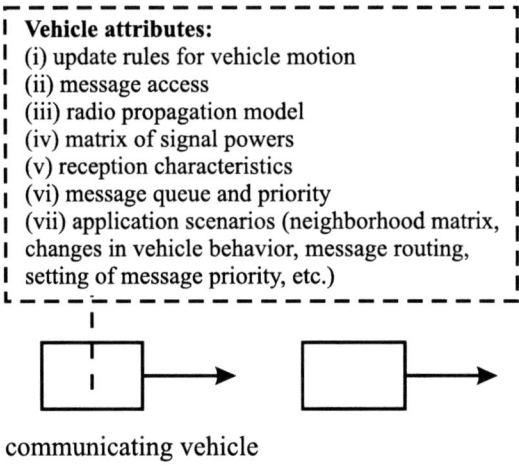

Fig. 9.12 Scheme of simulations of ad-hoc networks and traffic flow within a united network model. Taken from [54]

In this model, simulations of traffic flow and an ad-hoc vehicle communication network as well as applications of an ad-hoc vehicle network for traffic flow are performed in a united simulation network. In this simulation network (Fig. 9.12), each vehicle exhibits different *attributes* needed for vehicle motion, vehicle communications, and application scenarios. The vehicle attribute "update rules for vehicle motion" is given by a microscopic three-phase traffic flow model (Chap. 11).

If there are messages to be sent and the medium is free, the vehicle sends the message that has the highest priority and/or is the first one in the message queue in this vehicle. To prevent collisions between messages sent by different communicating vehicles, a message access method is usually applied. During a motion of a communicating vehicle in a traffic network calculated through the use of the update rules for vehicle motion, the vehicle attribute "message access" calculates vehicle access possibility for message sending at each time instant. As an example, we use here IEEE 802.11e basic access method [55–57]. No access is possible when medium is busy. After the medium has been free, in accordance with the IEEE 802.11e access method, there is a backoff procedure applied for each of the communicating vehicles independently of each other. At the end of the backoff procedure, a decision whether the medium is free or busy is made.

In the model (Fig. 9.12), to make the decision whether the medium is free or busy or else the vehicle has received a message or not, based on the vehicle attribute "radio propagation model", signal powers of messages sent by all other communicating vehicles are calculated. If a signal power is greater than a given threshold signal power denoted by P_{th} (model parameter), then this signal power of the associated message is stored into a "matrix of signal powers" of the vehicle:

- at each time instant, the matrix of signal powers of the vehicle contains signal powers of messages sent by other vehicles in ad-hoc network that are greater than the threshold signal power P_{th} at the location of the vehicle.

This threshold P_{th} is chosen to be much smaller than a carrier sense threshold (CSTh). The smaller P_{th} is chosen, the greater the accuracy of simulations of ad-hoc network performance, however, the longer the simulations run time. Signal reception characteristics (whether the medium is free or busy as well as whether the vehicle has received a message or not) are associated with an analysis of the matrix of signal powers, which is automatically made at each time instant for each communicating vehicle individually. The decision about signal collisions is further used for a study of ad-hoc network performance.

ID of sending vehicle	25	382	37	36	31
Distance between the receiving vehicle 33 and sending vehicle	234 m	345 m	300 m	70 m	562 m
Received signal power of the message sent at the location of the vehicle 33	-91 dBm	-95 dBm	-93 dBm	**-81 dBm**	-99 dBm

Fig. 9.13 Hypothetical example for matrix of signal powers

We consider an application of the matrix of signal powers for a hypothetical example[4] of a communicating vehicle with a vehicle-ID (identification number) 33 (Fig. 9.13). In this case, in the matrix of the vehicle 33 there are several signal powers of those messages sent at the time t by other vehicles in an ad-hoc network whose signal powers are greater than the threshold CSTh $= -96$ dBm at the location of the vehicle 33. However, only the signal power of a message sent by vehicle ID 36 that is equal to -81 dBm is greater than a signal receiving threshold RXTh $= -90$ dBm. The ratio between the power of this greatest signal power of a message sent by vehicle ID 36 and the sum of the powers of all other signals stored in the matrix

---

[4] In this example, we have used a well-known two-ray-ground radio propagation model with the communication range 200 m: $P(\gamma) = P_0/(\gamma/200)^2$, where γ is the distance between two communicating vehicles, $P_0 = 10^{-9}$ mW; RXTh $= -90$ dBm, CSTh $= -96$ dBm, SNR $= -6$ dBm.

9.5 Traffic Control based on Wireless Car Communication

is greater than the required signal-to-noise ratio (SNR) for the whole duration of the message. Thus in the matrix of signal powers the signal sent by vehicle 36 that is 70 m outside of the location of vehicle 33 could be considered to be received by vehicle 33.

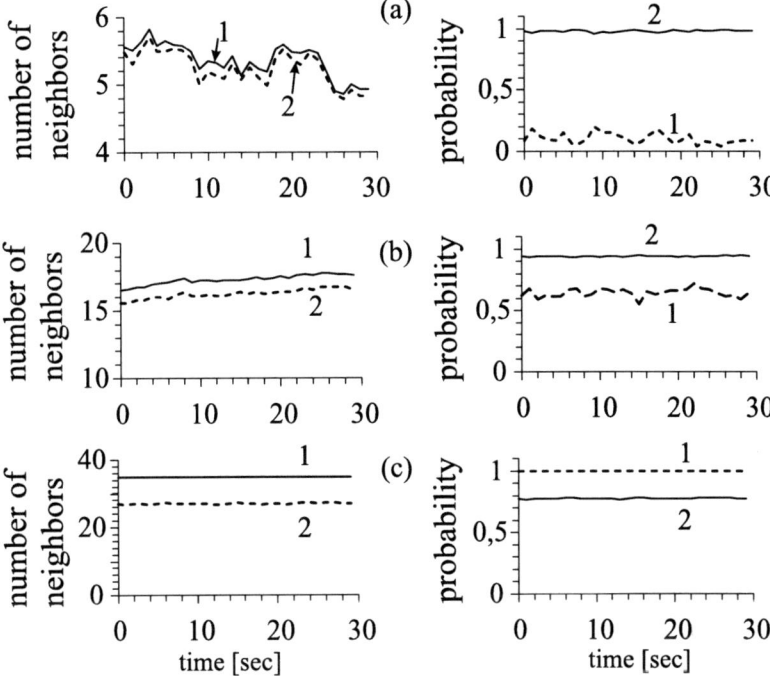

Fig. 9.14 Simulations of performance characteristics of neighbor table for different percentages of communicating vehicles 20% (a), 50% (b), 100% (c). Figures left: number of communicating neighbors for a vehicle within the communication range (curves 1) and the average number of communicating neighbors in the neighbor table (curves 2). Figures right: probabilities for one or more mistakes in the neighbor tables (curves 1) and for message receiving from one of the neighbors (curves 2). Taken from [54]

Based on an application, which should be simulated, in the model each communicating vehicle exhibits an attribute of message queue organization and individual message priority performance governed automatically. Because each communicating vehicle manages these features individually, this attribute can be chosen differently for various types of the communicating vehicles.

Each communicating vehicle exhibits an attribute "application scenario". This attribute governs the organization of all messages that are received and to be sent. Based on this attribute and the message context just received by the vehicle, the vehicle can change its behavior in traffic flow (e.g., the vehicle slows down or changes the lane, or else changes the route, etc.).

For an illustration of the model[5], we consider simulations of a matrix of neighbors for a communicating vehicle, also called a neighbor table. In this table, current vehicle locations, ID, and other vehicle characteristics (speed, acceleration, etc.) of all communicating vehicles are stored, which are in the communication range. The neighbor table can be used by message routing in the ad-hoc network and support of driver assistance systems. However, due to signal collisions not all messages are received and, therefore, some vehicles do not appear in the neighbor table. As a result, probability of such mistakes in the neighbor tables increases rapidly with the percentage of communicating vehicles (right, curves 1 in Fig. 9.14), whereas probability of message receiving only slightly decreases with the percentage of communicating vehicles (right, curves 2). At any percentage of communicating vehicles under consideration mistake probability is not equal to zero (right, curves 1).

9.5.2 Prevention of Traffic Breakdown at On-Ramp Bottleneck Through Vehicle Ad-Hoc Network

As mentioned in Sect. 9.4, we can assume that there can be the following two hypothetical possibilities to prevent traffic breakdown at an on-ramp bottleneck through changes in driver behavior of communicating vehicles:

(i) A decrease in the amplitude of disturbances on the main road occurring when vehicles merge from on-ramp onto the right lane of the main road. This decreases the probability of nucleus occurrence required for traffic breakdown.
(ii) An increase in probability of over-acceleration.

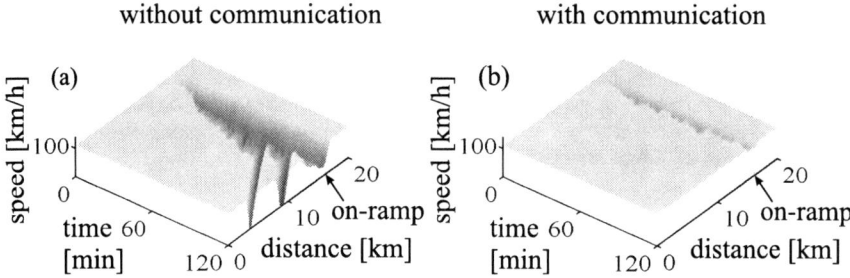

Fig. 9.15 Simulations of prevention of traffic breakdown at on-ramp bottleneck through vehicle communication: Speed in time and space without communication (a) and with vehicle communication (b). Taken from [54]

---

[5] In simulations presented, we have used $P_{th} = -116$ dBm, which allows us to have a good balance between accuracy and simulation time. Simulation results are changed in the range of about 1%, when instead of $P_{th} = -116$ dBm, the threshold $P_{th} = -126$ dBm has been used.

9.5 Traffic Control based on Wireless Car Communication 167

In simulations presented in Fig. 9.15, we assume that vehicles moving in the on-ramp lane send a message for neighbor vehicles moving in the right road lane when the vehicle intends to merge from the on-ramp onto the main road. We assume that the following vehicle in the right lane increases a time headway for the vehicle merging. Simulations show that in comparison with the case in which no vehicle communication is applied and traffic breakdown occurs (Fig. 9.15 (a)) this change in driver behavior of communicating vehicles decreases disturbances in free flow at the bottleneck. This results in the prevention of traffic breakdown (Fig. 9.15 (b)).

9.5.3 Influence of Ad-Hoc Vehicle Network on Congested Traffic Patterns

Here we consider a case in which flow rates upstream of a bottleneck are great and traffic control through the use of changes in driver behavior in free flow at the bottleneck discussed in Sect. 9.5.2 is not applied. In this case, traffic breakdown occurs at the bottleneck resulting in synchronized flow upstream of the bottleneck. Later, wide moving jams begin to emerge in the synchronized flow, i.e., an GP appears (Fig. 9.16 (a, b)). Nevertheless, as mentioned in Sect. 9.4, we can assume that there can be the following two hypothetical possibilities to prevent moving jam emergence in synchronized flow through changes in driver behavior of communicating vehicles moving in synchronized flow:

(i) A decrease in the amplitude of disturbances in synchronized flow upstream of the bottleneck. This decreases the probability of nucleus occurrence required for the emergence of wide moving jams.
(ii) A decrease in the density of synchronized flow upstream of the bottleneck. This decreases the critical speed required for the emergence of wide moving jams in synchronized flow. The lower the critical speed, the smaller the probability for the emergence of wide moving jams.

We assume that after synchronized flow has just occurred due to traffic breakdown at the bottleneck, communicating vehicles, which reach the synchronized flow, send priority messages about the speed reduction to vehicles moving in free flow upstream. Each message comprises a minimum space gap that should be maintained by vehicles while moving in the synchronized flow.

Due to this change in driver behavior incorporated in a three-phase traffic flow model, at the same flow rates upstream of the bottleneck as those in Fig. 9.16 (a, b) rather than the GP a widening synchronized flow pattern (WSP) is forming (Fig. 9.16 (c, d)). Whereas in the pinch region of the GP the mean space gap is 15 m, it is 25 m within the WSP. Due to the transformation of the GP into the WSP, two effects are achieved: (i) wide moving jams do not occur and (ii) the average speed within synchronized flow upstream of the bottleneck increases from about 40 km/h

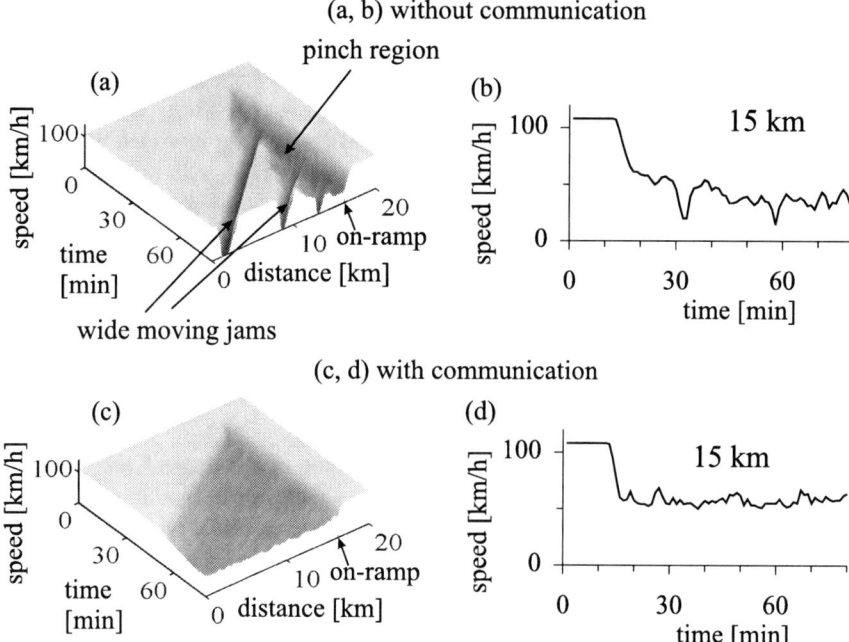

Fig. 9.16 Simulation of the effect of vehicle communication on congested patterns: Speed in time and space (a, c) and at a location $x = 15$ km (b, d) without communication (a, b) and with communication (c, d). Taken from [54]

within the GP to 60 km/h within the WSP. These effects can result in a considerable increase in the efficiency and safety of traffic.

9.6 Adaptive Cruise Control

Adaptive cruise control (ACC) is one of the ways to enhance driver comfort and safety in traffic (e.g., [58–60]). An ACC-vehicle measures the space gap g and speed difference between the preceding vehicle and ACC-vehicle $\Delta v = v_\ell - v$, where v is the ACC-vehicle speed, v_ℓ is the speed of the preceding vehicle. Based on the current values of g, v, and Δv, the ACC vehicle calculates a time headway τ between the ACC-vehicle and the preceding vehicle. For simplicity, we consider here only a case in which absolute values of Δv, the difference $\Delta \tau = \tau - \tau_{ACC}$, where τ_{ACC} is a desired time headway chosen by a driver, and acceleration (deceleration) of the preceding vehicle are not very great.

Then for an usual ACC the acceleration (deceleration) of the ACC-vehicle is given by a formula (e.g., [58–60]):

9.6 Adaptive Cruise Control

$$a_{ACC}(t) = K_1(g(t) - g_{ACC}(v(t))) + K_2 \Delta v(t) \quad (9.18)$$

where g_{ACC} is a desired space gap (Fig. 9.17) related to the desired time headway τ_{ACC}; K_1, K_2 are dynamic coefficients of ACC (K_1, $K_2 > 0$)[6].

The main objective of the ACC is to maintain the speed difference Δv close to zero and the space gap g close to g_{ACC}, i.e., the time headway should be close to the desired time headway τ_{ACC}. At higher speeds, $g_{ACC} = v\tau_{ACC}$, i.e., it is an increasing speed function (Fig. 9.17). When $\Delta v = 0$ and $g > g_{ACC}$, as follows from (9.18) the ACC-vehicle accelerates; otherwise, i.e., at $g < g_{ACC}$, the ACC-vehicle decelerates (Fig. 9.17).

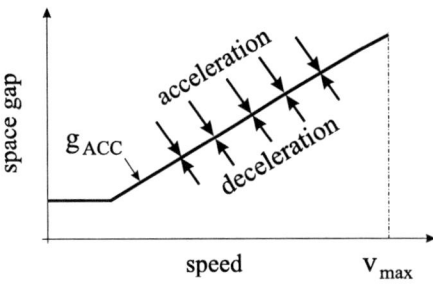

Fig. 9.17 Explanation of ACC: Qualitative speed dependence of desired space gap $g_{ACC}(v)$.

Traffic flow characteristics can be improved considerably through the use of ACC based on three-phase traffic theory [67]. In accordance with this theory (Sect. 3.2), when the space gap to the preceding vehicle is within a 2D-region in the space-gap–speed plane (dashed region in Fig. 9.18), a vehicle with such an ACC adapts its speed to the speed of the preceding vehicle without caring, what the precise space gap is [67]:

$$a(t) = K_{\Delta v} \Delta v(t) \quad \text{at } g_{safe} \leq g \leq G, \quad (9.19)$$

where G and g_{safe} are the synchronization and safe space gaps, respectively, $K_{\Delta v}$ is a dynamic coefficient ($K_{\Delta v} > 0$). At $g > G$ the ACC-vehicle accelerates, whereas at $g < g_{safe}$ the ACC-vehicle decelerates. In other words, outside of the 2D-region in the space-gap–speed plane formula (9.19) is not applied.

In contrast with the usual ACC (9.18), acceleration (deceleration) of a vehicle with the ACC based on three-phase traffic theory (9.19) does not depend on the space gap, i.e., on the time headway to the preceding vehicle at all. In other words, the ACC mode (9.19) does not maintain a desired time headway chosen by the driver. Moreover, the dynamic coefficient $K_{\Delta v}$ in the ACC mode (9.19) can be chosen

---

[6] The effect of ACC-vehicles (9.18) on traffic flow at different dynamic coefficients of ACC adaptation has been simulated with three-phase traffic flow models in [61–66]. Some results of these simulations can be found in Sect. 23.6 of the book [1].

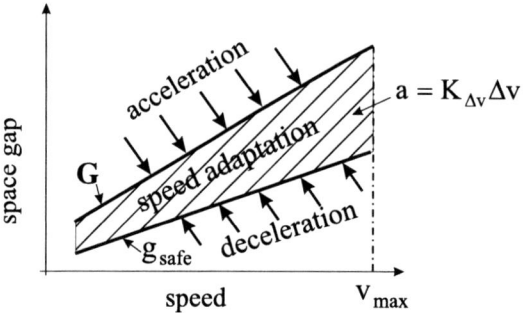

Fig. 9.18 Explanation of ACC in the framework of three-phase traffic theory [67]: A part of 2D-region of space gaps between a vehicle moving in accordance with the ACC mode (9.19) and the preceding vehicles in the space-gap–speed plane (dashed region).

to be considerably smaller than the dynamic coefficient K_2 in (9.18) for the usual ACC. This explains the following possible advantages of the ACC (9.19) based on three-phase traffic theory in comparison with the usual ACC (9.18):

- The removing of a conflict between the dynamic and comfortable ACC behavior. In particular, a much comfortable vehicle motion in congested traffic is possible.
- The reduction of fuel consumption and CO_2 emissions while moving in congested traffic.
- Because the ACC (9.19) leads to the reduction in speed changes in traffic flow, a sequence of such ACC-vehicles can prevent traffic breakdown at a bottleneck and wide moving jam emergence in synchronized flow.

References

1. B.S. Kerner, *The Physics of Traffic*, (Springer, Berlin, New York, 2004)
2. B.S. Kerner, H. Rehborn, M. Aleksić, A. Haug, Transp. Res. C **12**, 369–400 (2004)
3. H. Rehborn, S.L. Klenov, in *Encyclopedia of Complexity and System Science*, ed. by R.A. Meyers. (Springer, Berlin, 2009), pp. 9282–9302
4. B.S. Kerner, H. Rehborn, H. Kirschfink, German Patent DE 196 47 127 C2 (1998); US Patent US 5861820
5. B.S. Kerner, German Patent DE 199 40 957 C2 (1999)
6. B.S. Kerner, German Patent DE 199 44 075 C2 (1999); USA Patent US 2002045985; Japan Patent JP 2002117481
7. B.S. Kerner, M. Aleksić, U. Denneler, German Patent DE 199 44 077 C1 (1999)
8. H. Rehborn, J. Palmer, Straßenverkehrstechnik, No. 8, 463–470 (2008)
9. H. Rehborn, J. Palmer, Traf. Eng. & Cont. **49**, 261–266 (2008)
10. H. Rehborn, J. Palmer, in *Proceedings of IEEE Intelligent Vehicles Symposium*, (Eindhoven, 2008)
11. J. Palmer, H. Rehborn, L. Mbekeani, in *Proceedings of 15th World Congress on Intelligent Transport Systems*, (New York, 2008)
12. E.E. Stokes, Phil. Mag. **33**, 349–356 (1848)

13. B.S. Kerner, H. Rehborn, S.L. Klenov, M. Prinn, J. Palmer, German patent publication DE 102008003039 (2009)
14. E.D. Arnold, *Ramp Metering: A Review of the Literature*, (Virginia Transportation Research Council, USA, 1998)
15. A.D. May, *Traffic Flow Fundamentals*, (Prentice-Hall, Inc., New Jersey, 1990)
16. M. Papageorgiou, *Application of Automatic Control Concepts in Traffic Flow Modeling and Control*, (Springer, Berlin, New York, 1983)
17. E. Smaragdis, M. Papageorgiou, Transp. Res. Rec. **1856**, 74-86 (2003)
18. A. Hegyi, T. Bellemans, B. De Schutter, in *Encyclopedia of Complexity and System Science*, ed. by R.A. Meyers. (Springer, Berlin, 2009), pp. 3943–3964
19. M. Papageorgiou, A. Kotsialos, IEEE Trans. on ITS. **3**, 271-281 (2002)
20. M. Papageorgiou, H. Hadj-Salem, J.-M. Blosseville, Transp. Res. Rec. **1320**, 58-64 (1991)
21. M. Papageorgiou, H. Hadj-Salem, F. Middleham, Transp. Res. Rec. **1603**, 99–98 (1997)
22. M. Papageorgiou, J.-M. Blosseville, H. Hadj-Salem, Trans. Res. A. **24**, 361–370 (1990)
23. M. Papageorgiou, Y. Wang, E. Kosmatopoulos, I. Papamichail, Traf. Eng. & Cont. **48**, 271–276 (2007)
24. M. Papageorgiou, E. Kosmatopoulos, I. Papamichail, Y. Wang, IEEE Trans. on ITS. **9**, 360-365 (2008)
25. B.S. Kerner, Traf. Eng. & Cont. **48**, 28–35 (2007)
26. B.S. Kerner, Traf. Eng. & Cont. **48**, 68–75 (2007)
27. B.S. Kerner, Traf. Eng. & Cont. **48**, 114–120 (2007)
28. B.S. Kerner, IEEE Trans. on ITS **8**, 308–320 (2007)
29. B.S. Kerner, Transp. Res. Rec. **2088**, 80–89 (2008)
30. B.S. Kerner, in *Proceedings of the 83rd TRB Annual Meeting*, (Washington, DC, 2004), TRB Paper 04-3062
31. B.S. Kerner, Physica A **355**, 565–601 (2005)
32. B.S. Kerner, Transp. Res. Rec. **1999**, 30–39 (2007)
33. G. Noecker, D. Hermann, A. Hiller:, in *Proceedings of the 7th World Congress on Intelligent Transportation Systems*, (Torino, Italy 2000) Paper No. 2013
34. R.G. Herrtwich, G. Noecker:, in *Traffic and Granular Flow01*, ed. by M. Fukui, Y. Sugiyama, M. Schreckenberg, D.E. Wolf. (Springer, Berlin 2003), pp. 271–280
35. G. Noecker, K. Mezger, B.S. Kerner, in *Proceedings of Workshop FAS2005*, ed. by C. Stiller, M. Maurer. (University of Karlsruhe, Germany 2003), pp. 151–162
36. The INVENT Project. http://www.invent-online.de/
37. B.S. Kerner, Trans. Res. Rec. **1678**, 160–167 (1999); in *Transportation and Traffic Theory*, ed. by A. Ceder. (Elsevier Science, Amsterdam 1999), pp. 147–171; Networks and Spatial Economics, **1**, 35 (2001); Mathematical and Computer Modelling, **35**, 481–508 (2002); Phys. Rev. E **65**, 046138 (2002)
38. Dedicated Short Range Communications working group. http://grouper.ieee.org/groups/scc32/dsrc/index.html
39. The Fleetnet Project. http://www.fleetnet.de
40. The Now: Network on Wheels Project. http://www.network-on-wheels.de
41. Internet ITS Consortium. http://www.internetits.org
42. The WILLWARN Project. http://www.prevent-ip.org/en/
43. Vehicle safety communications consortium. http://www-nrd.nhtsa.dot.gov/pdf/nrd-12/CAMP3/pages/VSCC.htm
44. Car2Car Communication Consortium. http://www.car-to-car.org/
45. R. Schmitz, M. Torrent-Moreno, H. Hartenstein, W. Effelsberg, in *Proceeding of 29th Annual IEEE International Conference on Local Computer Networks*, (IEEE, Tampa, Florida, 2004), pp. 594–601
46. M. Torrent-Moreno, D. Jiang, H. Hartenstein, in *VANET'04: Proceedings of the 1st ACM International Workshop on Vehicular Ad Hoc Networks*, (Philadelphia, Pennsylvania, 2004), pp. 10–18

47. M. Torrent-Moreno, S. Corroy, F. Schmidt-Eisenlohr, H. Hartenstein, in *MSWiM'06: Proceedings of the 9th ACM international Symposium on Modeling Analysis and Simulation of Wireless and Mobile Systems*, (Terromolinos, Spain, 2007), pp. 68–77
48. F. Schmidt-Eisenlohr, M. Torrent-Moreno, J. Mittag, H. Hartenstein, in *WONS'07: Proceedings of the Fourth Annual Conference on Wireless on Demand Network Systems and Services*, (Obergurgl, Austria, 2007), pp. 50–58
49. D. Choffnes, F. Bustamante, in *Proceedings of the International Workshop on Vehicular Ad Hoc Networks*, (Cologne, Germany, 2005)
50. J. Maurer, T. Fgen, W. Wiesbeck, in *Proceedings of the International Workshop on Intelligent Transportation*, (Hamburg, Germany, March 2005).
51. Q. Chen, D. Jiang, V. Taliwal, L. Delgrossi, in *Proceedings of the International Workshop on Vehicular Ad Hoc Networks*, (Los Angeles, 2006)
52. First Simulation Workshop 2007. http://www.comesafety.org/index.php?id=34
53. B. Sklar, IEEE Communications Magazine, **35**, 90–100 (1997)
54. B. S. Kerner, S. L. Klenov, A. Brakemeier, arXiv: 0712.2711 (2007), available at http://arxiv.org/abs/0712.2711; in *Proceedings of IEEE IV 2008*, paper 35; in *Proceedings of the Fourth international Workshop on Vehicle-to-Vehicle Communication V2VCOM 2008*, (2008), pp. 57–63
55. IEEE Std.802.11-1999, Part 11: Wireless LAN Medium Access Control (MAC) and Physical Layer (PHY) specifications. IEEE Std.802.11, 1999 edition.
56. IEEE Std.802.11e/D4.4, Draft Supplement to Part 11: Wireless LAN Medium Access Control (MAC) and Physical Layer (PHY) specifications: Higher-speed Physical Layer the 5 GHz Band, IEEE Std.802.11a-1999, 1999.
57. IEEE Std.802.11a, Supplement to Part 11: Wireless LAN Medium Access Control (MAC) and Physical Layer (PHY) specifications: Medium Access Control (MAC) Enhancements for Quality of Service (QoS), June 2004.
58. S. Becker, M. Bork, H.T. Dorissen, G. Geduld, O. Hofmann, K. Naab, G. Nöcker, P. Rieth, J. Sonntag, in *Proceedings of the 1st World Congress on Applications of Transport Telematics and Intelligent Vehicle-Highway Systems*, (1994), pp. 1828–1843.
59. R. Müller, G. Nöcker, in *Proceedings of the Intelligent Vehicles '92 Symposium*, ed. by I. Masaki. (IEEE, Detroit, USA, 1992), pp. 173–178
60. G. Nöcker, in *Band Elektronik im Kraftfahrzeugwesen*, ed. by G. Walliser. Vol. 437, (Expert Verlag 1994), pp. 299–322 (ISBN 3-8169-1024-6).
61. B.S. Kerner, cond-mat/0309017 (2003), e-print in http://arxiv.org/abs/cond-mat/0309017
62. B.S. Kerner, in *Proceedings of the 10th World Congress on Intelligent Transport Systems*, Paper No. 2043 T, (Madrid, Spain, 2003)
63. L.C. Davis, Phys. Rev. E. **69**, 066110 (2004)
64. B.S. Kerner, in *Transportation and Traffic Theory. Proceedings of the 16th Inter. Sym. on Transportation and Traffic Theory*, ed. by H.S. Mahmassani. (Elsevier, Amsterdam, 2005), pp. 181–204
65. L.C. Davis, Physica A **368**, 541–550 (2006)
66. L.C. Davis, Physica A **379**, 274–290 (2007)
67. B.S. Kerner, German patent DE 10308256.5 (2004); USA patent US 20070150167 (2007) http://www.freepatentsonline.com/20070150167.html; German patent publications DE 102005017559 (2007), DE 102005017560 (2007), DE 102005033495 (2007), DE 102007008255 (2007), DE 102007008253 (2007), DE 102007008254 (2007), DE 102007008257 (2007), DE 102008023704 (2008)

Chapter 10
Earlier Theoretical Basis of Transportation Engineering: Fundamental Diagram Approach

10.1 Traffic Description and Control based on Fundamental Diagram of Traffic Flow

10.1.1 Fundamental Diagram of Traffic Flow

Beginning from the classic work of Greenshields [1], the fundamental diagram of traffic flow is the basis for earlier traffic flow theories and models as well as the basic methodology for empirical studies of measured traffic data (see, e.g., reviews and books [2–27]).

The fundamental diagram is a flow-rate–density relationship, i.e., a correspondence between a given vehicle density and the flow rate in traffic flow (Fig. 10.1 (a)). The fundamental diagram reflects the result of empirical observations that the greater the density, the lower the averaged speed in vehicular traffic (Fig. 10.1 (b)). Thus in the flow–density plane, the fundamental diagram should pass through the origin (when the density is zero so is the flow rate) and should have at least one maximum (Fig. 10.1 (a)). The fundamental diagram gives also connections between the average space gap between vehicles and the average speed (Fig. 10.1 (c)) as well as between link travel time (travel time on a link of a traffic network) and flow rate (Fig. 10.1 (d))[1]. The fundamental diagram is up to now the most often subject of traffic flow measurements and it is the teaching basis in transportation engineering.

However, *neither* empirical fundamental diagram *nor* any associated relationship between averaged traffic flow variables are associated with the empirical basis of three-phase traffic theory. In contrast, the empirical basis of this theory is the empirical phase definitions [S] and [J] as well as other empirical *spatiotemporal features* of traffic breakdown and resulting congested patterns found in measured traffic data (Part I).

---

[1] There are a huge number of publications devoted to measurements of flow–density (fundamental diagram), speed-density, speed–space-gap, and link-travel-time–flow relationships (see, e.g., a review by Hall in [13]).

B.S. Kerner, *Introduction to Modern Traffic Flow Theory and Control*,
DOI 10.1007/978-3-642-02605-8_10, © Springer-Verlag Berlin Heidelberg 2009

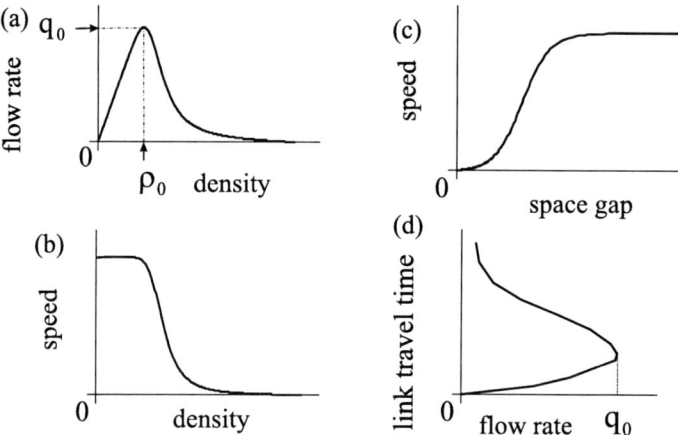

Fig. 10.1 Qualitative example of fundamental diagram: (a) Flow–density relationship (fundamental diagram). (b–d) The speed–density (b), speed–space-gap (c) and link-travel-time–flow (d) relationships associated with (a)

The main reason why empirical fundamental diagrams of traffic flow and other associated relationships between averaged traffic flow variables are *not* some empirical fundamentals of three-phase traffic theory is as follows:

- Vehicular traffic occurs in space and time. Thus to understand traffic, empirical *spatiotemporal* features of traffic breakdown and resulting congested patterns must be understood. However, most of these empirical spatiotemporal features are *averaged* in time and space and, therefore, *lost* in empirical fundamental diagrams of traffic flow and other associated relationships between averaged traffic flow variables.

Moreover, even if a fundamental diagram consists of several parts each of them is related to traffic data measured at different road locations (see, e.g., Fig. 2.6 in the review by Hall in [13]), it is *not* possible to reconstruct *spatiotemporal characteristics* of a congested pattern from the fundamental diagram:

- A solely analysis of different congested traffic states and phase transitions in the flow–density plane is not consistent with empirical features of congested traffic. This is because most of these spatiotemporal features are lost in such a congested traffic analysis.
- A classification of traffic states and phases based on such a solely traffic flow analysis in the flow–density plane (as well as in speed–density, space-gap–density planes, etc.), which is very often made in the literature, ignores the spatiotemporal nature of traffic and, therefore, it cannot be used for efficient engineering applications.

10.1 Traffic Description and Control based on Fundamental Diagram   175

- A classification of traffic states and phases should be made from an analysis of common empirical features of the spatiotemporal traffic dynamics (see [28] and Part I of the present book).

Probably, a difficulty in the understanding of the above criticism on the fundamental diagram of traffic flow is that there is *nothing wrong* in the fundamental diagram. This diagram, however, reflects only some *averaged characteristics* of traffic flow. Rather than a study of these averaged characteristics, the main objective of transportation engineering methods, which could be successfully applied for traffic flow management, should be the understanding of *spatiotemporal characteristics* of congested patterns; this is the main objective of this book. As abovementioned, these characteristics are *lost* in the fundamental diagram as well as in all other relationships between averaged traffic flow variables.

The latter critical comment is also related to a huge number of other approaches for the description of characteristics of congested traffic like optimal velocity functions, speed, flow rate and density correlation functions [29–31] as well as so-called cumulative N- and T-curves [15, 32]: as in the fundamental diagram approach, in these approaches important spatiotemporal characteristics of real measured traffic patterns needed for the understanding of traffic breakdown and resulting congested patterns are lost.

In the remaining of this chapter, based on a critical consideration of traffic flow theories and highway capacity definitions in the framework of the fundamental diagram of traffic flow, we will try to show and explain that and why

- the methodology of transportation engineering based on the fundamental diagram of traffic flow is *inconsistent* with *spatiotemporal measured traffic data* and, therefore, the methodology leads to traffic flow theories that cannot be used for efficient dynamic management and control in congested freeway networks.

10.1.2 Two Model Classes in Framework of Fundamental Diagram Approach

Earlier traffic flow models, which claim to explain and predict the onset of traffic congestion, are reviewed in [3–26]. All these models are associated with a so-called fundamental diagram approach to traffic flow modeling:

- An approach to traffic flow theories and models, which is based on the fundamental diagram of traffic flow, we call the fundamental diagram approach to traffic flow theory and modeling.

In 1955 Lighthill and Whitham [33] wrote in their classic work (see p. 319 in [33]): "... The fundamental hypothesis of the theory is that at any point of the road the flow (vehicles per hour) is a function of the concentration (vehicles per mile)..." (Fig. 10.1 (a)). The fundamental diagram for a traffic flow model means that in a hypothetical model limit case in which all vehicles move at time-independent speeds there is a single desired (or optimal) space gap.

The models within the framework of the fundamental diagram approach can be classified into two main classes. The first model class is associated with the classic Lighthill-Whitham-Richards (LWR) model introduced in 1955-1956 [33, 34] (see also references in [4–6, 8, 11, 13, 19, 20, 24, 25]). An example for this model class is a cell-transmission model of Daganzo [35]. The basic idea of the LWR model is that maximum flow rate (denoted by q_0 in Fig. 10.1 (a)), which is associated with the maximum point on the fundamental diagram, determines capacity of free flow at a bottleneck, i.e., the bottleneck capacity. Thus if the flow rate upstream of the bottleneck exceeds the bottleneck capacity, then traffic breakdown should occur.

The second model class is associated with the classic General Motors (GM) model of Herman, Gazis, Rothery, Montroll, and Potts introduced in 1959-1961 [36, 37]. The basic idea of the GM model approach is as follows. Beginning at a critical density, there is an instability of steady model states at the fundamental diagram. The instability is associated with a finite value of driver reaction time. Examples for the GM model class are the optimal velocity (OV) models by Newell [38], Whitham [39], and Bando *et al.* [40–42], Payne's macroscopic model and its variants [43,44], Wiedemann's psychophysical traffic flow model and its variants [9,45], Gipps's model [46], the Nagel-Schreckenberg (NaSch) cellular automata (CA) model and its variants [47, 48], the Krauß model [49, 50], the Intelligent Driver Model (IDM) of Treiber *et al.* [51] as well as a huge number of other traffic flow models (see below and references in [4–6, 8, 9, 11–14, 17–23, 26, 27]).

10.1.3 Validation of Theoretical Fundamental Diagram

Thus the fundamental diagram (Fig. 10.1 (a)) is the basic hypothesis for all earlier traffic flow theories models reviewed in [3–26]. However, empirical observations show that whereas shapes of fundamental diagrams of free flow are qualitatively the same at different road locations, in contrast, the shape of empirical fundamental diagrams of congested traffic depends qualitatively on the location at which data measurements have been performed within a congested pattern. An example of different fundamental diagram types associated with an GP at an on-ramp bottleneck is shown in Fig. 10.2 (see explanations in Chap. 15 of the book [28]).

This means that in a general case the validation of a theoretical fundamental diagram of a traffic flow model, which is very often used by researchers as an empirical basis for the fundamental diagram, cannot be performed. Depending on a road location various empirical fundamental diagram types exist. Moreover, if data, which are measured at many different locations within congested traffic, is used for averaging to one empirical fundamental diagram, then there is almost no possibility to find a relation of such a fundamental diagram to spatiotemporal features of real traffic patterns.

10.2 Lighthill-Whitham-Richards (LWR) Theory

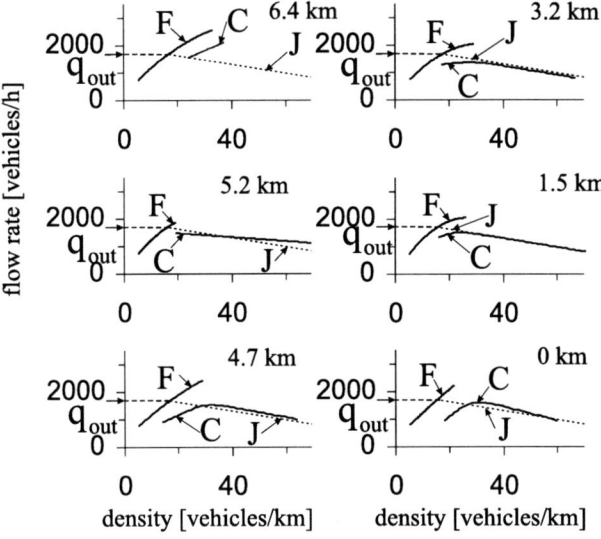

Fig. 10.2 Dependence of empirical fundamental diagram types of congested traffic at different locations within the GP shown in Fig. 5.10 (a). F – free flow. C – congested traffic. J – line *J* (dashed line). On-ramp bottleneck is at the location about 6.4 km. See explanations of these fundamental diagram types in Chap. 15 of Ref. [28]

10.2 Congested Traffic Description in the Framework of Lighthill-Whitham-Richards (LWR) Traffic Flow Theory

10.2.1 LWR Traffic Flow Theory

The basic idea of the classic LWR traffic flow theory is as follows: the maximum flow rate q_0 associated with the maximum point (ρ_0, q_0) at the fundamental diagram (Fig. 10.1 (a, d)) determines free flow capacity at a bottleneck denoted by $q_{\rm cap}$:

$$q_{\rm cap} = q_0. \tag{10.1}$$

If the sum of the flow rates in free flow upstream of the bottleneck reaches the maximum flow rate q_0 at the fundamental diagram (Fig. 10.1 (a)), i.e, it reaches the bottleneck capacity (10.1), then a further increase in the upstream flow rates must lead to congestion formation and upstream congestion propagation.

This hypothesis of the LWR theory means that congested traffic occurs only then, when the sum of the flow rates in free flow upstream exceeds the bottleneck capacity $q_{\rm cap}$ determined by (10.1). This conclusion of the LWR traffic flow theory about the reason for the onset of congestion is very often used as the definition of congested traffic. As we will see below, this hypothesis of the LWR theory about the onset of congestion as well as the associated congested traffic definition are inconsistent with empirical results.

In accordance with the LWR theory, traffic flow phenomena should be explained based on the law of conservation of the number of vehicles on the road

$$\frac{\partial \rho(x,t)}{\partial t} + \frac{\partial Q(\rho(x,t))}{\partial x} = 0, \qquad (10.2)$$

in which there is a relationship between the flow rate Q and density ρ

$$Q = Q(\rho) \qquad (10.3)$$

associated with the fundamental diagram of traffic flow. Here, x is a spatial coordinate in the direction of traffic flow, and t is time. The LWR model (10.2), (10.3) has discontinuous solutions in the form of shock-waves (Fig. 10.3) with a shock-wave velocity

$$v_s = \frac{Q(\rho_2) - Q(\rho_1)}{\rho_2 - \rho_1}, \qquad (10.4)$$

where the flow rates $Q(\rho_2)$, $Q(\rho_1)$ associated with (10.3) correspond to points in the flow–density plane that lie on the fundamental diagram; ρ_1 and ρ_2 are the densities upstream and downstream of the shock wave, respectively (Fig. 10.3). The shock wave can be represented in the flow–density plane by the line labeled by "shock" whose slope is equal to the shock velocity v_s (Fig. 10.3 (b)).

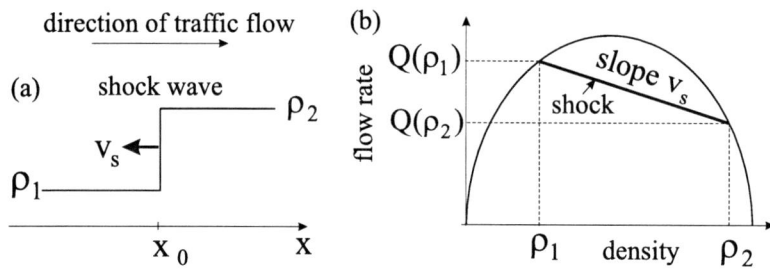

Fig. 10.3 Qualitative representation of a shock wave in space (a) and in the fundamental diagram with the line labeled by "shock" (b) [33]

10.2.2 Features of Onset of Congestion in LWR Theory

Because the LWR model (10.2), (10.3) has discontinuous solutions in the form of shock waves, its numerical simulations are usually based on one of finite differ-

10.2 Lighthill-Whitham-Richards (LWR) Theory

ence approximation methods for partial differential equations associated with the Godunov family methods. One of such discrete versions of the LWR model (10.2), (10.3) that is consistent with the LWR theory is a cell-transmission model of Daganzo [35].

In Daganzo's cell transmission model, a rectangular lattice with time spacing τ_{step} and space length (cell length) Δx is overlaid on the time–space plane; the x-coordinates represent the center of the cells into which a road has been discretized; t-coordinates are the times at which the cell vehicle densities are evaluated. Then the model reads as follows [35]:

$$\rho(x,t+\tau_{step}) = \rho(x,t) - [q(x+\Delta x/2, t+\tau_{step}/2) - q(x-\Delta x/2, t+\tau_{step}/2)]\tau_{step}/\Delta x, \quad (10.5)$$

$$q(x+\Delta x/2, t+\tau_{step}/2) = \min(S(\rho(x,t)), R(\rho(x+\Delta x,t))), \quad (10.6)$$

where $R(\rho)$ and $S(\rho)$ are non-increasing so-called *receiving* and non-decreasing so-called *sending* curves, respectively; these two monotonic curves $R(\rho)$ and $S(\rho)$ take values in the interval $[0, q_0]$: $R(\rho) = Q(\rho)$ at $\rho \geq \rho_0$ and $R(\rho) = q_0$ at $\rho < \rho_0$; $S(\rho) = Q(\rho)$ at $\rho \leq \rho_0$ and $S(\rho) = q_0$ at $\rho > \rho_0$, where (ρ_0, q_0) is the maximum point at the fundamental diagram (Fig. 10.1 (a)).

Based on numerical simulations of Daganzo's cell-transmission model (10.5), (10.6) on a road with an on-ramp bottleneck[2], common qualitative features of traffic breakdown and resulting congested patterns of the LWR theory at the bottleneck can be shown (Fig. 10.4) [52–54]. Numerical simulations presented in Fig. 10.4 are made at a given flow rate in free flow on the main road upstream of the bottleneck q_{in} and different on-ramp inflow rates q_{on}.

If $q_{on} = 0$, the speed and density are spatially homogeneous. When the flow rate q_{on} increases beginning from zero (points 1–3 in Fig. 10.4 (a, b)), the sum of the flow rates $q_{on} + q_{in}$ increases too as long as the condition

$$q_{on} + q_{in} < q_0 \quad (10.7)$$

is satisfied. In this case, no moving shock wave occurs.

There is a critical flow rate to the on-ramp $q_{on} = q_{on}^{(d)}$ at which the flow rate $q_{on} + q_{in}$ is equal to the maximum flow rate on the fundamental diagram:

$$q_{in} + q_{on}^{(d)} = q_0. \quad (10.8)$$

When the flow rate q_{on} increases further, i.e.,

$$\Delta q = q_{in} + q_{on} - q_0 > 0, \quad (10.9)$$

then a shock wave of lower speed and greater density propagating upstream appears (Fig. 10.4 (c)).

Independent of the choice in q_{in} (but $q_{in} < q_0$), the absolute value of the shock-wave velocity $|v_s|$ increases continuously beginning from *zero* (Fig. 10.4 (b)), when

---

[2] A bottleneck model used for simulations can be found in [54].

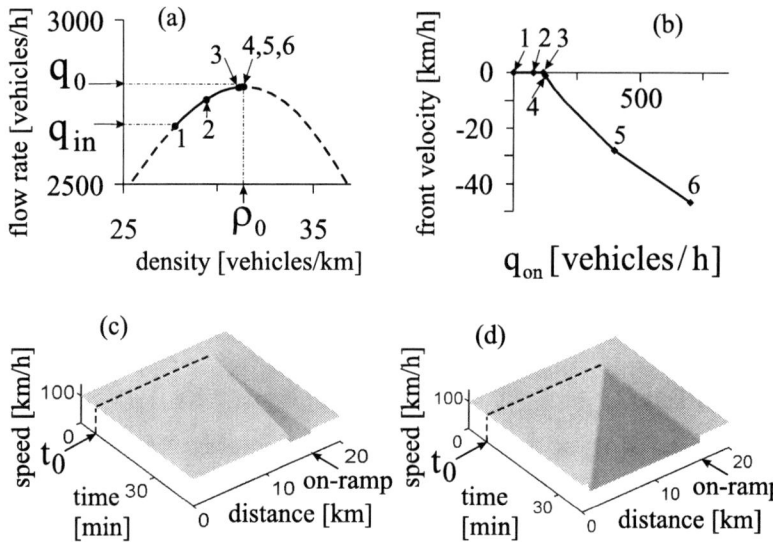

Fig. 10.4 Traffic breakdown in the LWR theory: (a, b) The flow rate and density on main road downstream of bottleneck (a) and shock velocity v_s (b) at different on-ramp inflow rates q_{on} (points 1-6); dashed curve in (a) is a part of the fundamental diagram used in simulations in the vicinity of the maximum point (ρ_0, q_0). (c, d) Propagation of shock-waves under condition (10.9) for $q_{on} = 160$ (c) and 400 vehicles/h (d); at $t = t_0$ the on-ramp inflow is switched on. Simulations of Daganzo's cell-transmission model at on-ramp bottleneck made in [53, 54]

q_{on} first reaches and then exceeds the critical flow rate $q_{on}^{(d)}$ in (10.8). We find that if $\Delta q \to 0$, then $|v_s| \to 0$. For example, at $\Delta q = 3$ vehicles/h ($q_{on} = 120$ vehicles/h), the shock wave has the velocity $v_s \approx -0.5$ km/h (point 4 in Fig. 10.4 (b)).

The greater q_{on}, specifically, the greater Δq (10.9), the greater the absolute value of the shock wave velocity $|v_s|$ (Fig. 10.4 (b–d)). The flow rate downstream of the bottleneck, which is equal to q_0 under condition (10.8), remains approximately to be equal to q_0, when q_{on} increases (points 4–6 in Fig. 10.4 (a)).

A summary of features of the LWR theory that are inconsistent with empirical results (Chaps. 3 and 5) is as follows:

1. In the LWR theory, there is *no* induced traffic breakdown at a bottleneck.
2. In the LWR theory, there is *no* hysteresis effect associated with congested pattern emergence and dissolution in empirical observations.
3. In the LWR theory, the absolute value of the velocity $|v_s|$ increases continuously beginning from zero, when q_{on} first reaches and then exceeds the critical flow rate $q_{on}^{(d)}$ (Fig. 10.4 (b)) associated with condition (10.8). This is qualitatively different from the behavior of the upstream front velocity of a congested pattern at a bottleneck observed in real measured traffic data: when the flow rate q_{on} increases and traffic breakdown occurs, rather than a continuous increase in

$|v_{\mathrm{s}}|$ beginning from zero with the increase in q_{on} a wave of synchronized flow occurs abruptly and propagates upstream with a *great* velocity.

4. In empirical observations, the flow rate downstream of the bottleneck at which breakdown occurs in a realization (day) is often smaller than flow rates observed in free flow downstream of the bottleneck before breakdown occurrence. This cannot be explained by the LWR theory.
5. In empirical data, after traffic breakdown has occurred at a bottleneck, wide moving jams can emerge spontaneously in synchronized flow upstream of the bottleneck, i.e., an GP emerges spontaneously. Independent of the density, *no* spontaneous moving jam emergence can be found in the LWR theory.
6. The classic LWR-formula for shock wave velocity (10.4) cannot be applied for a correct calculation of the front velocities of a wide moving jam while the jam propagates through synchronized flow. Thus instead of the LWR formula (10.4), one should use the classic Stokes's formula for a shock wave velocity derived by Stokes in 1848 [55]:

$$v_{\mathrm{s}} = \frac{q_2 - q_1}{\rho_2 - \rho_1}, \tag{10.10}$$

where q_2, ρ_2 and q_1, ρ_1 are the flow rate and density downstream and upstream of a shock wave, respectively. The fundamental difference between the LWR formula (10.4) and Stokes's formula (10.10) is that there is *no* given relationship between the density and flow rate in Stokes's formula (10.10). For these reasons, rather than the LWR formula (10.4) in the model ASDA for tracking and prediction of wide moving jam propagation (Sect. 9.1), Stokes's formula (10.10) has been applied.

For these reasons, the LWR theory and the associated traffic flow models (like cell-transmission models) cannot be used for reliable analysis of freeway traffic control and dynamic traffic management strategies.

10.2.3 Discussion of The Term "Shock Waves in Traffic Flow"

As shown above, the LWR shock wave theory cannot be used for the description of empirical spatiotemporal features of traffic congested patterns found in real measured traffic data. However, because this theory is the basic theory in the teaching of many generations of transportation engineers, the term *shock wave in traffic flow* is automatically associated with the LWR traffic flow theory.

This is one of the reasons why we do not use this term in three-phase traffic theory. Another important reason for this is as follows.

- In real traffic, a variety of diverse congested traffic patterns are observed whose spatiotemporal features are qualitatively different. To distinguish these patterns, different terms for them are needed.

For example, as shown in Chap. 2 the downstream front of a *wide moving jam* exhibits the characteristic jam feature [J] to propagate through bottlenecks while maintaining the mean velocity of the jam downstream front. In contrast, if a *narrow moving jam* propagating upstream in free flow reaches a bottleneck at which metastable free flow has been before, the jam is caught at the bottleneck. In the frame of the LWR theory, narrow and wide moving jams are both *shock waves*. Therefore, instead of the single term *shock wave in traffic flow* we use different specific terms for congested traffic patterns with qualitatively different spatiotemporal features, like narrow and wide moving jams, synchronized flow patterns, general patterns, etc. (see Chaps. 2–7).

10.3 Traffic Breakdown Description through Free Flow Instability in General Motors (GM) Model Class

10.3.1 GM Model

As long ago as 1958–1961, Herman, Montroll, Potts, Gazis, Rothery [36, 37] and Kometani and Sasaki [56, 57] assumed that the onset of congestion is associated with an instability of free flow. This instability is related to a finite reaction time of drivers. This driver reaction time is responsible for an over-deceleration effect discussed in Sect. 3.2.3: If a vehicle begins to decelerate unexpectedly, then owing to the finite driver reaction time the following vehicle starts deceleration with a delay. When the time delay is long enough, the driver of the following vehicle decelerates stronger than it is needed to avoid collisions. As a result, the speed of the following vehicle becomes lower than the speed of the preceding vehicle. If this over-deceleration effect is realized for the following drivers, a wave of vehicle speed reduction appears and increases in amplitude over time. This instability should occur when the vehicle density on the fundamental diagram exceeds some critical density.

In the car-following GM model of Herman, Montroll, Potts, Gazis, Rothery [36, 37] a driver reaction time denoted by τ_0 is explicitly used in vehicle deceleration (acceleration) $a(t+\tau_0)$: the vehicle reacts with the time delay τ_0 on any changes in the space gap to the preceding vehicle $g(t)$ and the speed difference $\Delta v(t) = v_\ell(t) - v(t)$ between the speed of the preceding vehicle $v_\ell(t)$ and the vehicle speed $v(t)$. As a result, the GM model reads as follows [37]:

$$a(t+\tau_0) = \frac{\Delta v(t)[v(t+\tau_0)]^{m_1}}{T_0[g(t)+d]^{m_2}}, \quad (10.11)$$

$$a(t) = \frac{dv(t)}{dt}, \quad (10.12)$$

where d is the vehicle length; T_0, m_1, m_2 are constants. By integrating Eqs. (10.11), (10.12) at $m_1 < 1$, $m_2 > 1$, one gets model solutions for steady states related to the fundamental diagram [37]

10.3 General Motors (GM) Model

$$V_0(\rho) = v_0[1 - (\rho/\rho_{max})^{(m_2-1)}]^{(1-m_1)^{-1}}, \quad (10.13)$$

where v_0 is constant, $\rho_{max} = 1/d$.

One of the mathematical realizations of the GM model class proposed by Nagatani and Nakanishi [58] and further developed by Lubashevky et al. [59] reads as follows

$$\frac{da}{dt} = \frac{f(v(t), v_\ell(t), g(t)) - a(t)}{\tau_0}. \quad (10.14)$$

In both models [37, 58, 59], steady state model solutions in the flow–density plane lie on the fundamental diagram.

The idea of the GM model (10.11) with a driver reaction time is incorporated either explicitly or implicitly in many other traffic flow models reviewed in [4–7, 9, 11–14, 16–27].

10.3.2 Newell's Optimal Velocity (OV) Model and its Variants

At $m_1 = 0$ the GM model is associated with Newell's optimal velocity (OV) model [38] (see also the review [22])

$$v(t + \tau_0) = V(g(t)). \quad (10.15)$$

In this case, the OV function $v = V(g)$ and the associated fundamental diagram $Q(\rho) = \rho V_0(\rho)$ are determined from (10.15) at time-independent space gap g and speed v:

$$v = V(g). \quad (10.16)$$

In the Bando et al. OV model [40–42]

$$\frac{dv(t)}{dt} = \frac{V(g(t)) - v(t)}{\tau_0} \quad (10.17)$$

it is assumed that

$$V(g) = 0.5v_0[\tanh(\beta_0(g - g_0)) + \gamma_0], \quad (10.18)$$

β_0, γ_0, v_0, τ_0, g_0 are model parameters.

There are many other traffic flow models associated with a generalization of OV models, which can be written in the form (e.g., [20])

$$\frac{dv(t)}{dt} = \phi(v(t), v_\ell(t), g(t)), \quad (10.19)$$

where the OV function $v = V(g)$ and the associated fundamental diagram $Q(\rho) = \rho V_0(\rho)$ are determined from (10.19) at time-independent space gap g and speeds v, v_ℓ as well as $v = v_\ell$:

$$\phi(V, V, g) = 0. \quad (10.20)$$

In the models (10.19), a driver reaction time is used only implicitly in a traffic flow model. An example of these models is the IDM of Treiber *et al.* [51] with

$$\phi = c_1[1 - (v/v_0)^{\delta_0} - (g^*(v,v_\ell)/g)^2], \tag{10.21}$$

$$g^*(v,v_\ell) = g_1 + c_2(v/v_0)^{1/2} + \tau_1 v + v(v - v_\ell)/2(a_0 b_0)^{1/2}, \tag{10.22}$$

where c_1, c_2, δ_0, g_1, v_0, τ_1, a_0, and b_0 are model parameters.

These models exhibit qualitatively the same features of free flow instability and resulting wide moving jam formation as those firstly found in [60,61] from an analysis of a version of Payne's macroscopic traffic flow model.

10.3.3 Payne's Macroscopic Traffic Flow Model and its Variants

Payne's model reads as follows [43, 44]:

$$\frac{\partial \rho(x,t)}{\partial t} + \frac{\partial (\rho(x,t)v(x,t))}{\partial x} = 0, \tag{10.23}$$

$$\frac{\partial v(x,t)}{\partial t} + v \frac{\partial v(x,t)}{\partial x} = \frac{V_0(\rho) - v(x,t)}{\tau_0} - \frac{c_0^2}{\rho} \frac{\partial \rho(x,t)}{\partial x}, \tag{10.24}$$

where the speed–density relationship $V_0(\rho)$ determines the fundamental diagram $Q(\rho) = \rho V_0(\rho)$; τ_0, c_0 are model parameters. To avoid solutions with speed discontinuities, Kühne introduced [62] a viscosity term $\mu \partial^2 v(x,t)/\partial x^2$ in the equation (10.24) of Payne's model. In [60], the equation (10.24) of Payne's model has been rewritten as the Navier-Stokes equation:

$$\frac{\partial v(x,t)}{\partial t} + v \frac{\partial v(x,t)}{\partial x} = \frac{V_0(\rho) - v(x,t)}{\tau_0} - \frac{c_0^2}{\rho} \frac{\partial \rho(x,t)}{\partial x} + \frac{\mu}{\rho} \frac{\partial^2 v(x,t)}{\partial x^2} \tag{10.25}$$

with the speed–density relationship

$$V_0(\rho) = c_3[1 + [\exp[(\rho/\rho_{\max}) - c_4]/c_5]^{-1} - c_6], \tag{10.26}$$

c_i, $i = 3,4,5,6$ are model parameters. There are a variety of other Navier-Stokes-like and gas-kinetic non-local traffic flow models, which, as Payne's model, consist of the vehicle balance and velocity equations (see models of Cremer [12] and Papageorgiou [16] as well as other macroscopic models discussed in Sect. D of the review by Helbing [20]). We may call all these macroscopic traffic flow models as Payne-like models. This is because these models show qualitatively the same features of free flow instability and resulting wide moving jam formation in free flow as those found in [60,61,63] for the version of Payne's model (10.23), (10.25), (10.26).

In the Aw-Rascle macroscopic model [64], instead of Payne's equation (10.24), the following equation has been introduced:

$$\frac{\partial v(x,t)}{\partial t} + v\frac{\partial v(x,t)}{\partial x} = \frac{V_0(\rho) - v(x,t)}{\tau_0} + \rho\frac{dP}{d\rho}\frac{\partial v(x,t)}{\partial x}, \quad (10.27)$$

where $P(\rho)$ is a density function. When the same fundamental diagram is chosen in this model as that in Payne's model, then the Aw-Rascle macroscopic model (10.23), (10.27) [64] exhibits also qualitatively the same features of free flow instability and resulting congested pattern formation [65] as those in Payne-like models and other models of the GM model class found earlier in [60, 61, 63, 66, 67].

10.3.4 Wiedemann's Psychophysical Traffic Flow Model and its Variants

In Wiedemann's psychophysical model [9], following the preceding vehicle a driver changes acceleration (deceleration), if some thresholds called "action points" are crossed. Some of these action points are associated with the following driver behavioral assumption made in Wiedemann's model and in its variants [9, 45].

If the absolute value $|\Delta v|$ of the speed difference $\Delta v = v_\ell - v$ in car-following is smaller than a threshold value, then a driver is *not able to recognize* whether the space gap to the preceding vehicle increases or decreases over time. In other words, it is assumed that at small values of $|\Delta v|$ a driver is not able to recognize whether she/he is slower or faster than the preceding vehicle. In a car-following in which the speed of the preceding vehicle is a time-independent and under a assumption that Δv is a very small negative value (because the driver accelerates initially with a small acceleration), the driver is not able to recognize that she/he is faster than the preceding vehicle as long as an action point associated with a safe space gap is crossed that depends on the vehicle speed. After the driver reaches the safe space gap, she/he decelerates with a small deceleration; however, as long as Δv is a small enough value, the driver is not able to recognize whether she/he is slower than the preceding vehicle. The driver recognizes this only after she/he crosses another action point associated with a greater space gap than the safe space gap. At this action point the driver accelerates with a small acceleration up to the safe space gap, then the driver decelerates with the small deceleration, and so on.

This driver behavioral assumption of Wiedemann's model and its variants, which is probably valid for *free flow*, contradicts a driver behavioral assumption of three-phase traffic theory discussed in Sect. 3.2.1 in which is assumed that in *synchronized flow* independent of the speed difference Δv a driver *does recognize* whether she/he is slower or faster than the preceding vehicles; therefore, when the preceding vehicle moves in synchronized flow at a time-independent speed, then the speed adaptation effect leads to car-following in the framework of three-phase traffic theory in which the driver follows the preceding vehicle at $\Delta v = 0$ and any space gap to the preceding vehicle within the space gap range (3.12) (Sect. 3.2.2).

Thus rather than the 2D *steady* states of synchronized flow of three-phase traffic theory (Fig. 3.3 (a)), under condition that the preceding vehicle moves at a time-independent speed, in Wiedemann's model some 2D *dynamic* states of traffic flow appear, which are associated with changes in driver behavior in the action points.

It should be stressed that rather than this 2D dynamic vehicle motion occurring in car-following of Wiedemann's model with small values of $|\Delta v|$ and the associated small vehicle deceleration and acceleration, the following driver behavioral assumption made in Wiedemann's model and its variants determines features of traffic breakdown in the models [9,45]: When the speed difference Δv is initially a large enough negative value and the driver crosses an action point at which she/he must decelerate, then as in other models of the GM model class the driver decelerates with a time delay (called also "dead" time).

10.3.5 Nagel-Schreckenberg Cellular Automata Traffic Flow Model

A different mathematical description for a driver time delay within the framework of the GM model class has been introduced by Nagel and Schreckenberg [47]. In the Nagel-Schreckenberg (NaSch) cellular automata (CA) model, in which the time and space are discrete values, the driver time delay is described through the use of model driver fluctuations. The discrete time is $t = n\tau_{\text{step}}$, $n = 0, 1, 2, ..$; τ_{step} is the time step and the road is divided into cells of a finite length (see a history of CA traffic flow model development in review [27]). In the initial version of the NaSch CA model the cell length has been chosen to be equal the vehicle length d [47]. The update rules for vehicle motion in the NaSch CA model can be written as follows

$$v_{n+1} = \max(0, \min(v_{\text{free}}, v_n + 1, g_n) - \xi_n), \quad (10.28)$$

$$x_{n+1} = x_n + v_{n+1}, \quad (10.29)$$

$$\xi_n = \begin{cases} 1 & \text{for } r \leq p, \\ 0 & \text{otherwise}, \end{cases} \quad (10.30)$$

$r = \text{rand}(0,1)$ denotes a random number uniformly distributed between 0 and 1; $p < 1$; time and space are in the units of τ_{step} and d, respectively.

To understand the NaSch CA model, firstly note that g_n determines in (10.28) a safe speed: when $v_n > g_n$, then a driver decelerates at time step $n + 1$ because from (10.28) it follows that

$$v_{n+1} \leq g_n. \quad (10.31)$$

This prevents vehicle collisions. Note that the safe speed determined through the space gap g was first introduced by Pipes [68].

Secondly, a case

$$v_{n+1} = v_n + 1 \quad (10.32)$$

means that the vehicle accelerates at time step $n + 1$. However, in the NaSch CA model due to model fluctuations with probability p this acceleration does not occur

10.3 General Motors (GM) Model

and instead of (10.32) from (10.28) it follows that the vehicle maintains its speed:

$$v_{n+1} = v_n. \tag{10.33}$$

Thus in this case, model fluctuations simulate a time delay in vehicle acceleration; this time delay is equal to

$$\tau_{\text{del}}^{(a)} = \frac{\tau_{\text{step}}}{1-p}. \tag{10.34}$$

Thirdly, let us assume that the vehicle should decelerate at time step $n+1$ that occurs if $v_n > g_n$. Then without taking fluctuations into account from (10.28) the condition $v_{n+1} = g_n$ would be satisfied. However, due to model fluctuations with probability p we find $v_{n+1} = \max(0, g_n - 1)$, i.e., the vehicle decelerates stronger than is needed for safety conditions. This is the main idea of the GM model for the over-deceleration effect, which should explain traffic flow instability.

In car-following models like the GM model and different kinds of OV models, *each* of the vehicles exhibits the same microscopic time delays, which are determined by deterministic rules of model motion (for the models of identical vehicles that are considered here). In contrast, in the NaSch CA model these driver time delays are simulated through the use of random model fluctuations. As a result, in the NaSch CA model these driver time delays are described as "collective effects" that occur *on average* in traffic flow.

One of the advantages of the NaSch CA model (10.28) is very fast computer simulation times of traffic flow in large traffic networks. A disadvantage of this model in comparison with deterministic traffic flow models is very great (non-realistic) fluctuations of vehicle speed.

10.3.6 Krauß's Stochastic Traffic Flow Model

The problem of large speed fluctuations of the NaSch CA model has been solved in Krauß's stochastic microscopic model [49,50], which uses the same ideas for mathematical description of driver delay times in vehicle acceleration and deceleration as those in the NaSch CA model. Krauß's model can be written as follows [49,50]

$$v_{n+1} = \max(0, \min(v_{\text{free}}, v_n + a_{\max}\tau_{\text{step}}, v_n^{(\text{safe})}) - \xi_n), \tag{10.35}$$

$$x_{n+1} = x_n + v_{n+1}\tau_{\text{step}}, \tag{10.36}$$

$\xi_n = a_{\max}\tau_{\text{step}}r$, $r = \text{rand}(0,1)$, $a_{\max}$ is the maximum acceleration, a discrete time $t = n\tau_{\text{step}}$, $n = 0, 1, 2, ..$, a safe speed $v_n^{(\text{safe})} = v^{(\text{safe})}(g_n, v_{\ell,n})$ in (10.35) is a solution of Gipps's equation [46]

$$v^{(\text{safe})}\tau_{\text{safe}} + X_{\text{d}}(v^{(\text{safe})}) = g_n + X_{\text{d}}(v_{\ell,n}), \tag{10.37}$$

where τ_{safe} is a safe time headway,

$$X_{\mathrm{d}}(u) = b\tau_{\mathrm{step}}^2 \left(\alpha\beta + \frac{\alpha(\alpha-1)}{2} \right) \qquad (10.38)$$

is the distance traveled by the vehicle with an initial speed u at a time-independent deceleration b until it comes to a stop, in (10.38) $\alpha = \lfloor u/b\tau_{\mathrm{step}} \rfloor$ and $\beta = u/b\tau_{\mathrm{step}} - \alpha$ are the integer and the fractional part of $u/b\tau_{\mathrm{step}}$, respectively; $\lfloor z \rfloor$ denotes the integer part of a real number z.

10.3.7 Traffic Breakdown resulting from Free Flow Instability in GM Model Class: Wide Moving Jam Emergence

What types of congested traffic patterns should occur due to free flow instability in the GM model class? This question has been answered by Kerner and Konhäuser in 1993–1994 from their numerical study of a version of Payne's macroscopic model (10.23), (10.25), (10.26) [60, 61]: As a result of the instability of free flow at the vehicle density that is greater than a critical density denoted by $\rho_{\mathrm{cr}}^{(J)}$ associated with a critical flow rate $q_{\mathrm{cr}}^{(J)}$, wide moving jams emerge *spontaneously* in free flow (F→J transition for short).

It has been found that within the flow range and associated density range [60]

$$q_{\mathrm{out}} \leq q < q_{\mathrm{cr}}^{(J)} \quad (\rho_{\mathrm{min}} \leq \rho < \rho_{\mathrm{cr}}^{(J)}), \qquad (10.39)$$

where q_{out} and ρ_{min} are the flow rate and density in free flow formed in the outflow of a wide moving jam (Sect. 2.5), free flow is *metastable* with respect to an F→J transition. This means that under condition (10.39) small enough disturbances in an initial homogeneous free flow decay; however, if a nucleus required for the F→J transition, i.e., a great enough local disturbance appears in the free flow, this disturbance grows leading to the occurrence of a wide moving jam (Fig. 10.5 (a)).

If the initial free flow density is only slightly smaller than the critical density $\rho_{\mathrm{cr}}^{(J)}$, then the F→J transition is associated with a "boomerang" behavior of the disturbance (Fig. 10.5 (a)). This boomerang effect found out in [60, 69] is as follows: firstly, the disturbance propagates downstream in free flow; then the disturbance comes to a stop strongly growing in its amplitude; as a result, the disturbance begins to propagate upstream; finally, a wide moving jam is forming that propagates upstream, i.e., the jam propagates through the location at which the initial disturbance has occurred. As found later in [63], the same boomerang effect occurs at an on-ramp bottleneck (Fig. 10.5 (b)). Indeed, due to a disturbance in free flow that permanently exists at the bottleneck, a wide moving jam emerges spontaneously at the bottleneck in metastable free flow.

It must be stressed that the term *boomerang effect* used in [70, 71] has a sense *only* when traffic breakdown occurs without influence of a bottleneck. Otherwise, if a bottleneck is the reason for traffic breakdown and the upstream congestion propagation, then this well-known and usual way of the congestion occurrence has noth-

10.3 General Motors (GM) Model

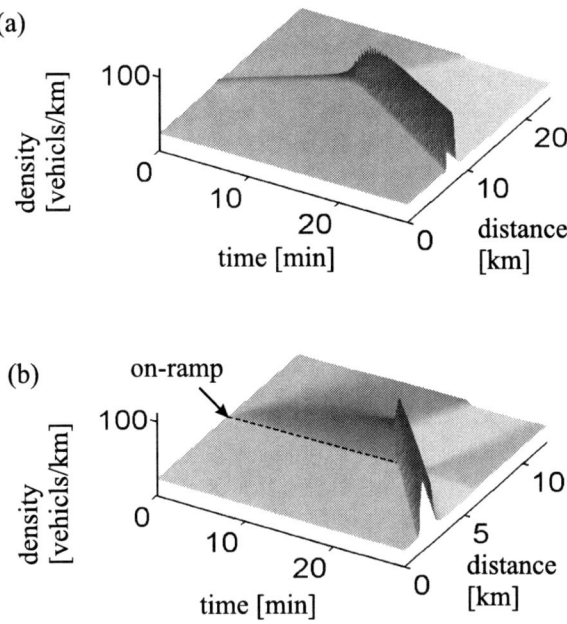

Fig. 10.5 "Boomerang" behavior of the growing disturbance in metastable free flow with respect to wide moving jam formation [60, 63, 69]. Vehicle density development in space and time. Simulations of a Payne-like model (10.23), (10.25), (10.26): (a) Homogeneous road. (b) On-ramp bottleneck

ing to do with the *boomerang* effect. Recall that in Fig. 7.7 (a, b) of Sect. 7.2 we have presented and discussed a sequence of F→S→J transitions, specifically MSP formation at an off-ramp bottleneck (F→S transition) with the subsequent transformation of the MSP into a wide moving jam (S→J transition) that occur between the off-ramp bottleneck and an upstream on-ramp bottleneck. The same measured traffic data[3] as that shown in Fig. 7.7 (b, c) has been explained in [70, 71] by the boomerang effect, i.e., "growing perturbations on a homogeneous freeway section without on- and off-ramps" (caption to Fig. 1 of [70]) leading to an F→J transition without bottlenecks. As follows from the discussion of this data made in Sects. 7.2.1 and 7.2.3

---

[3] It should be noted that for the same road section of the freeway A5-South in Germany the origin of the axis for road locations was chosen differently in [70, 71] and in [28] (see section schemes shown in Fig. 2 of [71] and Fig. 2.1 of [28], respectively). For this reason, the location $x = 0$ km in Fig. 7.7 is related to the location $x = 465.1$ km in Figures of [70, 71]. Road detectors on this road section are labeled by S1, S2, ..., in Fig. 2 of [71] and the same detectors are labeled, respectively, by D1, D2, ..., in Fig. 2.1 of [28].

- the conclusion of [70, 71] that in this empirical data the boomerang effect is observed as well as the resulting conclusion about the observation of a direct spontaneous F→J transition are *invalid*.

This invalid conclusion about the boomerang effect, which occurs without bottlenecks [70], and the resulting incorrect statement that in the empirical data shown in Fig. 7.7 (b, c)[4] a direct spontaneous F→J transition is observed [70, 71] can be explained by the ignoring in [70] of the empirical result [28] that an F→S transition occurs at the off-ramp bottleneck *upstream* of the off-ramp due to lane changing of vehicles going to the off-ramp.

It should be noted that at the same flow rate on the main road upstream of the bottleneck as that used for the simulation of the boomerang effect shown in Fig. 10.5 (b) and a greater flow rate to the on-ramp a sequence of wide moving jams emerges spontaneously (Fig. 10.6) [20, 51, 63, 66, 67].

The conclusion of Ref. [60, 63] that an instability of free flow leads to an F→J transition is the *general* one for all models of the GM model class, which shows this instability (e.g., [19–22]). In particular, F→J transitions at the bottleneck for the NaSch CA model (Fig. 10.6 (e, f)) and for Wiedemann's model (Fig. 10.6 (g, h)) show qualitatively the same features of spontaneous wide moving jam emergence as those in a version of the Payne-like model of Ref. [60, 63, 67] (Fig. 10.6 (b, c)).

This general result of the models of the GM model class [20–23, 51, 60, 61, 63, 66, 67, 70, 72] (Fig. 10.6) that free flow instability, which should explain traffic breakdown, leads to an F→J transition is inconsistent with real measured traffic data:

- Rather than the F→J transition, in real free flow an F→S transition governs traffic breakdown at a bottleneck (Sect. 3.1.3).
- As a result, the traffic flow models with free flow instability of the GM model class reviewed in [4–6, 9, 11–14, 18–23, 26] cannot be used for a description of traffic breakdown at the bottleneck as observed in measured traffic data.

10.3.8 Wide Moving Jam Propagation

This critical conclusion about the GM model class does *not* concern characteristic parameters of wide moving jams (Sect. 2.5) firstly found out in the Kerner-Konhäuser theory of wide moving jam propagation [60] and later incorporated in a huge number of different traffic flow models (see, e.g., the review by Helbing [20]). The empirical evidence of the characteristic parameters of wide moving jam propagation have been found in [73].

To explain the characteristic parameters of wide moving jam propagation, the mean time delay in vehicle acceleration $\tau_{\text{del, jam}}^{(a)}$ at the downstream front of a wide moving jam (Sect. 2.5.1) should be longer than the time delay in vehicle acceleration $\tau_{\text{del}}^{(a)}$ at higher speeds in traffic flow. This is now known as a slow-to-start rule

---

[4] This critical conclusion is also related to the analysis of all other data measured on this road section and shown in Figs. 1 and 2 of [70] and Figs. 3 and 4 of [71].

10.3 General Motors (GM) Model

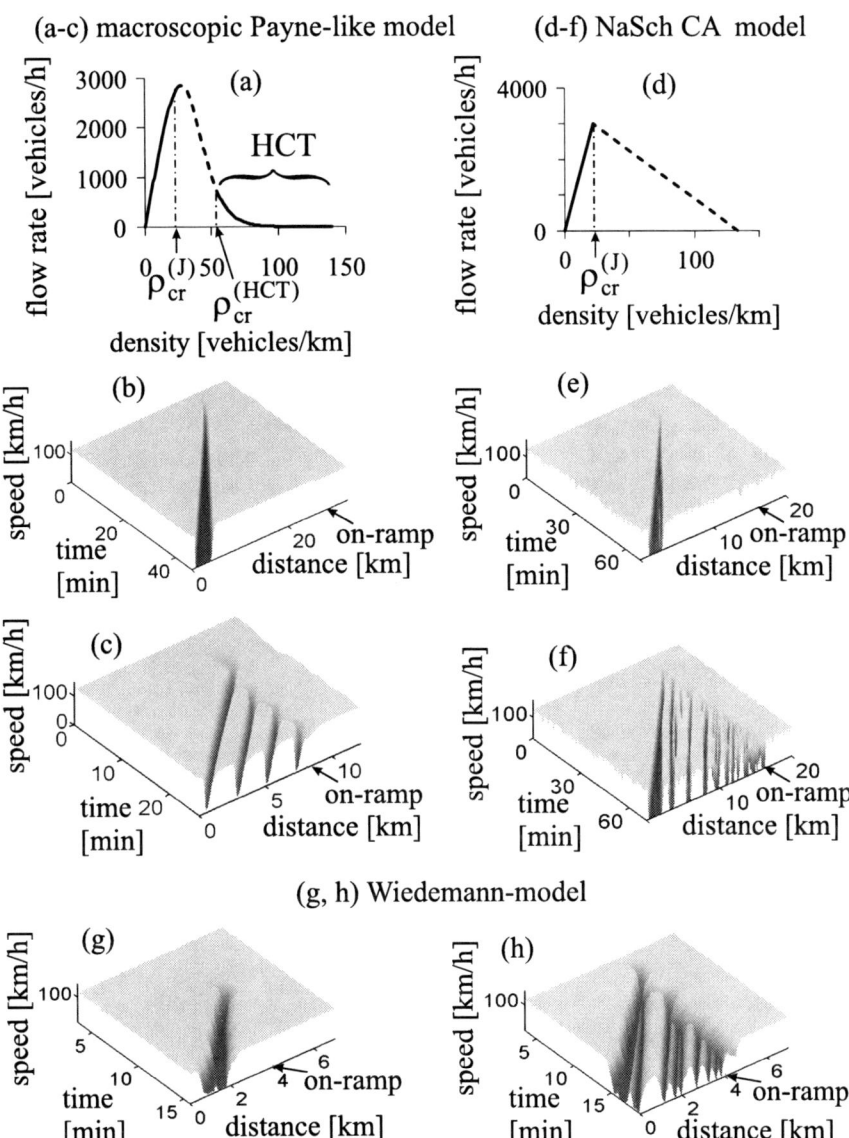

Fig. 10.6 Traffic breakdown at on-ramp bottleneck in the GM model class with free flow instability [20, 51, 63]: Wide moving jam formation in free flow in a Payne-like model of Ref. [60] (a–c), in the Nagel-Schreckenberg (NaSch) CA model (d–f), and in Wiedemann's model (g, h). (a, d) Fundamental diagrams of the Payne-like model (a) and the NaSch CA model (d); dashed parts of the diagrams are related to unstable states. (b, c, e, f–h) Average speed in space and time. Taken from [54]

in traffic flow modeling [48, 74]. In the theory of wide moving jam propagation derived in [60, 61] from a numerical study of the macroscopic model (10.23), (10.25), (10.26), the slow-to-start rule has been simulated through the use of the fundamental diagram (10.26) in which there is a wide density range in a neighborhood of the jam density within which $V_0(\rho) \approx 0$ (Fig. 10.6 (a)). As a result, vehicles could not almost accelerate from the initial speed $v = 0$ within the jam before the density decreases considerably, i.e., the mean space gap between vehicles increases. Thus only after a relatively long mean time delay in vehicle acceleration $\tau_{\text{del, jam}}^{(a)}$ could vehicles escape from the jam.

Rather than the implicit simulation of the slow-to-start rule through the use of a special form of the fundamental diagram discussed above, another mathematical idea of simulation of slow-to-start rule firstly introduced by Takayasu and Takayasu [74] has been used in a further development of the NaSch CA model [48]. Recall that in the NaSch CA model probability of fluctuations p simulates the time delay in acceleration in accordance with (10.34). Thus to simulate the slow-to-start rule, this probability has been taken considerably greater at $v = 0$ than at $v > 0$ [48]:

$$p(v_n) = \begin{cases} p_1 \text{ for } v_n = 0, \\ p_2 \text{ for } v_n > 0, \end{cases} \tag{10.40}$$

where p_1, p_2 are constants, $p_1 > p_2$.

10.3.9 Why Wide Moving Jams do not emerge spontaneously in Empirical Free Flow at Bottleneck

To explain the critical conclusion about the GM model class made in Sect. 10.3.7, we consider the following hypothesis of three-phase traffic theory [75–78]:

- At any density of free flow at which an F→J transition and an F→S transition are possible, a nucleus required for the F→S transition is considerably smaller than that required for the F→J transition.

To understand this hypothesis, we should compare two driver behavioral effects in free flow: the speed adaptation (Sect. 3.2.2) and over-deceleration effects. As the speed adaptation effect, the over-deceleration effect occurs also within a local speed disturbance within which the speed is lower and the density is greater than in an initial free flow.

According to the over-deceleration effect, a driver approaching a slower preceding vehicle decelerates stronger than is required to avoid collisions. As a result, the speed of the driver becomes *lower* than the speed of the preceding vehicle. This can occur with a great probability if an initial space gap is small enough. If all following vehicles move initially also at small enough space gaps, then due to over-deceleration they decelerate stronger than is required to avoid collisions. Then the speed of each following vehicle decreases up to zero – wide moving jam emerges.

Thus for the nucleation of a wide moving jam in free flow, most of the drivers should decelerate to a lower speed than the speed of the associated preceding vehicle.

According to the speed adaptation effect, however, a driver, who moves initially in free flow, while approaching a slower preceding vehicle, begins to decelerate within a synchronization space gap. The synchronization space gap is great enough. For this reason, the driver will not necessarily decelerate to a lower speed than the speed of the preceding vehicle; this driver has enough time to compensate the driver's reaction time and to adjust the speed to that of the preceding vehicle. Thus the necessary condition for an F→J transition – the over-deceleration effect, which can occur only at small enough initial space gaps between vehicles following each other, is much harder to satisfy than the necessary condition for an F→S transition – the speed adaptation effect, which occurs within the synchronization gap. This driver behavior – speed adaptation within the synchronization gap explains the above hypothesis of three-phase traffic theory and empirical observations in which the F→S transition governs traffic breakdown.

10.3.10 Homogeneous Congested Traffic

In Payne-like macroscopic models (e.g., [60, 67]), the Aw-Rascle model [65], OV models (e.g., [41, 42]), the IDM [20, 51] as well as some other traffic flow models (see references in [20]), the density region at the fundamental diagram, within which traffic flow is unstable, can be limited at greater densities: flow states at the fundamental diagram are unstable *only* within the density range (dashed part of the fundamental diagram in Fig. 10.6 (a)):

$$\rho_{cr}^{(J)} < \rho < \rho_{cr}^{(HCT)}. \quad (10.41)$$

In other words, within the density range

$$\rho_{cr}^{(HCT)} < \rho \leq \rho_{max} \quad (10.42)$$

homogeneous model states of congested traffic are stable with respect to small amplitude fluctuations.

- These model states of congested traffic upstream of an on-ramp bottleneck in which the speed, density, and flow rate are *homogeneous in space and time-independent* have been found out in model simulations by Helbing *et al.* [20, 66] and called *homogeneous congested traffic* (HCT) [20, 51, 51, 66, 67] (Fig. 10.7).
- In accordance with these traffic flow models [20, 66, 67], the more the density in HCT exceeds $\rho_{cr}^{(HCT)}$, the more stable is HCT with respect to non-homogeneous speed disturbances. This means that
  - the smaller the flow rate and the lower the speed within HCT, the more homogeneous in space and time should be HCT at a highway bottleneck and

– independent of how great density, low speed, and small flow rate are within HCT there should be *no* flow interruption within HCT, i.e., HCT is associated with non-interrupted traffic flow.

These model features of HCT [20, 66, 67][5] are inconsistent with empirical observations in which very *non-homogeneous* and complex in space and time congested patterns are observed, when the vehicle density in congested traffic is great enough, respectively, the average flow rate in congested traffic is small enough. An empirical example of such a congested pattern has been shown in Fig. 7.12.

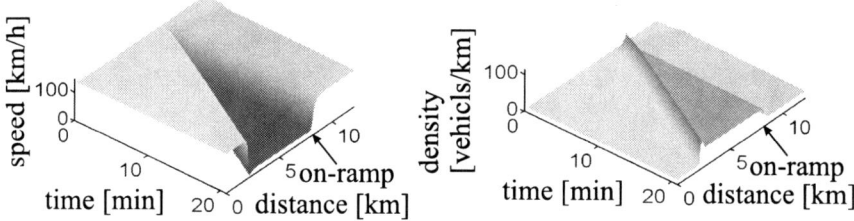

Fig. 10.7 Simulations of HCT at on-ramp bottleneck: Speed (figures left) and vehicle density (right) in space and time. Payne-like model (10.23), (10.25), (10.26) of Ref. [60]

In contrast with empirical non-homogeneous congested patterns observed at a great enough vehicle density in congested traffic (Fig. 7.12), in Ref. [51, 70, 71] congested traffic states have been published, which should prove the existence of HCT in measured traffic data. Based on the measured data used in Ref. [70], let us show that this empirical proof is invalid.

In Fig. 10 of Ref. [70], spatiotemporal speed distributions within two congested patterns are shown to be homogeneous during congested pattern existence. The average speed within the associated patterns is very low. It must be stressed that these results of Ref. [70] have been derived with an adaptive smoothing method of data processing discussed in [70], i.e., with *processed data sets*. In contrast, our Fig. 10.8 shows *real unprocessed raw measured* data for one of these congested patterns related to Fig. 10 (a) of Ref. [70].

To explain real measured data shown in Fig. 10.8, we should note that already in raw unprocessed data there is a large error in the average speed, when very low speeds are measured; if speeds v of all vehicles that have passed a detector during a 1-min interval are within the range $0 < v < 20$ km/h, then the road computer sets the average speed to 10 km/h. Only if no vehicle passes a detector during a 1-min interval the speed (and flow rate) is zero. This explains why in the speed data shown

---

[5] It must be noted that HCT is *not* a general result of traffic flow models of the GM model class. No HCT appears independent of the density, for example, in the NaSch CA model or in Krauß's model: beginning from the critical density $\rho_{cr}^{(J)}$, all states of congested traffic at the fundamental diagram in these models are also unstable up to the jam density (Fig. 10.6 (d)).

10.3 General Motors (GM) Model 195

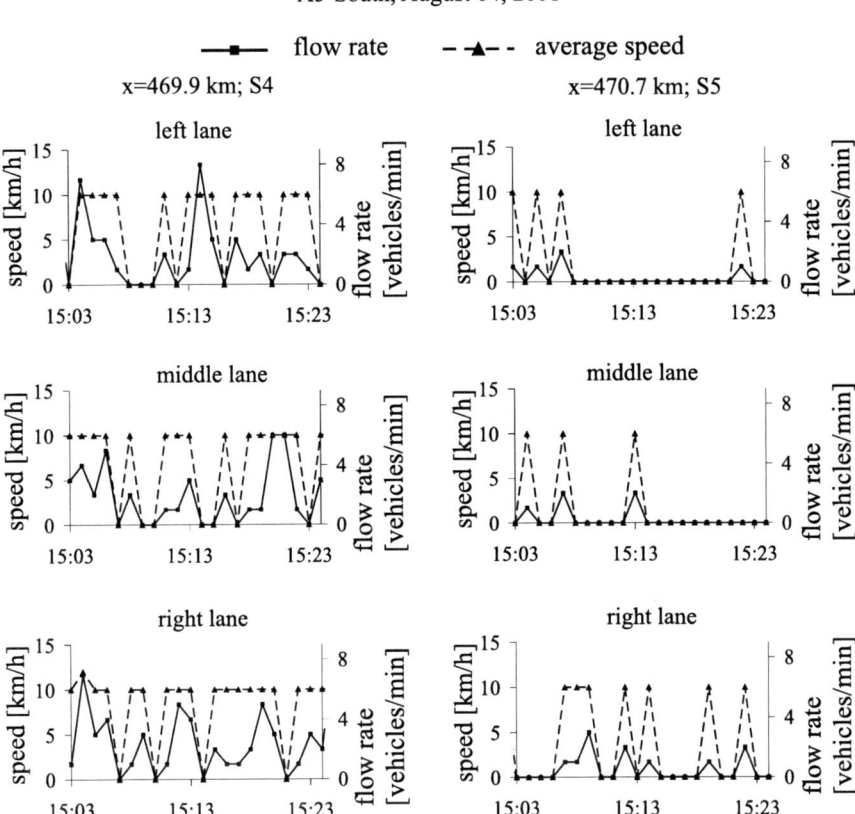

Fig. 10.8 Measured unprocessed flow rate (solid curves) and speed (dashed curves) at two detector locations within a congested pattern that has been presented in Fig. 10 (a) of Ref. [70] as "homogeneous congested traffic" (HCT). The locations $x = 469.9$ km for the detector S4 and $x = 470.7$ km for the detector S5 are chosen in accordance with the freeway section sketch shown in Fig. 6 of Ref. [70]. The raw data is 1-minute averaged data. Taken from [54, 79]

in Fig. 10.8 there are mostly two speed values, zero and 10 km/h. Only when average speeds are higher than 30 km/h, the speed can be used in deciding whether the speed distribution is really a homogeneous one or not.

We see that for low speeds based on this data *no* conclusion about features of a spatiotemporal pattern can be made from an analysis of speeds *only*, as made in [70, 71]. This critical conclusion is independent of a method used for the further processing of the data.

Rather than the average speed, the flow rate is measured with a sufficient accuracy at any density. We can see from the flow rate distribution shown in Fig. 10.8 that there are extremely complex spatiotemporal flow rate changes both in space and time between zero and 8 vehicles/min. This explains that in contrast with the statement of Ref. [70], the congested pattern is extremely non-homogeneous in space and time. Note that features of such complex spatiotemporal congested patterns occurring at heavy bottlenecks caused for example by accidents or bad weather condition have been discussed in Sect. 7.2.5.

Thus in reality the congested pattern shown in Fig. 10 (a) of Ref. [70] has no relation with HCT. The same critical conclusion can be made about the congested pattern shown in Fig. 10 (b) of Ref. [70] and in Fig. 6 of [71], which should be another "empirical" example of HCT.

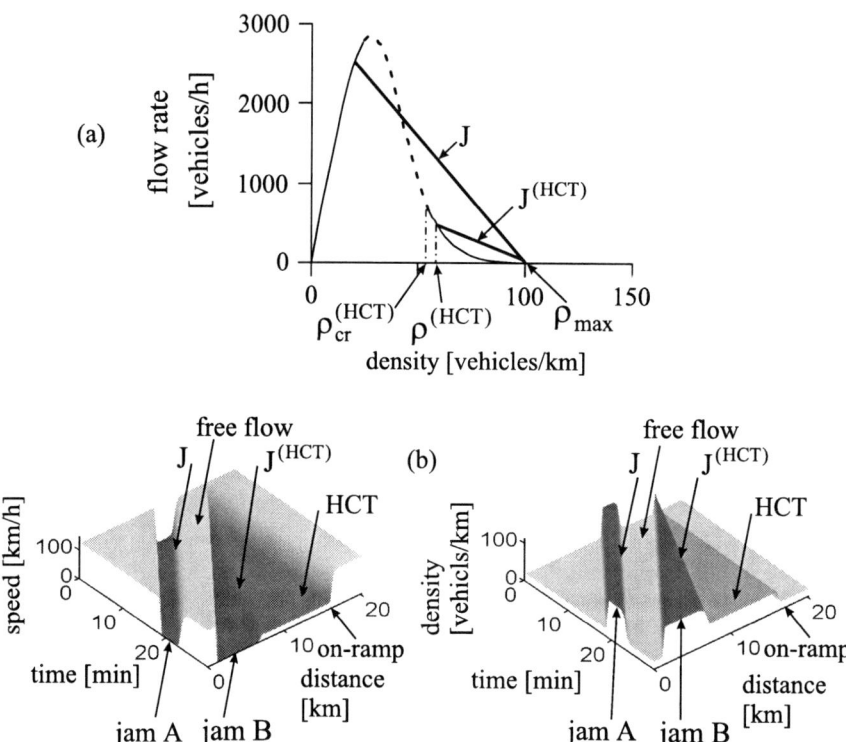

Fig. 10.9 Simulations of HCT model solutions and wide moving jam propagation: (a) Fundamental diagram with HCT taken from Fig. 10.6 (a) together with the lines J and $J^{(\mathrm{HCT})}$ that represent in the flow–density plane the downstream fronts of two wide moving jams denoted by "jam A" and "jam B," respectively in (b). (b) Propagation of two wide moving jams when downstream of the jams either free flow ("jam A") or HCT ("jam B") occur, respectively; speed (left) and density (right) in space and time. Payne-like model (10.23), (10.25), (10.26) of Ref. [60]

Another point of the criticism of the HCT model solutions is as follows. In all known empirical observations, the downstream front of a wide moving jam propagates through any dense congested traffic and bottlenecks while maintaining the mean velocity of the front v_{g} (Sect. 2.5). In other words, the empirical feature [J] of wide moving jams is *independent* of the state of traffic flow downstream of the jam. However, traffic flow models with the HCT model solutions [20,51,66,70,71] *cannot* satisfy this very important characteristic empirical feature of traffic.

To illustrate this, we consider features of wide moving jam propagation in these models with an example of the fundamental diagram shown in Fig. 10.9 (a). We find that the characteristic jam feature [J] in these models is satisfied when free flow is downstream of the jam (wide moving jam labeled by "jam A" in Fig. 10.9) [60]. In contrast, the jam feature [J] does *not* remain when an HCT model solution is downstream of a wide moving jam (the jam labeled by "jam B"). This because rather than the line J, the downstream front of the "jam B" propagates with a negative velocity $v_g^{(\mathrm{HCT})}$ that is associated with a line $J^{(\mathrm{HCT})}$ in the flow–density plane between the state within the jam with the jam density $\rho_{\max}$ and a point at the fundamental diagram for an HCT solution (with a density $\rho^{(\mathrm{HCT})}$ in Fig. 10.9 (a)). As a result, the absolute value of the downstream front velocity of the "jam A" $|v_g|$ is considerably greater than the one for the "jam B" $|v_g^{(\mathrm{HCT})}|$. Moreover, the greater the density $\rho^{(\mathrm{HCT})}$ of the HCT, the smaller $|v_g^{(\mathrm{HCT})}|$. As abovementioned, this is inconsistent with measured traffic data of wide moving jam propagation. Studying traffic flow models with HCT model solutions with various fundamental diagrams we can make the following general conclusion:

- HCT model solutions of Ref. [20,51,66,70,71] are not consistent with the empirical feature [J] of wide moving jam propagation through any dense congested traffic while maintaining the mean velocity of the jam front. For this reason, the features of these HCT model solutions have no sense for real traffic flow.

10.3.11 Oscillating Congested Traffic

Due to the existence of HCT model solutions, the traffic flow models mentioned in Sect. 10.3.10 exhibit also model solutions called *oscillating congested traffic* (OCT) [20,51,66,67,70–72] (Fig. 10.10). OCT appears in a neighborhood of the critical density $\rho_{\mathrm{cr}}^{(\mathrm{HCT})}$ for an instability of HCT: when the density in HCT decreases and it approaches the critical density $\rho_{\mathrm{cr}}^{(\mathrm{HCT})}$, then due to HCT instability, OCT occurs. Thus OCT model solutions result from the existence of HCT model solutions in these models, i.e., as HCT, OCT model solutions have no relation to real traffic flow.

It must be noted that by a choice of the flow rate q_{on} one can find model solutions with a "spatial combination" of HCT and OCT [20,51,70,71] (Fig. 10.11). In these model solutions, the HCT is just upstream of the bottleneck, whereas due to an instability of the HCT further upstream of the bottleneck the HCT transforms into an OCT of a large amplitude. In other words, this spatial combination of the HCT (near the bottleneck) with the OCT (further upstream of the bottleneck) looks like an GP found in empirical observations [75].

A model solution with a "spatial combination" of HCT and OCT (Fig. 10.11) presented *solitary from other model solutions* and *regardless of its features* is often

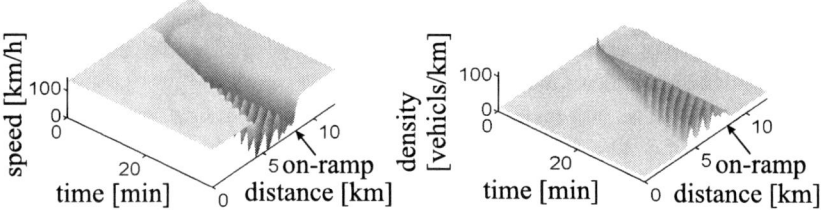

Fig. 10.10 Simulations of OCT at on-ramp bottleneck: Speed (figures left) and vehicle density (right) in space and time. Payne-like model (10.23), (10.25), (10.26) of Ref. [60]

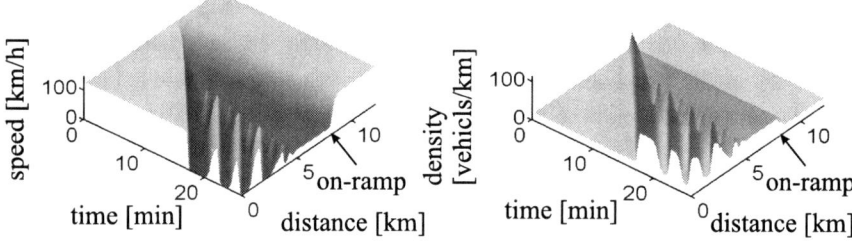

Fig. 10.11 Simulations of model solution with a spatial combination of HCT and OCT at on-ramp bottleneck: Speed (figures left) and vehicle density (right) in space and time. Payne-like model (10.23), (10.25), (10.26) of Ref. [60]

used [20, 51, 70] as a "proof" that traffic flow models with HCT and OCT are able to describe the GP found in real measured traffic data: the HCT and OCT shown in Fig. 10.11 are associated in [20, 51, 70, 71] with the pinch region of synchronized flow and region of wide moving jams of the GP, respectively.

The main question to the model solution with a spatial combination of HCT and OCT (Fig. 10.11) is whether this model solution exhibits the fundamental empirical features of GPs? These GP features are as follows [28]:

- The greater the bottleneck strength, *the greater* the mean frequency of moving jam emergence within the GP and *the smaller* the mean width of the region of synchronized flow upstream of the bottleneck (the smaller the mean width of the pinch region) of the GP.

To prove that the model solution with a spatial combination of HCT and OCT (Fig. 10.11) does *not* exhibit these empirical GP features, we consider this model solution under increase in the flow rate to the on-ramp q_{on} (Fig. 10.12). We find that

- the greater the flow rate to the on-ramp q_{on} (i.e., the greater the bottleneck strength), *the smaller* the mean frequency of moving jam emergences and *the greater* the mean width of the region of HCT (Fig. 10.12).

This is inconsistent with real measured traffic data.

10.3 General Motors (GM) Model

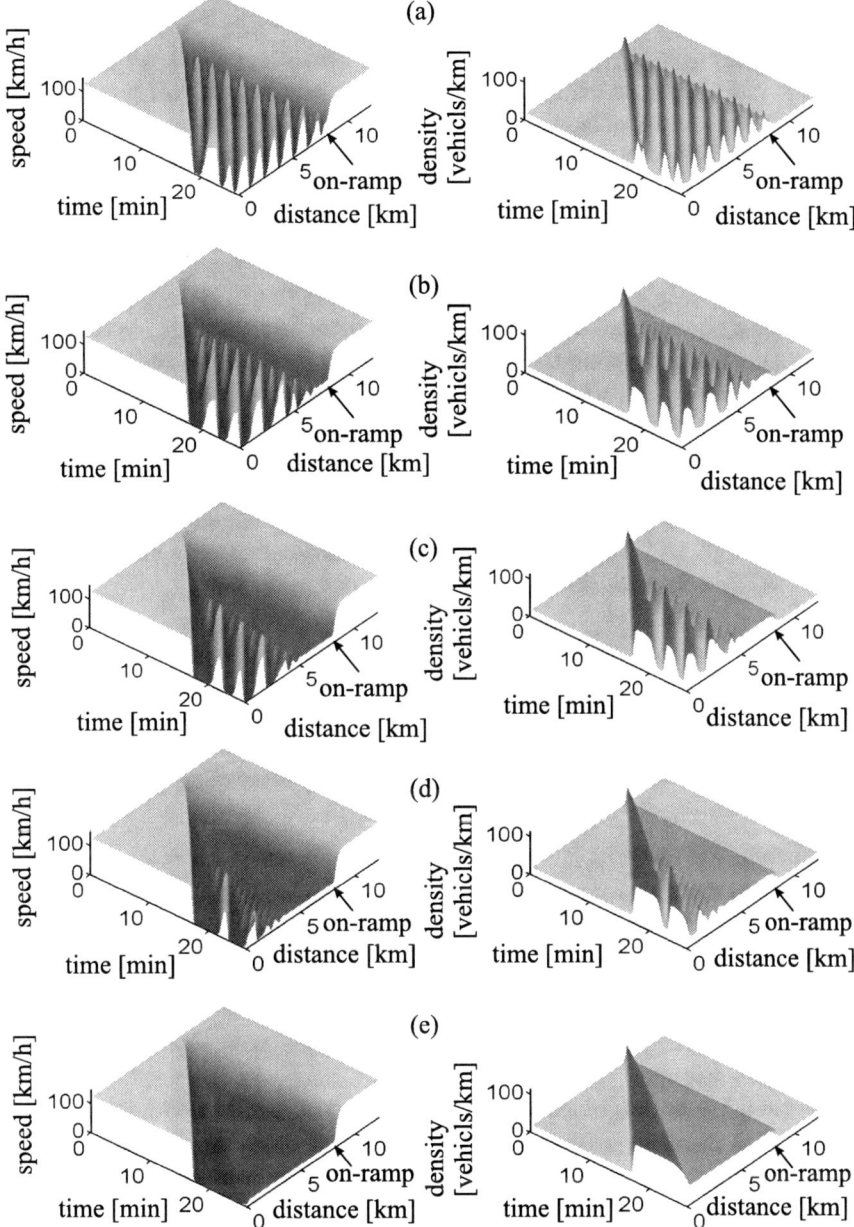

Fig. 10.12 Simulations of model solutions with a spatial combination of HCT and OCT at on-ramp bottleneck at a given flow rate q_{in} and different q_{on}: q_{on} = 180 (a), 247 (b), 266 (c), 281 (d), and 324 (e) vehicles/h. Speed (figures left) and vehicle density (right) in space and time. Figure (c) is taken from Fig. 10.11. Payne-like model (10.23), (10.25), (10.26) of Ref. [60]

10.3.12 Diagrams of Congested Patterns

In accordance with criticisms of the GM model class made above, we should also mention that congested pattern diagrams, which describe different types of congested traffic patterns in these models [20,21,51,63,66,67,72,80], have no relation to real traffic. This criticism includes also earlier results of the author *et al.* [63,80] made in the context of the fundamental diagram approach.

- The main point of the criticism of any congested pattern diagram at a bottleneck in the framework of the GM model class firstly derived from numerical simulations by Helbing *et al.* [66] and later developed in [20,21,51,67,70,72] is that at a great enough flow rate in free flow on the main road upstream of the bottleneck a transition from free flow to a congested pattern is associated with an F→J transition, i.e., with wide moving jam emergence in free flow at the bottleneck (Fig. 10.6). In contrast with this model result, in real traffic the onset of congestion is governed by an F→S transition (Chap. 3).
- In addition, at a great bottleneck strength (in particular, a great flow rate to the on-ramp), OCT and HCT model solutions appear in the diagrams of some of these models [20,51,66,67,70]. However, as shown above features of the OCT and HCT model solutions have no relation to real traffic flow.

10.4 Common Features of earlier Traffic Flow Models

10.4.1 Models Combining LWR and GM Approaches

Considering wide moving jam emergence in the GM model class, we have above assumed that the critical density $\rho_{\text{cr}}^{(J)}$ for the instability leading to F→J transitions is related to free flow, specifically, the condition

$$\rho_{\text{cr}}^{(J)} < \rho_0 \qquad (10.43)$$

is satisfied, where the density ρ_0 is associated with the maximum flow rate at the fundamental diagram.

However, parameters of many traffic flow models of the GM model class can also be chosen in the way that the critical density $\rho_{\text{cr}}^{(J)}$, at which steady speed states on the fundamental diagram become unstable as the density increases, is greater than the density ρ_0 (Fig. 10.13 (a, b)):

$$\rho_{\text{cr}}^{(J)} > \rho_0. \qquad (10.44)$$

Thus by changing of model parameters in some of the models of the GM model class, a model can exhibit qualitatively different conditions for congested traffic occurrence depending on whether condition (10.43) (Fig. 10.6), or condition

10.4 Common Features of earlier Traffic Flow Models

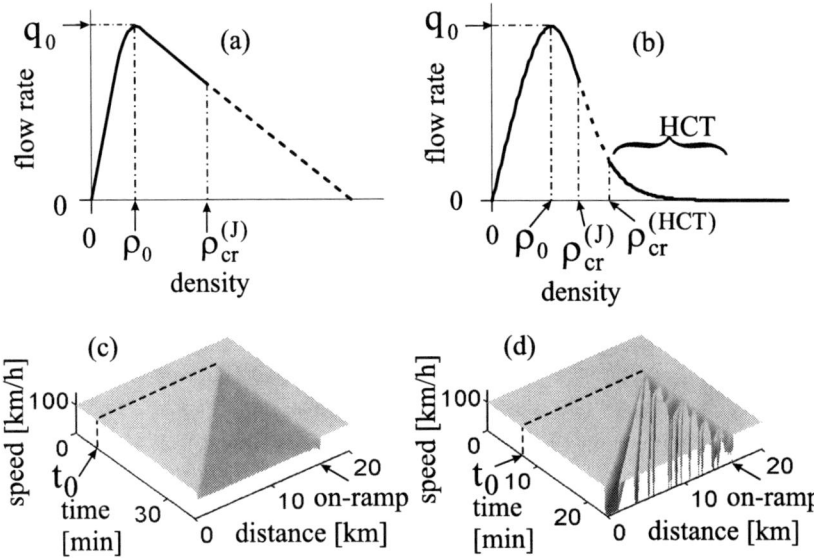

Fig. 10.13 Traffic breakdown and congested patterns at on-ramp bottleneck in models combining the LWR and GM approaches under condition (10.44): (a, b) Two types of fundamental diagrams; dashed parts show unstable flow states. (c) Propagation of shock waves under condition (10.9) that is the same as that in Fig. 10.4 (d). (d) Spontaneous moving jam emergence in dense traffic in (c) when the on-ramp inflow rate increases and the density approaches the critical density $\rho_{\rm cr}^{\rm (J)}$. At $t = t_0$ on-ramp inflow is switched on. Taken from [52]

(10.44) (Fig. 10.13) is satisfied, or else at any density there cannot be model instability leading to wide moving jam emergence at all, i.e., when the model exhibits qualitatively the same features as those of the LWR model (Fig. 10.4). This conclusion is related to all earlier traffic flow models that the author knows (see, e.g., [4–27, 37, 38, 40–51, 56–67]).

In particular, when the shape of the fundamental diagram is chosen, then such transitions from condition (10.43) to condition (10.44) and, finally, to the LWR model class occur in the OV model (10.17) by a gradual decrease in the model parameter τ_0, in a version of Payne's model (10.23), (10.25), (10.26) by a gradual increase in the model parameter c_0, in the IDM (10.21), (10.22) by a gradual increase in the model parameter c_1.

10.4.2 Summary of Features of earlier Traffic Flow Models

The above consideration of earlier traffic flow models in the framework of the fundamental diagram hypothesis shows that stability features of steady state model solutions determine most of the qualitative features of the onset of congestion and resulting congested patterns. For this reason, it turns out that mathematically very different traffic flow models, which exhibit qualitatively the same stability features of steady speed states at the fundamental diagram, show also qualitatively the same features of traffic congestion. This leads to the traffic flow model classification made above and to the following conclusions about common features of earlier traffic flow modeling approaches:

(i) Most of the traffic flow models used by traffic researches are based on fundamental diagram hypothesis. These models can be classified into two main classes. The first model class is associated with the classic LWR theory. The basic idea of the LWR theory is that the maximum flow rate associated with the maximum point at the fundamental diagram determines free flow capacity at a bottleneck. Thus if the flow rate upstream of the bottleneck exceeds the capacity, then traffic breakdown should occur. The second model class is associated with the basic idea of the GM model: beginning at a critical density there is an instability of free flow caused by a driver time delay.

(a) The LWR theory and all traffic flow models based on this theory *cannot* show and predict the fundamental empirical features of traffic breakdown as well as empirical spontaneous moving jam emergence.

(b) The instability of free flow beginning at a critical vehicle density incorporated in models and theories in the context of the GM model approach leads to an F→J transition. In contrast, in empirical observations, traffic breakdown is associated with an F→S transition. Thus the models *cannot* explain traffic breakdown observed in real traffic flow.

(ii) These traffic flow models are basic traffic flow models for simulations of freeway control and dynamic management strategies. However, we have to conclude that the related simulations of the control and dynamic management strategies *cannot* predict many of the freeway traffic phenomena that would occur through the use of a dynamic management strategy.

(iii) It must be stressed that the above criticism of the earlier traffic flow modeling approaches does not diminish the following **achievements** of these approaches and associated models, which are also used in three-phase traffic theory and the related three-phase traffic flow models:

(a) In accordance with empirical results, there are characteristic parameters of wide moving jam propagation (Sect. 2.5), which have been firstly discovered in [60] and later incorporated in a huge number of traffic flow models of the GM model class [13, 19–23]. One of these characteristic parameters is that the downstream front of wide moving jams propagates on average steadily along a road.

(b) Classic ideas of traffic flow theories introduced and developed within the fundamental diagram approach about different driver time delays, various mathematical descriptions of driver acceleration and deceleration as well as safety conditions are also very important elements used in three-phase traffic flow models, which overcome drawbacks of the earlier modeling approaches to traffic congestion (item (i) and (ii)). In particular, pertinent pioneering ideas, which were introduced in earlier models and theories by Herman, Montroll, Potts, Rothery, Gazis [36, 37], Kometani and Sasaki [56], Newell [38], Gipps [46], Payne [43,44], Nagel, Schreckenberg, Schadschneider, and co-workers [47,48], Bando, Sugiyama, and colleagues [41], Takayasu and Takayasu [74], Krauß et al. [49], Nagatani and Nakanishi [58] and by many other groups (see references in [4–6, 8, 9, 11–14, 16–23, 27]), are also very important elements of three-phase traffic models discussed in Chap. 11.

10.5 Empirical Tests of earlier Traffic Flow Models

Obviously, traffic flow theories must be based on real behavior of drivers in traffic, and their solutions should show phenomena observed in traffic flow. For this reason, an empirical test of a traffic flow theory is of great importance. There are several approaches to perform such a test:

(i) Vehicle speeds, time headways, and accelerations measured in a car-following experiment are compared with that, which a model shows (e.g., [4]).

(ii) Spatiotemporal evolution of speed (or density) associated with a processing of real empirical data made for example through the use of adaptive smoothing methods are compared with results of numerical simulations of congested patterns [51,70,71].

(iii) The empirical fundamental diagram for traffic flow (flow–density relationship) associated with measurements made at a road location are compared with a theoretical fundamental diagram for traffic flow used in a model or that results from the model (e.g., [29–31, 83–91]).

(iv) Time headway distributions, optimal velocity (OV) functions, and some other single vehicle characteristics that are measured at a road location or result from data aggregation are compared with the associated model results (e.g., [29–31]).

(v) Traffic variables (e.g., flow rate and average speed) are measured at three or more different freeway locations. Traffic data measured at two road locations, which are the farthest upstream and downstream, are used as upstream and downstream boundary conditions for a model, respectively. The model calculates spatiotemporal distributions of traffic variables between these two locations. The distributions should correspond to empirical data measured at the locations, which have not been used as the boundary conditions for the modeling (e.g., [12,92,94,95]).

In all these cases, model parameters are chosen to have the best agreement with the associated empirical data.

As explained above, past traffic flow theories and models reviewed in [13,17–22, 26] cannot explain and reproduce empirical features of traffic breakdown. Nevertheless, some of these models can show a good correspondence with empirical data in the above empirical test approaches (i)–(v).

To explain this, we should note that empirical data used in the tests (i)–(iv) does not contain important empirical features of spatiotemporal traffic dynamics considered in Part I. To study these fundamental features (traffic breakdown and resulting spatiotemporal congested patterns), real unprocessed measured data, which should include spatiotemporal distribution of synchronized flow within a congested pattern (see Sect. 2.4.11 of the book [28]), should be studied.

In particular, due to smoothing and/or selection of data in the empirical test (ii) [51,70,71], important characteristics of congested patterns can get lost resulting in invalid analysis and classification of empirical congested traffic states. A characteristic example of such invalid analysis of empirical features of traffic congestion and the associated test of simulation results made in [51,70,71] has been considered in Sect. 10.3.10 (Fig. 10.8). Another characteristic example of the analysis of empirical features of traffic congestion made in [70,71], which results in the incorrect statement about the empirical observation of the boomerang effect, i.e., a direct spontaneous F→J transition, has been discussed in Sect. 10.3.7.

The fundamental diagram, OV functions as classified in [29–31], time headway distributions, and other macroscopic and single vehicle characteristics of traffic flow used up to now in empirical tests (iii) and (iv) are associated with an averaging of spatiotemporal traffic pattern characteristics. For this reason, important features of spatiotemporal congested patterns and phase transitions in traffic flow are lost in these characteristics. Thus in contrast with [29–31] it is not justified to use these macroscopic and microscopic characteristics as the solely empirical basis for a decision whether a traffic flow model can describe real traffic flow or not.

These critical conclusions explain why the NaSch CA models [47,48] and their further developments including a CA model with "comfortable driving" [96] show empirical fundamental diagrams, empirical OV functions and time headway distributions satisfactory [31], even though these CA models [47,48,96] as found in [97] (see explanations to simulation results of the NaSch CA model with "comfortable driving" [96] presented in Figs. 20–23 of Ref. [97]) cannot show and predict the main empirical spatiotemporal features of phase transitions and synchronized flow.

Moreover, as explained in [93], a criterion for the definitions of synchronized flow and wide moving jam of Ref. [29–31], which is based on a comparison of the flow–density correlations within empirical data associated with congested traffic, is invalid. As shown in [81,93,98], this is associated with a large systematic error in calculations of the vehicle density within wide moving jams made in Ref. [29–31] from empirical data (for a more detailed consideration see Ref. [93]). For this reason, empirical tests of the CA traffic flow models made in Ref. [29–31] are also invalid.

In addition, it should also be noted that in empirical tests of the CA traffic flow models made in Ref. [29–31], simulations of spatiotemporal congested patterns shown by the CA traffic flow models are made for a spatially homogeneous road, i.e., the road without bottlenecks. In contrast, measured data used for these tests are related to congested traffic occurring due to bottlenecks. It should be noted that the traffic dynamics at a bottleneck makes the greatest influence on spatiotemporal congested patterns occurring in simulations performed at the bottleneck [28]. This means that for an adequate empirical test of traffic flow models rather than simulations on the homogeneous road [29–31], model simulations of spatiotemporal congested patterns should be made at bottlenecks [81, 82].

In empirical test (v), firstly congested traffic is measured at a road location. Then measured traffic variables associated with this congested traffic are used at the downstream boundary of a traffic flow model. In other words, at this boundary traffic variables are *given* as time functions associated with congested traffic measured at the road location related to this model boundary. As a result, in simulations congested traffic given at the model boundary propagates further upstream. If the characteristic features of wide moving jam propagation can be shown by the model (as explained above, this is the case for many models of the GM model class [20–22]), then the downstream jam front velocity can be chosen close to an empirical one. For this reason, an approximate correspondence between model and some empirical traffic variable functions is possible [92, 94, 95], although the models [20–22] cannot show and cannot reproduce empirical traffic breakdown and many resulting empirical spatiotemporal congested patterns.

This is because the test (v) is inconsistent with the "open character" of non-linear dynamic process, "traffic". As mentioned, in this test congested traffic is often given at the farthest downstream boundary of a freeway network model. This is not the case for real traffic, in which congestion occurs spontaneously within a real freeway network, mostly at a bottleneck. This bottleneck cannot be the farthest downstream boundary of this network. To simulate the open traffic process adequately with real traffic, vehicles should leave freely the farthest downstream boundary of a network model. This means that free flow conditions should be given at the farthest downstream boundary of the network model.

10.6 Applications of Highway Capacity Definitions in Transportation Engineering

In each field of science, the choice of a term for a phenomenon observed in measured data should reflect features of the phenomenon. It can turn out that the measured data has been understood many years later after the phenomenon has been firstly observed and, consequently, the term chosen does not reflect the empirical features of this phenomenon. Unfortunately, this is very often the case in traffic science and transportation engineering. This is because that measured data for the most of macroscopic spatiotemporal phenomena have been understood only recently [28].

This can explain why invalid terms can lead (and they do lead) to invalid methods for dynamic freeway traffic management, control, and dynamic traffic assignment widely used in transportation engineering.

10.6.1 Highway Capacity as a Particular Fixed or Stochastic Value

Traffic breakdown at a bottleneck limits a highway capacity in free flow at the bottleneck (Chap. 4). In the *Highway Capacity Manual*, highway capacity is defined as the maximum flow rate downstream of a bottleneck that can be observed in free flow. As mentioned in Sect. 2.2.3, recently Elefteriadou *et al.* [99] have found out that highway capacity exhibits a probabilistic behavior: At a given flow rate traffic breakdown can occur but it should not necessarily occur. Thus on one day traffic breakdown occurs, however, on another day at the same flow rates traffic breakdown is not observed. Moreover, Persaud *et al.* found [100] that empirical probability of traffic breakdown at a bottleneck is an increasing flow rate function (Fig. 2.6).

These empirical findings have led to the definition of highway capacity as a stochastic variable (stochastic or probabilistic capacity) [99, 101, 103, 103, 104]. As in the former capacity definition, in this probabilistic (stochastic) highway capacity definition is assumed that there is a *particular highway capacity*. However, in contrast with the definition of highway capacity as a fixed (deterministic) value, a particular probabilistic (stochastic) highway capacity cannot be exactly known at a given time instant, specifically, the capacity is known with some probability only. This probability is associated with probability of traffic breakdown at the bottleneck. The probability as a flow rate function can be found based on observations of many days (and years) when traffic breakdown occurred at the bottleneck.

Thus independent of whether highway capacity is defined either as a fixed (deterministic) or probabilistic (stochastic) value, in these capacity definitions it is assumed that at each time instant there is a *particular highway capacity* of free flow at the bottleneck.

- If the existence of the particular highway capacity is assumed, it can further be assumed that the capacity can be used as a *control parameter* for dynamic traffic management and control methods.

In contrast, in three-phase traffic theory is assumed that there are *the infinite number of highway capacities* of free flow at the bottleneck (Chap. 4).

In this context, the following question seems to be reasonable:

- Why highway capacity of free flow at the bottleneck cannot be defined as a particular either fixed or stochastic value?

Naturally, one can *define* the highway capacity as a particular value, however, only *as long as* the capacity definition is not applied for traffic control, dynamic traffic assignment, dynamic routing, and for other dynamic management methods and strategies. This is because the use of a particular (fixed or stochastic) highway capacity as a control parameter in dynamic traffic management and control methods is inconsistent with the fundamental empirical feature of traffic breakdown at the bottleneck (Chap. 3):

- There can be either spontaneous or induced traffic breakdown at the same bottleneck.

10.6 Applications of Highway Capacity Definitions

Indeed, let us assume that there is a particular (fixed or stochastic) highway capacity of free flow at a bottleneck. Then at any flow rate in free flow downstream of the bottleneck that is smaller than the particular capacity at a time instant *no* traffic breakdown could be induced at the bottleneck. This is inconsistent with empirical observations in which traffic breakdown is induced at the bottleneck (for example, through the propagation of a moving synchronized flow pattern shown in Fig. 3.2) at the flow rate at which free flow is observed at the bottleneck.

- Highway capacity of free flow at the bottleneck *cannot* depend on whether there is a congested pattern, which has occurred outside of the bottleneck and independent of the bottleneck existence, or not.

The contradiction between the feature of highway capacity associated with an induced traffic breakdown at the bottleneck and the definition of the capacity as a particular (fixed or stochastic) value can be solved by the following assumption of three-phase traffic theory (Chap. 4):

- The infinite number of the flow rates in free flow downstream of the bottleneck at which traffic breakdown *can be induced* at the bottleneck are the *infinite number* of highway capacities of free flow at the bottleneck.

Note that the minimum and maximum capacities, which limit the range of the infinite number of highway capacities of free flow at the bottleneck, do not depend on whether there is a congested pattern that propagates through the bottleneck, or not:

- If the flow rate in free flow downstream of the bottleneck is smaller than the minimum capacity, then *no* traffic breakdown is possible at the bottleneck: Traffic breakdown does not occur at the bottleneck, even if a localized congested pattern (e.g., a wide moving jam or an MSP) has passed through the bottleneck.
- If the flow rate in free flow downstream of the bottleneck is greater than the maximum capacity, then traffic breakdown does occur at the bottleneck. This traffic breakdown at the bottleneck does not depend on whether a congested pattern propagates through the bottleneck, or not.

Summarizing results of this discussion of highway capacity definitions, we can make the following conclusion. The qualitative difference between two approaches to the highway capacity definition as a *particular stochastic* value or as the *infinite number* of values is as follows:

- The definition of highway capacity as a particular stochastic value assumes that at each time instant there is a highway capacity of free flow at the bottleneck but we do not know its value; this is probably because the capacity value depends on traffic parameters, like weather, the percentage of long vehicles, etc., which change randomly over time in real traffic. If the existence of such a particular stochastic capacity is assumed, we could further assume that through measurements of traffic parameters the capacity can approximately be estimated as a time-function and, therefore, be used as a control parameter for dynamic traffic management and control methods.

- In contrast, the definition of highway capacity as the infinite number of values assumes that even when each of the traffic parameters were *known and time-independent*, there are the infinite number of highway capacities between the minimum and maximum capacities[6].

10.6.2 Capacity Drop

Just after traffic breakdown has occurred, the discharge flow rate can be smaller than the pre-discharge flow rate (Sect. 2.2.2). This decrease in the flow rate in free flow downstream of the bottleneck just after traffic breakdown has occurred has been called *capacity drop* [105].

The term *capacity drop* assumes that there is a particular capacity of free flow at the bottleneck [5,7,99,101,102,104]. However, in accordance with the fundamental empirical features of traffic breakdown discussed in Sect. 3.1.3, in three-phase traffic theory there are the infinite number of capacities of free flow at a bottleneck and, therefore, the term *capacity drop* is not used in this theory.

10.6.3 Capacity Definitions based on earlier Traffic Flow Models

There can be two definitions of highway capacity, i.e, capacity of free flow at a bottleneck associated with the two model classes, the LWR and GM model classes.

In the LWR model, capacity of free flow at a bottleneck (bottleneck capacity) is equal to the flow rate q_{cap} downstream of the bottleneck (10.1): if the sum of flow rates upstream of the bottleneck exceeds q_{cap}, which is the maximum flow rate at the fundamental diagram, then traffic breakdown should occur leading to congested traffic upstream of the bottleneck.

Note that a random choice of q_{cap} (to satisfy probabilistic features of traffic breakdown, see Sect. 10.6.1) changes *none* of qualitative features of the onset of congestion in the LWR theory shown in Fig. 10.4; as explained in Sect. 10.2, these LWR theory features are, however, inconsistent with the fundamental empirical features of traffic breakdown. Thus even under assumption about a random character of q_{cap} in the capacity definition (10.1), the definition is not associated with empirical results. Therefore, dynamic traffic management methods based on this capacity definition are also not consistent with real traffic features [107–111].

In the GM model class, the maximum bottleneck capacity can be defined to be equal to the flow rate downstream of the bottleneck at which free flow instability occurs in a traffic flow model. However, as shown in Sect. 10.3 in this model class the instability leads to an F→J transition. This is inconsistent with the empirical fact that traffic breakdown is governed by an F→S transition. For this reason, the

---

[6] Note that the theoretical basis of ANCONA on-ramp metering method (Sect. 9.2) is consistent with this capacity definition.

definition of free flow capacity made within the framework of the GM model class does not satisfy the empirical features of highway capacity (Sect. 3.1.3).

10.6.4 Transportation Engineering and Highway Capacity

Most known methods and models for dynamic traffic management like on-ramp metering, speed limit control, dynamic traffic assignment, etc. are based on the basic assumption that there is a particular (deterministic or stochastic) highway capacity of free flow at a bottleneck (Sect. 10.6.1). In particular, on-ramp metering methods try to control on-ramp inflow making the flow rate downstream of the bottleneck as close as possible to highway capacity. In many methods for dynamic traffic assignment is assumed that between a link travel time and a link flow rate there is a relationship characterized by highway capacity.

In these and other methods of dynamic traffic management, there are many approaches for a choice of highway capacity based on an analysis of measured data. However, the critical consideration made above about highway capacity as a particular value (Sect. 10.6.1) leads to the subsequent critical conclusion that independent of the choice of highway capacity for a dynamic traffic management method, the method is inconsistent with the fundamental empirical features of traffic breakdown.

10.6.5 ALINEA On-Ramp Metering Method

Traffic flow models discussed in the previous section are standard ones for validation of freeway traffic control, dynamic traffic management, and dynamic traffic assignment. Because these models cannot show empirical features of traffic breakdown, the related simulations of freeway control and dynamic management strategies based on these models cannot predict many of freeway traffic phenomena that would occur through the use of a control strategy. In this section, this critical comment, which can be applied for all freeway traffic flow control strategies' simulations that the author knows of, is illustrated for the well-known ALINEA method of Papageorgiou *et al.* for feedback on-ramp metering [112–118]. There are at least two reasons for the choice of the ALINEA method: (i) Various ALINEA strategies are used in many real on-ramp metering installations. (ii) The theoretical basis of ALINEA does clearly follow from the LWR theory.

10.6.5.1 Theoretical Basis of ALINEA

The ALINEA control rule, which determines the flow rate of vehicles that can merge onto the main road from the on-ramp $q_{\text{on}}^{(\text{cont})}$ at each time t_s, $t_s = T_{\text{av}}s$, $s = 1, 2, ...$, i.e., during the time interval $t_s \leq t < t_{s+1}$ reads as follows [112–115]:

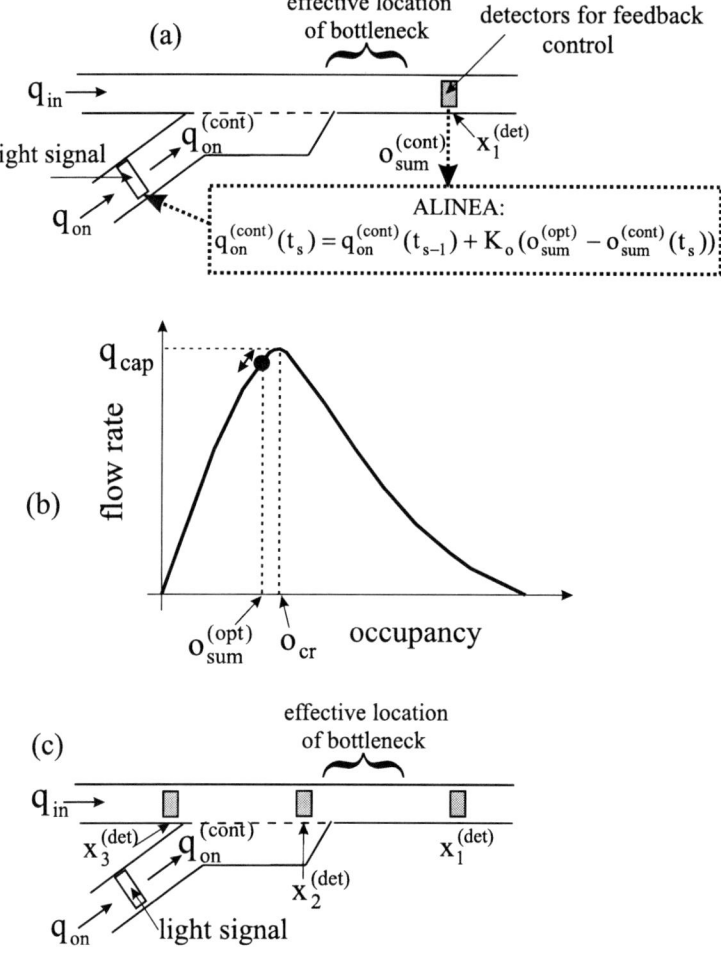

Fig. 10.14 Explanation of ALINEA [112–118]: (a) Scheme of ALINEA application. (b) Fundamental diagram. (c) Three possible locations of feedback control detectors

$$q_{on}^{(cont)}(t_s) = q_{on}^{(cont)}(t_{s-1}) + K_o(o_{sum}^{(opt)} - o_{sum}^{(cont)}(t_s)), \qquad (10.45)$$

where T_{av} (e.g., T_{av}= 1 min) is the averaging time interval for data measured by feedback control detector (labeled "detector for feedback control" in Fig. 10.14 (a)); $o_{sum}^{(opt)}$ is a chosen optimal (target) occupancy; $o_{sum}^{(cont)}(t_s)$ is occupancy measured via the feedback control detector and averaged during the time interval $t_{s-1} < t \leq t_s$; K_o is constant. To measure the occupancy $o_{sum}^{(cont)}(t_s)$ associated with the sum of the flow rates $q_{in} + q_{on}$ and to avoid the influence of disturbances due to vehicle merging within the on-ramp merging region, the feedback control detector in ALINEA of

10.6 Applications of Highway Capacity Definitions

Ref. [112–114] is located at some distance downstream of the end of the on-ramp merging region (detector location $x_1^{(\text{det})}$ in Fig. 10.14 (a)).

The theoretical basis of ALINEA [112–114, 116–118] is associated with the LWR theory [8, 33] that the capacity downstream of the bottleneck q_{cap} is found from (10.1), i.e., the capacity is related to the maximum point of the fundamental diagram for downstream free flow at the bottleneck (Fig. 10.14 (b)). Following [116], we denote the critical occupancy associated with this capacity q_{cap} by o_{cr} (Fig. 10.14 (b)). In accordance with these assumptions [8,33,116], traffic breakdown should not occur as long as downstream occupancy $o_{\text{sum}}^{(\text{cont})}$ does not exceed critical occupancy o_{cr} (Fig. 10.14 (b)). Thus in (10.45) the condition

$$o_{\text{sum}}^{(\text{opt})} < o_{\text{cr}} \quad (10.46)$$

should be satisfied [116].

To prevent congestion emergence at the bottleneck at the highest possible throughput downstream of the bottleneck with the ALINEA rule (10.45), the optimal occupancy $o_{\text{sum}}^{(\text{opt})}$ should be chosen as close as possible to o_{cr} [116]:

$$\frac{o_{\text{cr}} - o_{\text{sum}}^{(\text{opt})}}{o_{\text{sum}}^{(\text{opt})}} \ll 1. \quad (10.47)$$

Through the use of ALINEA, occupancy $o_{\text{sum}}^{(\text{cont})}(t_s)$ measured downstream of the bottleneck can change over time, however, this occupancy should be in a neighborhood of $o_{\text{sum}}^{(\text{opt})}$ (labeled by black point in Fig. 10.14 (b)) [116]. One might indeed conclude that ALINEA strategy ensures the highest possible throughput downstream of the bottleneck and it maintains free flow at the bottleneck.

However, this theoretical basis of ALINEA – traffic breakdown at the bottleneck does not occur as long as condition (10.46) is satisfied (Fig. 10.14 (b)) – is inconsistent with the empirical features of traffic breakdown (Sect. 3.1.3).

10.6.5.2 Traffic Breakdown under ALINEA Application

For simulations of the ALINEA rule (10.45), we choose an optimal occupancy $o_{\text{sum}}^{(\text{opt})}$ that satisfies conditions (10.46) and (10.47). In this case, when the flow rate to the on-ramp q_{on} increases slowly (Fig. 10.15 (a)), traffic breakdown occurs under ALINEA application (Fig. 10.15 (b)). Later, an GP is formed upstream of the bottleneck. Thus ALINEA cannot prevent traffic breakdown and upstream propagation of a congested pattern occurring due to traffic breakdown. This failure of ALINEA to prevent traffic breakdown is explained based on empirical features of traffic breakdown as follows. In an initially free flow downstream of the bottleneck, traffic breakdown occurs with a probability that is greater than zero within a flow rate range (3.25), i.e.,

$$q_{\text{th}}^{(B)} \leq q_{\text{sum}} \leq q_{\text{max}}^{(\text{free B})}, \quad (10.48)$$

and the associated occupancy range in free flow downstream of the bottleneck satisfying conditions (Fig. 10.16)

$$o_{th}^{(B)}(q_{th}^{(B)}) \leq o_{sum}^{(opt)} \leq o_{cr}(q_{max}^{(free\ B)}). \quad (10.49)$$

Probability of traffic breakdown is equal to zero only when the flow rate downstream of the bottleneck satisfies condition (4.4), respectively the associated occupancy satisfies condition

$$o_{sum}^{(opt)} < o_{th}^{(B)}. \quad (10.50)$$

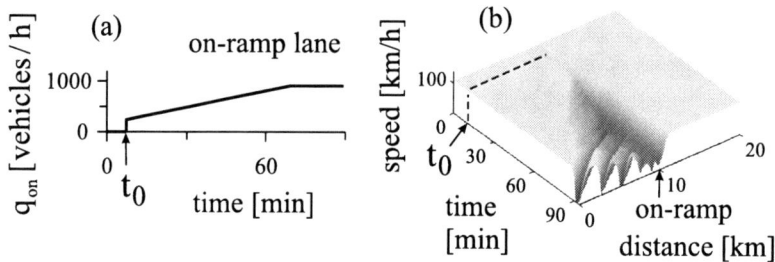

Fig. 10.15 ALINEA cannot prevent traffic breakdown and upstream congestion propagation. At $t = t_0$ on-ramp inflow is switched on. Taken from [28, 106]

This is a qualitative difference between theoretic basis of the ALINEA rule (10.45) discussed above (Fig. 10.14 (b)) [112–114, 116, 117] and three-phase traffic theory (Fig. 10.16). In three-phase traffic theory, under conditions (10.46), (10.47) for ALINEA performance the probability of traffic breakdown at the bottleneck is close to one (Sect. 3.3). This explains the result presented in Fig. 10.15 (b): firstly, traffic breakdown occurs under ALINEA application, then emergent congestion propagates upstream that can affect other upstream bottlenecks in a freeway network.

The above simulations of ALINEA method (10.45) have been made for locations of feedback control detector downstream of the effective bottleneck location as initially introduced in Ref. [112–114]. In this case, ALINEA can reliably perform only at the target occupancy satisfying condition (10.50), when traffic breakdown probability is zero. However, in this case no on-ramp metering is needed.

10.6.5.3 Influence of Feedback Detector Location on ALINEA Performance

An influence of a location of feedback control detector on ALINEA performance in the whole possible detector location range (Fig. 10.14 (c))

$$x_3^{(det)} \leq x \leq x_1^{(det)}, \quad (10.51)$$

10.6 Applications of Highway Capacity Definitions

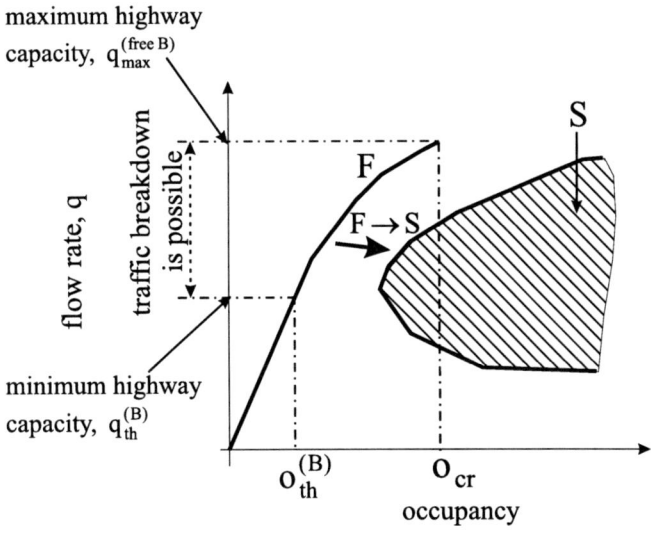

Fig. 10.16 Explanation of failure of ALINEA to prevent traffic breakdown. Taken from [28, 106]

i.e., between a road location $x_1^{(\mathrm{det})}$ downstream of the effective bottleneck location and a location $x_3^{(\mathrm{det})}$ upstream of the bottleneck have been studied in Ref. [28, 106].

When feedback control detector in ALINEA is downstream of the beginning of the on-ramp merging region, however, the detector is located *upstream* of the effective bottleneck location (specifically upstream of the road location at which traffic breakdown can firstly be measured in a neighborhood of the on-ramp bottleneck) ($x_2^{(\mathrm{det})}$ in Fig. 10.14 (c)) [118], then, as stressed in [28, 106], such an ALINEA application can suppress traffic congestion, i.e., maintain free flow at the bottleneck[7].

The same conclusion has been made in [109, 110] about an ALINEA application called UP-ALINEA [116] in which the detector is upstream of the beginning of the on-ramp merging region, i.e., upstream of the on-ramp bottleneck ($x_3^{(\mathrm{det})}$ in Fig. 10.14 (c)).

However, these applications of ALINEA and UP-ALINEA suppress congestion at the bottleneck only at the expense of extremely rapid growth of the vehicle queue at the light signal and the associated growth of travel time (waiting time) [28, 106]. This conclusion can be explained as follows[8]. The ALINEA rule (10.45) is a *linear* controller. In contrast, traffic breakdown exhibits a *discontinues* character, i.e., it is fundamentally a *non-linear* effect. This explains why independent of the choice of a location for feedback control detector and target occupancy (within the range

---

[7] For a more detailed consideration of this ALINEA application, see Sect. 3.3 of [106].

[8] As shown in [108], other ALINEA variants like ALINEA/Q [116] do not solve this problem: ALINEA/Q cannot prevent traffic breakdown with subsequent congested pattern upstream propagation.

(10.49)), the ALINEA method is inconsistent with the non-linear character of traffic breakdown and non-linear features of resulting traffic congestion.

10.7 Comparison of Feedback On-Ramp Metering Methods

Here, based on numerical simulations we discuss briefly results of a comparison of ANCONA (Sect. 9.2) with ALINEA and UP-ALINEA applications in which feedback control detector is located *upstream* of the effective bottleneck location (detector locations $x_3^{(\mathrm{det})} \leq x \leq x_2^{(\mathrm{det})}$ in Fig. 10.14 (c)). Therefore, these ALINEA and UP-ALINEA applications can suppress traffic congestion, i.e., maintain free flow at the bottleneck[9].

A comparison of ANCONA with ALINEA in which feedback control detector is located *upstream* of the effective bottleneck location, however, downstream of the beginning of the on-ramp merging region ($x_2^{(\mathrm{det})}$ in Fig. 10.14 (c)) is shown in Fig. 10.17. We can see that the waiting time at light signal in ANCONA (curve 1 in Fig. 10.17 (b)) is considerably shorter that that is in ALINEA (curve 2).

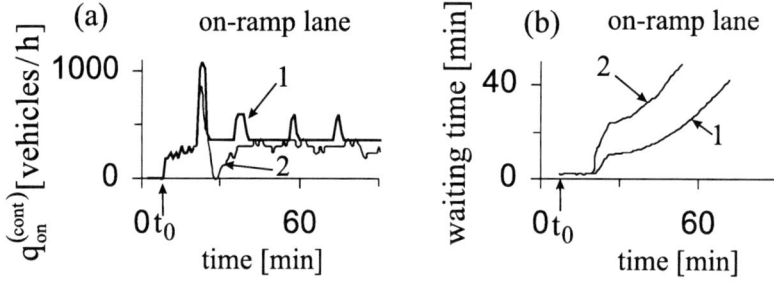

Fig. 10.17 Comparison of ANCONA with ALINEA at detector location $x_2^{(\mathrm{det})}$ in Fig. 10.14 (c) at the same flow rates q_{in}, q_{on} and bottleneck model in both methods. In (a, b), curves 1 for ANCONA and curves 2 for ALINEA. At $t = t_0$ on-ramp inflow is switched on. Taken from [28]

In UP-ALINEA [116], feedback control detector is upstream of the bottleneck ($x_3^{(\mathrm{det})}$ in Fig. 10.14 (c)) as that in ANCONA (Fig. 9.7). In this case, UP-ALINEA can also suppress any congested pattern and maintain free flow at the bottleneck (Fig. 10.18) [108]. However, UP-ALINEA exhibits a considerably longer waiting time at light signal than is the case under ANCONA application during the same time interval (curve 1 for ANCONA and curve 2 for UP-ALINEA in Fig. 10.18 (d)). Thus UP-ALINEA suppresses congestion at the bottleneck only at the expense of extremely rapid growth of the vehicle queue at the light signal and the associated growth of waiting time.

---

[9] A more detailed comparison of ANCONA with ALINEA at different locations of control detector and values of optimal occupancy can be found in [28, 106, 109, 110].

10.7 Comparison of Feedback On-Ramp Metering Methods

The crucial difference between queue length growths at light signal in the on-ramp lane over time under ANCONA applications (curves 1 in Figs. 10.17 (b) and 10.18 (d)) and applications of ALINEA or UP-ALINEA (curves 2) can be explained as follows. Because in the case under consideration feedback control detector is upstream of the effective bottleneck location, ALINEA and UP-ALINEA suppress any congested pattern at the bottleneck (Fig. 10.18 (b)). At chosen great enough flow rates $q_{\rm in}$, $q_{\rm on}$ used in simulations at which an GP is formed without on-ramp metering (Fig. 9.8), congestion suppression via ALINEA and UP-ALINEA can be achieved only if these methods decreases the average flow rate of vehicles that merge onto the main road from the on-ramp to very small values (curves 2 for $q_{\rm on}^{(\rm cont)}(t)$ in Figs. 10.17 (a) and 10.18 (c)).

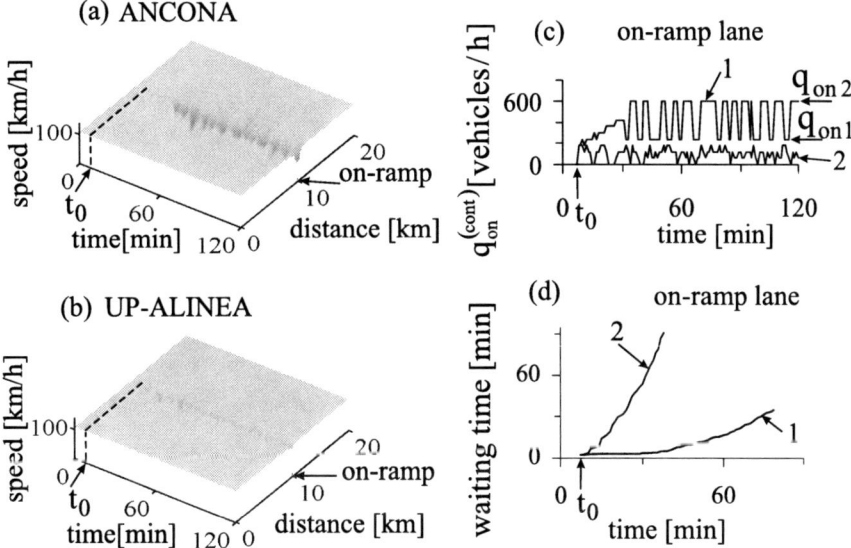

Fig. 10.18 Comparison of ANCONA with UP-ALINEA at the same flow rates $q_{\rm in}$, $q_{\rm on}$ and bottleneck model in both methods. (a, b) Speed in space and time for ANCONA (a) and UP-ALINEA (b); at $t = t_0$ on-ramp inflow is switched on. (c, d) Time dependences of the flow rate $q_{\rm on}^{(\rm cont)}$ (c) and waiting time at light signal in the on-ramp lane (d); curves 1 for ANCONA and curves 2 for UP-ALINEA. Taken from [109]

In contrast, ANCONA is based on the fact that traffic breakdown is associated with an F→S transition at the bottleneck. It is also taken into account that through the use of ANCONA a return phase transition from synchronized flow to free flow (S→F transition) can be achieved at the bottleneck. This S→F transition can lead to synchronized flow dissolution. ANCONA decreases the flow rate $q_{\rm on}^{(\rm cont)}$ only after traffic breakdown has occurred at the bottleneck. This leads to a considerably greater $q_{\rm on}^{(\rm cont)}$ (curves 1 in Figs. 10.17 (a) and 10.18 (c)) in comparison with ALINEA and UP-ALINEA (curves 2). This explains why the vehicle waiting time at the light signal under ANCONA application (curves 1 in Figs. 10.17 (b) and 10.18 (d)) is also considerably shorter than under ALINEA and UP-ALINEA applications (curves 2). For this reason, simulations show [106, 109] that ANCONA exhibits considerably greater throughputs than ALINEA and UP-ALINEA.

Thus ANCONA exhibits the following benefits in comparison with ALINEA and UP-ALINEA methods:

(a) Greater throughputs on the main road and on-ramp.
(b) Considerably shorter vehicle waiting times at light signal in the on-ramp lane.
(c) The upstream propagation of congestion does not occur even if a congested pattern occurs at the bottleneck: the congested pattern is spatially localized in the vicinity of the bottleneck.

References

1. B.D. Greenshields, in *Highway Research Board Proceedings*, **14**, pp. 448–477 (1935)
2. M. Koshi, M. Iwasaki, I. Ohkura, in *Proc. 8th International Symposium on Transportation and Traffic Theory*, ed. by V.F. Hurdle. (University of Toronto Press, Toronto, Ontario, 1983), pp. 403
3. D. Drew, *Traffic Flow Theory and Control*, (NY: McGraw Hill, New York, 1968)
4. W. Leutzbach, *Introduction to the Theory of Traffic Flow*, (Springer, Berlin, 1988)
5. A.D. May, *Traffic Flow Fundamentals*, (Prentice-Hall, Inc., New Jersey, 1990)
6. F.A. Haight, *Mathematical Theories of Traffic Flow*, (Academic Press, New York, 1963)
7. *Highway Capacity Manual 2000*, (National Research Council, Transportation Research Board, Washington, D.C., 2000)
8. C.F. Daganzo, *Fundamentals of Transportation and Traffic Operations*, (Elsevier Science Inc., New York, 1997)
9. R. Wiedemann, *Simulation des Verkehrsflusses*, (University of Karlsruhe, Karlsruhe, 1974)
10. I. Prigogine, R. Herman, *Kinetic Theory of Vehicular Traffic*, (American Elsevier, New York, 1971)
11. G.B. Whitham, *Linear and Nonlinear Waves*, (Wiley, New York, 1974)
12. M. Cremer, *Der Verkehrsfluss auf Schnellstrassen*, (Springer, Berlin, 1979)
13. N.H. Gartner, C.J. Messer, A.K. Rathi (editors), *Traffic Flow Theory: A State-of-the-Art Report*, (Transportation Research Board, Washington DC, 2001)
14. D.C. Gazis, *Traffic Theory*, (Springer, Berlin, 2002)
15. G.F. Newell, *Applications of Queuing Theory*, (Chapman Hall, London, 1982)
16. M. Papageorgiou, *Application of Automatic Control Concepts in Traffic Flow Modeling and Control*, (Springer, Berlin, New York, 1983)
17. M. Brackstone, M. McDonald, Transportation Research F **2**, 181 (1998)
18. D.E. Wolf, Physica A **263**, 438–451 (1999)
19. D. Chowdhury, L. Santen, A. Schadschneider, Physics Reports **329**, 199 (2000)
20. D. Helbing, Rev. Mod. Phys. **73**, 1067–1141 (2001)
21. T. Nagatani, Rep. Prog. Phys. **65**, 1331–1386 (2002)
22. K. Nagel, P. Wagner, R. Woesler, Oper. Res. **51**, 681–716 (2003)
23. R. Mahnke, J. Kaupužs, I. Lubashevsky, Phys. Rep. **408**, 1–130 (2005)
24. N. Bellomo, V. Coscia, M. Delitala, Math. Mod. Meth. App. Sc. **12**, 1801–1843 (2002)
25. B. Piccoli, A. Tosin, in *Encyclopedia of Complexity and System Science*, ed. by R.A. Meyers. (Springer, Berlin, 2009), pp. 9727–9749
26. H. Rakha, P. Pasumarthy, S. Adjerid, Transportation Letters, **1**, 95–110 (2009)
27. S. Maerivoet, B. De Moor, Phys. Rep. **419**, 1–64 (2005)

28. B.S. Kerner, *The Physics of Traffic*, (Springer, Berlin, New York, 2004)
29. L. Neubert, L. Santen, A. Schadschneider, M. Schreckenberg, Phys. Rev. E **60**, 6480–6490 (1999)
30. W. Knospe, L. Santen, A. Schadschneider, M. Schreckenberg, Phys. Rev. E **65**, 056133 (2002)
31. W. Knospe, L. Santen, A. Schadschneider, M. Schreckenberg, Phys. Rev. E **70**, 016115 (2004)
32. M.J. Cassidy, J.R. Windover, Transportation Res. Rec. **1484**, 73–79 (1995); W.H. Lin, C.F. Daganzo, Transp. Res. **31A**, 141–155 (1997); J.-C. Muñoz, C.F. Daganzo, in *Transportation and Traffic Theory in the 21st Century*, ed. by M.A.P. Taylor. (Elsevier Science, Amsterdam 2002), pp. 441–461
33. M.J. Lighthill, G.B. Whitham, Proc. Roy. Soc. A **229**, 281–345 (1995)
34. P.I. Richards, Oper. Res. **4**, 42–51 (1956)
35. C.F. Daganzo, Trans. Res. B **28**, 269–287 (1993)
36. R. Herman, E.W. Montroll, R.B. Potts, R.W. Rothery, Oper. Res. **7**, 86–106 (1959)
37. D.C. Gazis, R. Herman, R.W. Rothery, Oper. Res. **9**, 545–567 (1961)
38. G.F. Newell, Oper. Res. **9**, 209–229 (1961)
39. G.B. Whitham, Proc. R. Soc. London A **428**, 49 (1990)
40. M. Bando, K. Hasebe, A. Nakayama, A. Shibata, Y. Sugiyama, Jpn. J. Appl. Math. **11**, 203–223 (1994)
41. M. Bando, K. Hasebe, A. Nakayama, A. Shibata, Y. Sugiyama, Phys. Rev. E **51**, 1035–1042 (1995)
42. M. Bando, K. Hasebe, A. Nakayama, A. Shibata, Y. Sugiyama, J. Phys. I France **5**, 1389–1399 (1995)
43. H.J. Payne, in *Mathematical Models of Public Systems*, ed. by G.A. Bekey. Vol. 1, (Simulation Council, La Jolla, 1971)
44. H.J. Payne, Tran. Res. Rec. **772**, 68 (1979)
45. H.-T. Fritzsche, Transportation Engineering Contribution **5**, 317 (1994)
46. P.G. Gipps, Trans. Res. B **15**, 105–111 (1981)
47. K. Nagel, M. Schreckenberg, J. Phys. (France) I **2**, 2221–2229 (1992)
48. R. Barlović, L. Santen, A. Schadschneider, M. Schreckenberg, Eur. Phys. J. B **5**, 793–800 (1998)
49. S. Krauß, P. Wagner, C. Gawron, Phys. Rev. E **55**, 5597–5602 (1997)
50. S. Krauß, PhD thesis, DRL-Forschungsbericht 98-08 (1998). http://www.zaik.de/~paper
51. M. Treiber, A. Hennecke, D. Helbing, Phys. Rev. E **62**, 1805–1824 (2000)
52. B.S. Kerner, S.L. Klenov, J. Phys. A: Math. Gen. **39**, 1775–1809 (2006); 7605
53. B.S. Kerner, in *Transportation Research Trends*, ed. by P.O. Inweldi. (Nova Science Publishers, Inc., New York, USA, 2008), pp. 1–92
54. B.S. Kerner, in *Encyclopedia of Complexity and System Science*, ed. by R.A. Meyers. (Springer, Berlin, 2009), pp. 9302–9355
55. E.E. Stokes, Phil. Mag. **33**, 349–356 (1848)
56. E. Kometani, T. Sasaki, J. Oper. Res. Soc. Jap. **2**, 11 (1958)
57. E. Kometani, T. Sasaki, Oper. Res. **7**, 704–720 (1959)
58. T. Nagatani, K. Nakanishi, Phys. Rev. E **57**, 6415–6421 (1998)
59. I. Lubashevsky, P. Wagner, R. Mahnke, Eur. Phys. J. B **32**, 243–247 (2003)
60. B.S. Kerner, P. Konhäuser, Phys. Rev. E **50**, 54–83 (1994)
61. B.S. Kerner, P. Konhäuser, Phys. Rev. E **48**, 2335–2338 (1993)
62. R. Kühne, in *Highway Capacity and Level of Service*, ed. by U. Brannolte. (A.A. Balkema, Rotterdam, 1991), pp. 211
63. B.S. Kerner, P. Konhäuser, M. Schilke, Phys. Rev. E **51**, 6243–6246 (1995)
64. A. Aw, M. Rascle, SIAM J. Appl. Math. **60**, 916–938 (2000)
65. C.F. Tanga, R. Jiang, Q.S. Wu, Physica A **377**, 641–650 (2007)
66. D. Helbing, A. Hennecke, M, Treiber, Phys. Rev. Lett. **82**, 4360–4363 (1999)
67. H.Y. Lee, H.-W. Lee, D. Kim, Phys. Rev. E **59**, 5101–5111 (1999)
68. L.A. Pipes, J. Appl. Phys. **24**, 274–287 (1953)
69. B.S. Kerner, P. Konhäuser, bild der wissenschaft, Heft 11, 86–89 (1994)

70. M. Schönhof, D. Helbing, Transp. Sc. **41**, 135–166 (2007)
71. M. Schönhof, D. Helbing, Transp. Rec. B **43**, 784–797 (2009)
72. P. Berg, A. Woods, Phys. Rev. E **64**, 035602(R) (2001)
73. B.S. Kerner, H. Rehborn, Phys. Rev. E **53**, R1297–R1300; R4275–R4278 (1996)
74. M. Takayasu, H. Takayasu, Fractals **1**, 860–866 (1993)
75. B.S. Kerner, Phys. Rev. Lett. **81**, 3797–3400 (1998)
76. B.S. Kerner, in *Proceedings of the 3rd Symposium on Highway Capacity and Level of Service*, ed. by R. Rysgaard, Vol 2, (Road Directorate, Ministry of Transport – Denmark, 1998), pp. 621–642
77. B.S. Kerner, Trans. Res. Rec. **1678**, 160–167 (1999)
78. B.S. Kerner, Physics World **12**, 25–30 (August 1999); in *Transportation and Traffic Theory*, ed. by A. Ceder. (Elsevier Science, Amsterdam, 1999), pp. 147–171
79. B.S. Kerner, J. Phys. A: Math. Theor. **41**, 215101 (2008); 369801 (2008)
80. B.S. Kerner, P. Konhäuser, M. Schilke, Phys. Lett. A **215**, 45–56 (1996)
81. B.S. Kerner, S.L. Klenov, A. Hiller, Non. Dyn. **49**, 525–553 (2007)
82. B.S. Kerner, S.L. Klenov, A. Hiller, J. Phys. A: Math. Gen. **39**, 2001–2020 (2006)
83. J.-B. Lesort (editor), *Transportation and Traffic Theory. Proceedings of the 13th International Symposium on Transportation and Traffic Theory*, (Elsevier Science Ltd, Oxford, 1996)
84. A. Ceder (editor), *Transportation and Traffic Theory. Proceedings of the 14th International Symposium on Transportation and Traffic Theory*, (Elsevier Science Ltd, Oxford, 1999)
85. M.A.P. Taylor (editor), *Transportation and Traffic Theory in the 21st Century. Proceedings of the 15th International Symposium on Transportation and Traffic Theory*, (Elsevier Science Ltd, Amsterdam, 2002)
86. D.E. Wolf, M. Schreckenberg, A. Bachem (editors), *Traffic and Granular Flow. Proceedings of the International Workshop on Traffic and Granular Flow*, (World Scientific, Singapore, 1995)
87. M. Schreckenberg, D.E. Wolf (editors), *Traffic and Granular Flow' 97. Proceedings of the International Workshop on Traffic and Granular Flow*, (Springer, Singapore, 1998)
88. D. Helbing, H.J. Herrmann, M. Schreckenberg, D.E. Wolf (editors), *Traffic and Granular Flow' 99*, (Springer, Heidelberg, 2000)
89. M. Fukui, Y. Sugiyama, M. Schreckenberg, D.E. Wolf (editors), *Traffic and Granular Flow' 01*, (Springer, Heidelberg, 2003)
90. S.P. Hoogendoorn, S. Luding, P.H.L. Bovy, M. Schreckenberg, D.E. Wolf (editors), *Traffic and Granular Flow' 03*, (Springer, Heidelberg, 2005)
91. A. Schadschneider, T. Pöschel, R.D. Kühne, M. Schreckenberg, D.E. Wolf (editors), *Traffic and Granular Flow' 05. Proceedings of the International Workshop on Traffic and Granular Flow*, (Springer, Berlin, 2007)
92. E. Brockfeld, R.D. Kühne, A. Skabardonis, P. Wagner, Trans. Res. Rec. **1852**, 124–129 (2003)
93. B.S. Kerner, in *Encyclopedia of Complexity and System Science*, ed. by R.A. Meyers. (Springer, Berlin, 2009), pp. 9355–9411
94. Y. Wang, M. Papageorgiou, A. Messmer, in *Proceedings of the 83rd Annual Transportation Research Board Meeting*, TRB Paper No. 04-3429. (TRB, Washington DC, 2004)
95. E. Brockfeld, R.D. Kühne, P. Wagner, in *Proceeding of the Transportation Research Board 84th Annual Meeting*, TRB Paper No. 05-2152. (TRB, Washington, DC, 2005)
96. W. Knospe, L. Santen, A. Schadschneider, M. Schreckenberg, J. Phys. A: Math. Gen. **33**, L477–L485 (2000)
97. B.S. Kerner, S.L. Klenov, D.E. Wolf, J. Phys. A: Math. Gen. **35**, 9971–10013 (2002)
98. B.S. Kerner, S.L. Klenov, A. Hiller, H. Rehborn, Phys. Rev. E **73**, 046107 (2006)
99. L. Elefteriadou, R.P. Roess, W.R. McShane, Transp. Res. Rec. **1484**, 80–89 (1995)
100. B.N. Persaud, S. Yagar, R. Brownlee, Transp. Res. Rec. **1634**, 64–69 (1998)
101. M. Lorenz, L. Elefteriadou Trans. Res. Cir. **E-C018**, 84–95 (2000)
102. W. Brilon, J. Geistefeld, M. Regler, in *Transportation and Traffic Theory. Proceedings of the 16th Inter. Sym. on Transportation and Traffic Theory*, ed. by H.S. Mahmassani. (Elsevier, Amsterdam, 2005), pp. 125–144

References

103. W. Brilon, H. Zurlinden, Straßenverkehrstechnik, Heft 4, 164 (2004)
104. W. Brilon, M. Regler, J. Geistefeld, Straßenverkehrstechnik, Heft 3, 136 (2005)
105. F.L. Hall, K. Agyemang-Duah, Transp. Res. Rec. **1320**, 91–98 (1991)
106. B.S. Kerner, Physica A **355**, 565–601 (2005)
107. B.S. Kerner, Traf. Eng. & Cont. **48**, 28–35 (2007)
108. B.S. Kerner, Traf. Eng. & Cont. **48**, 68–75 (2007)
109. B.S. Kerner, Traf. Eng. & Cont. **48**, 114–120 (2007)
110. B.S. Kerner, IEEE Trans. on ITS, **8**, 308–320 (2007)
111. B.S. Kerner, Transp. Res. Rec. **1999**, 30–39 (2007)
112. M. Papageorgiou, J.-M. Blosseville, H. Hadj-Salem, Trans. Res. A. **24**, 361–370 (1990)
113. M. Papageorgiou, H. Hadj-Salem, J.-M. Blosseville, Transp. Res. Rec. **1320**, 58–64 (1991)
114. M. Papageorgiou, H. Hadj-Salem, F. Middleham, Transp. Res. Rec. **1603**, 99–98 (1997)
115. M. Papageorgiou, A. Kotsialos, IEEE Trans. on ITS. **3**, 271-281 (2002)
116. E. Smaragdis, M. Papageorgiou, Transp. Res. Rec. **1856**, 74–86 (2003)
117. M. Papageorgiou, Y. Wang, E. Kosmatopoulos, I. Papamichail, Traf. Eng. & Cont. **48**, 271–276 (2007)
118. M. Papageorgiou, E. Kosmatopoulos, I. Papamichail, Y. Wang, IEEE Trans. on ITS. **9**, 360–365 (2008)

Chapter 11
Three-Phase Traffic Flow Models

11.1 Overview of Three-Phase Traffic Flow Models

Three phase traffic theory is a qualitative theory. Many different mathematical three-phase traffic flow models can be developed in the framework of three-phase traffic theory. The first microscopic three-phase traffic flow model that can reproduce empirical features of traffic breakdown and resulting spatiotemporal congested patterns was developed by Kerner and Klenov in 2002 [1]. Some months later, Kerner, Klenov, and Wolf developed a cellular automata (CA) three-phase traffic flow model (KKW CA model) [2]. Later other traffic flow models in the framework of three-phase traffic theory have been developed: Davis [3] as well as Kerner and Klenov [4] have proposed different three-phase microscopic deterministic models; Jiang and Wu [5], Lee *et al.* [6], and Gao *et al.* [7] have developed various CA three-phase traffic flow models; Laval [8] and Hoogendoorn *et al.* [9] have developed macroscopic three-phase traffic flow models. Features of phase transitions and congested patterns that these models exhibit are similar to those found in 2002 in the Kerner-Klenov stochastic and KKW CA models. Recent simulation results in the framework of three-phase traffic theory can be found in Ref. [10–23][1].

In this chapter, we consider different classes of three-phase traffic flow models:

(i) An **a**cceleration **t**ime **d**elay model [4] (ATD model for short) in which rules of vehicle motion are *deterministic* ones.

(ii) A *stochastic* three-phase traffic flow model, in which a variety of driver time delays is simulated through the use of model fluctuations [1, 28].

(iii) The KKW CA model [2].

---

[1] Note that the use of only some of the hypotheses of three-phase traffic theory in a traffic flow model does not ensure that the model can show empirical phase transitions in vehicular traffic. An example is Colombo's continuum model [24] that consists of Eq. (10.2) in which dependence $Q(\rho)$ is associated with the fundamental hypothesis of three-phase traffic theory (Fig. 5.2) [25–27]. However, solutions of Colombo's model [24] can show none of the empirical features of F→S and S→J transitions because in this model other hypotheses of three-phase traffic theory needed for the description of phase transitions in real traffic flow have not been taken into account.

However, rather than a detailed description of these models and their solutions, we focus only on the incorporation of hypotheses of three-phase traffic theory about 2D-region of steady states of synchronized flow and the over-acceleration effect into these models[2].

This allows us in next Chap. 12 to show that by the ignoring of these hypotheses, the ATD model transforms into a deterministic model of the GM model class (Sect. 10.3.1), the stochastic three-phase traffic flow model transforms into Krauß's model (Sect. 10.3.6), and the KKW CA model transforms into the Nagel-Schreckenberg (NaSch) CA model (Sect. 10.3.5). On the one hand, this will show the linking between three-phase traffic flow models and earlier traffic flow models; on the other hand, this will explain why these and other earlier traffic flow models and theories have failed in explaining of empirical features of traffic breakdown and many resulting congested patterns.

It must be noted that in addition to the incorporation of the hypotheses of three-phase traffic theory, in the three-phase traffic flow models many ideas about simulations of the driver behavior introduced in earlier traffic flow models have also been used. In particular, simulations of the over-deceleration effect due to a driver time delay firstly introduced by Herman, Gazis, Kometani, Sasaki, Newell *et al.* [30–35] (Sect. 10.3.1), simulations of driver time delays through the use of models fluctuations introduced by Nagel and Schreckenberg [36], simulations of slow-to-start-rules introduced by Takayasu and Takayasu [37], Kerner and Konhäuser [38], Barlović *et al.* [39] as well as simulations of safe speed in traffic flow introduced by Gipps [40] and Krauß *et al.* [41] are also very important elements of these three-phase traffic flow models.

11.2 Deterministic Acceleration Time Delay Three-Phase Traffic Flow Model

11.2.1 Rules of Vehicle Motion

The Kerner-Klenov ATD model for identical vehicles and drivers reads as follows [4]:

$$\frac{dx}{dt} = v, \quad (11.1)$$

$$\frac{dv}{dt} = a, \quad (11.2)$$

---

[2] A detailed consideration of these models and their solutions, which explain the fundamental empirical features of traffic breakdown as well as all known empirical features of resulting congested patterns, can be found in [4] for the ATD model, in [28, 29] for the stochastic model, in [2] for the KKW CA model, and in Ref. [23] in which a theory of traffic congestion at heavy bottlenecks has been presented.

11.2 Deterministic Acceleration Time Delay Three-Phase Traffic Flow Model

$$\frac{da}{dt} = \begin{cases} (a^{(\text{free})} - a)/\tau_{\text{del}}(v, a) & \text{at } g > G \text{ and } g > g_{\max}^{(\text{jam})}, \\ (a^{(\text{syn})} - a)/\tau_{\text{del}}(v, a) & \text{at } g \leq G \text{ and } g > g_{\max}^{(\text{jam})}, \\ (a^{(\text{jam})} - a)/\tau_{\text{del}}(v, a) & \text{at } 0 \leq g \leq g_{\max}^{(\text{jam})}, \end{cases} \quad (11.3)$$

where a is a vehicle acceleration (deceleration); x is the vehicle space co-ordinate; $g = x_\ell - x - d$ is a space gap to the preceding vehicle, d is the vehicle length; the lower index ℓ marks variables related to the preceding vehicle; $g_{\max}^{(\text{jam})}$ is the maximum space gap within a wide moving jam (Fig. 11.1 (a)); $G(v, v_\ell)$ is a dynamic synchronization gap (Table 11.1); τ_{del} is a driver time delay; $a^{(\text{free})}$, $a^{(\text{syn})}$, and $a^{(\text{jam})}$ are desirable vehicle accelerations (deceleration) in the free flow, synchronized flow, and wide moving jam phases, respectively, which are found from the condition

$$a^{(\text{phase})} = \min(\max(\tilde{a}^{(\text{phase})}, a_{\min}), a_{\max}, a_{\text{safe}}), \quad (11.4)$$

the superscript "phase" in (11.4) means either "free", or "syn", or else "jam" for the related traffic phase; $a_{\min}$ and $a_{\max}$ ($a_{\min} < 0$, $a_{\max} \geq 0$) are, respectively, the minimum and maximum accelerations for cases in which there are no safety restrictions. In (11.4), functions $\tilde{a}^{(\text{free})}$, $\tilde{a}^{(\text{syn})}$, and $\tilde{a}^{(\text{jam})}$ associated with driver acceleration (deceleration) within the related traffic phase – free flow, or synchronized flow, or else wide moving jam – are determined as follows:

$$\tilde{a}^{(\text{free})}(g, v, v_\ell) = A(V(g) - v) + K\Delta v, \quad (11.5)$$

$$\tilde{a}^{(\text{syn})}(g, v, v_\ell) = A\min(V(g) - v, 0) + K\Delta v, \quad (11.6)$$

$$\tilde{a}^{(\text{jam})}(v) = -K^{(\text{jam})}v, \quad (11.7)$$

a_{safe} is a vehicle deceleration related to safety requirements taken as

$$a_{\text{safe}} = \frac{A_{\text{safe}}}{\tau_{\text{safe}}}(g - g_{\text{safe}}) + K_{\text{safe}}\Delta v, \quad (11.8)$$

where

$$\Delta v = v_\ell - v \quad (11.9)$$

is the speed difference between the speed of the preceding vehicle v_ℓ and the vehicle speed v; g_{safe} is a safe space gap, $g_{\text{safe}} = v\tau_{\text{safe}}$, τ_{safe} is a safe time headway; A, K, $K^{(\text{jam})}$, A_{safe}, K_{safe} are dynamic coefficients; $V(g)$ is a gap-dependent speed (Fig. 11.1 (a) and Table 11.1).

In Eqs. (11.3), the driver time delay $\tau_{\text{del}}(v, a)$ is defined as follows:

$$\tau_{\text{del}} = \begin{cases} \tau_{\text{reac}} & \text{at } a_{\text{safe}} < \min(0, \max(\tilde{a}^{(\text{phase})}, a_{\min}), a), \\ \tilde{\tau}_{\text{del}} & \text{otherwise}. \end{cases} \quad (11.10)$$

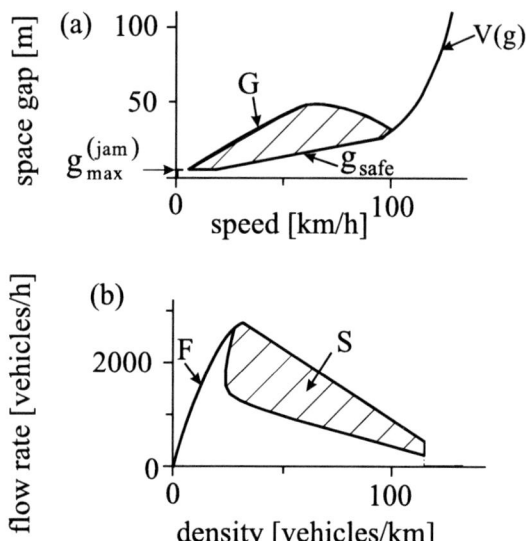

Fig. 11.1 Model steady states for ATD model: (a) In the space-gap–speed plane. (b) In the flow–density plane. 2D-dashed regions are related to steady states of synchronized flow. Taken from [4]

Table 11.1 Functions and parameters of ATD model as well as on-ramp bottleneck model used in simulations [4]

Functions
$G(v, v_\ell) = v\max(0, \tau_G(v) + \kappa(v, v_\ell)(v - v_\ell))$, $\kappa(v, v_\ell) = \begin{cases} \kappa^{(acc)} & \text{at } v < v_\ell, \\ \kappa^{(dec)} & \text{at } v \geq v_\ell, \end{cases}$
$\tau_G(v) = \tau_G^{(0)}(1 - 0.85(v/V_0)^2)$; $K(v, v_\ell) = \begin{cases} K^{(acc)} & \text{at } v < v_\ell, \\ K^{(dec)} & \text{at } v \geq v_\ell, \end{cases}$
$K^{(dec)}(v) = K_1^{(dec)}(1 - \lambda(v)) + K_2^{(dec)}\lambda(v)$, $\lambda(v) = (1 + \exp((v/v_c - 1)/\varepsilon))^{-1}$;
$V(g) = V_0 \tanh(\frac{g+g_0}{V_0 T})$; $\tau_{del}^{(d.\ inter)}(v) = \begin{cases} 0.5 \text{ s at } v \geq v_c, \\ 0.7 \text{ s otherwise}, \end{cases}$ $\tau_{del}^{(a.\ inter)}(v) = \begin{cases} 0.57 \text{ s at } v \geq v_c, \\ 0.87 \text{ s otherwise;} \end{cases}$
$A_{safe}(v_\ell) = A_{safe}^{(0)} \tau_{safe}(\tau_{safe} + v_\ell/(2b_{safe}))^{-1}$,
$K_{safe}(v_\ell) = A_{safe}^{(0)}(T_0 + v_\ell/(2b_{safe}))(\tau_{safe} + v_\ell/(2b_{safe}))^{-1}$.
Model of on-ramp bottleneck (Fig. 11.2)
vehicle merging rules: $g^+ > \gamma \hat{v} \tau_{safe}$, $g^- > \gamma v^- \tau_{safe}$;
the vehicle speed after merging: $\hat{v} = \min(v^+, v + \Delta v_r^{(1)})$;
superscripts + and − are related to the preceding and
trailing vehicles in the target lane on the main road, respectively.
Parameters
$V_0 = 120$ km/h, $T = 0.9$ s, $g_0 = 2$ m, $d = 7.5$ m, $A = 0.5$ s^{-1}, $K^{(acc)} = 0.8$ s^{-1},
$K^{(jam)} = 1$ s^{-1}, $K_1^{(dec)} = 0.95$ s^{-1}, $K_2^{(dec)} = 0.48$ s^{-1}, $v_c = 15$ m/s, $\varepsilon = 0.15$, $\tau_G^{(0)} = 2.5$ s,
$\kappa^{(acc)} = 0.5$ s^2/m, $\kappa^{(dec)} = 0.55$ s^2/m, $g_{max}^{(jam)} = 0.95$ m, $\tau_{del}^{(d)} = 1$ s, $\tau_{del}^{(a)} = 0.75$ s,
$\tau_{reac} = 0.4$ s, $a_{max} = 1$ m/s^2, $a_{min} = -1$ m/s^2, $A_{safe}^{(0)} = 1.25$ s^{-1}, $b_{safe} = 2$ m/s^2,
$\tau_{safe} = 1$ s, $T_0 = 0.42$ s, $L_m = 300$ m, $L_r = 500$ m, $\gamma = 0.22$, $\Delta v_r^{(1)} = 8$ m/s, $v_{free\ on} = 90$ km/h.

11.2 Deterministic Acceleration Time Delay Three-Phase Traffic Flow Model

Here, τ_{reac} is a driver reaction time that must be taken into account in the cases when the driver should decelerate unexpectedly to avoid collisions; $\tilde{\tau}_{\text{del}}$ is the time delay in other traffic situations, which is chosen different depending on whether the vehicle accelerates or decelerates:

$$\tilde{\tau}_{\text{del}} = \begin{cases} \tilde{\tau}_{\text{del}}^{(a)} & \text{at } a > 0, \\ \tilde{\tau}_{\text{del}}^{(d)} & \text{at } a \leq 0. \end{cases} \quad (11.11)$$

In turn, $\tilde{\tau}_{\text{del}}^{(a)}$ and $\tilde{\tau}_{\text{del}}^{(d)}$ in (11.11) depend on the acceleration (deceleration) a:

$$\tilde{\tau}_{\text{del}}^{(a)} = \begin{cases} \tau_{\text{del}}^{(a)} & \text{at } a < a^{(\text{phase})}, \\ \tau_{\text{del}}^{(a,\ \text{inter})} & \text{otherwise}, \end{cases} \quad (11.12)$$

$$\tilde{\tau}_{\text{del}}^{(d)} = \begin{cases} \tau_{\text{del}}^{(d)} & \text{at } a \geq a^{(\text{phase})}, \\ \tau_{\text{del}}^{(d,\ \text{inter})} & \text{otherwise}. \end{cases} \quad (11.13)$$

The sense of these mean driver time delays is as follows:

- $\tau_{\text{del}}^{(d,\ \text{inter})}$ is the mean driver time delay in interruption or reduction of driver deceleration that corresponds to situations in which a driver decelerates currently but wants either to stop the deceleration or to reduce it. The time delay is responsible for the over-deceleration effect discussed in Sect. 5.2.2.
- $\tau_{\text{del}}^{(d)}$ is the mean driver time delay in deceleration occurring in driving situations in which the driver starts to decelerate or wants to decelerate harder in cases in which the driver approaches a region of a lower speed downstream.
- $\tau_{\text{del}}^{(a)}$ is the mean time delay in acceleration occurring in driving situations in which the driver starts to accelerate or wants to increase vehicle acceleration. To satisfy the slow-to-start rule, in the ATD model the mean driver time delay in acceleration at the downstream front of a wide moving jam $\tau_{\text{del, jam}}^{(a)}$ is longer than $\tau_{\text{del}}^{(a)}$ at higher speeds.
- $\tau_{\text{del}}^{(a,\ \text{inter})}$ is the mean driver time delay in reduction of acceleration that corresponds to situations in which the driver accelerates currently but wants either to stop vehicle acceleration or to reduce it.

In the ATD model, at any vehicle speed the driver reaction time τ_{reac} is shorter than the time delay in interruption or reduction of driver deceleration $\tau_{\text{del}}^{(d,\ \text{inter})}(v)$. In turn, $\tau_{\text{del}}^{(d,\ \text{inter})}$ is shorter than the safe time headway τ_{safe}, i.e.,

$$\tau_{\text{safe}} > \tau_{\text{del}}^{(d,\ \text{inter})}(v) > \tau_{\text{reac}}. \quad (11.14)$$

Taking into account (5.4), we can write relations between some important driver time characteristics of the ATD model for lower synchronized flow speeds as follows:

$$\tau_G > \tau_{\text{del, jam}}^{(a)} > \tau_{\text{safe}} > \tau_{\text{del}}^{(d,\ \text{inter})}(v) > \tau_{\text{reac}}. \quad (11.15)$$

From (11.3)–(11.7) we can see that if the condition

$$g > G \qquad (11.16)$$

is satisfied, then a vehicle moves in accordance with the rules for free flow; under condition

$$g_{\max}^{(\mathrm{jam})} < g \leq G, \qquad (11.17)$$

the vehicle moves in accordance with the rules for synchronized flow; when

$$g \leq g_{\max}^{(\mathrm{jam})}, \qquad (11.18)$$

the vehicle decelerates within a wide moving jam.

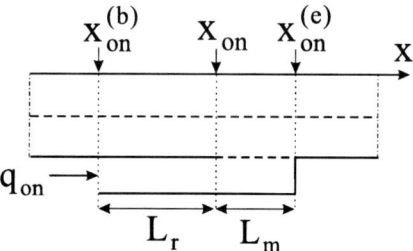

Fig. 11.2 Model of on-ramp bottleneck on two-lane road. Taken from [29]

A model of an on-ramp bottleneck used in simulations of the ATD model (Fig. 11.2) consists of two parts: (i) A part of the on-ramp lane of length L_r ($x_{\mathrm{on}}^{(b)} \leq x \leq x_{\mathrm{on}}$) in which vehicles move in accordance with the model (11.1)–(11.13), however, at another maximum speed $v_{\mathrm{free\ on}}$. (ii) A merging region of length L_m ($x_{\mathrm{on}} < x \leq x_{\mathrm{on}}^{(e)}$) in which vehicles, which move in the on-ramp lane in accordance with (11.2)–(11.13), can merge from the on-ramp lane onto the main road; formulae for this vehicle merging and model parameters are given in Table 11.1.

11.2.2 Speed Adaptation Effect

If the condition

$$g > G \qquad (11.19)$$

is satisfied and $\tilde{a}^{(\mathrm{free})} < a_{\max}$, a_{safe}, then from (11.3), (11.4) it follows that the vehicle accelerates (labeled by "acceleration" in Fig. 11.3) with acceleration found from the equation

$$\frac{da}{dt} = (\tilde{a}^{(\mathrm{free})} - a)/\tau_{\mathrm{del}}, \qquad (11.20)$$

where $\tilde{a}^{(\mathrm{free})}$ is given by (11.5).

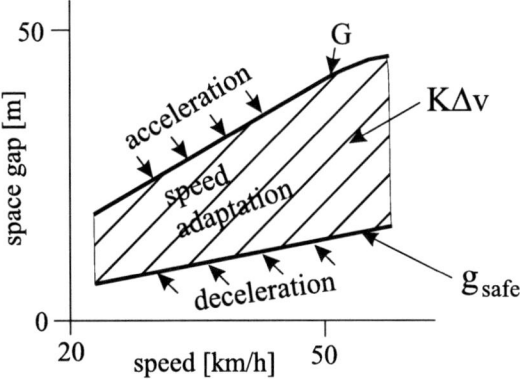

Fig. 11.3 Explanation of speed adaptation in ATD model. A part of steady states of synchronized flow for ATD model at lower speeds taken from Fig. 11.1 (a)

In contrast, if the condition

$$g < g_{\mathrm{safe}} \qquad (11.21)$$

is satisfied, the vehicle is not slower than the preceding vehicle, and $a_{\mathrm{safe}} < \max(\tilde{a}^{(\mathrm{free})}, a_{\mathrm{min}})$, then from (11.3), (11.4) it follows that the vehicle decelerates (labeled by "deceleration" in Fig. 11.3) with deceleration found from the equation

$$\frac{da}{dt} = (a_{\mathrm{safe}} - a)/\tau_{\mathrm{del}}, \qquad (11.22)$$

where a_{safe} is given by (11.8).

A different vehicle behavior we find within a 2D-region of steady states of synchronized flow (2D-dashed region in Fig. 11.3) that satisfy conditions

$$g_{\mathrm{safe}} < g \leq G. \qquad (11.23)$$

Indeed, at vehicle speeds v in synchronized flow, which are lower than the speed $V(g)$, formula (11.6) yields

$$\tilde{a}^{(\mathrm{syn})} = K\Delta v, \qquad (11.24)$$

where $K > 0$. If in (11.24) $a_{\mathrm{min}} < K\Delta v < a_{\mathrm{max}}$, a_{safe}, then from (11.3), (11.4), (11.24) we find that the vehicle acceleration (deceleration) is found from the equation

$$\frac{da}{dt} = (\tilde{a}^{(\mathrm{syn})} - a)/\tau_{\mathrm{del}}, \qquad (11.25)$$

where $\tilde{a}^{(\mathrm{syn})}$ is given by (11.24). Thus when a driver cannot pass the preceding vehicle moving at a synchronized flow speed, the driver adapts its speed to the speed of the preceding vehicle at any space gap within the space gap range (11.23), i.e., without caring what the precise space gap to the preceding vehicle is. When $\Delta v \neq 0$, then as follows from (11.24), the vehicle accelerates when it is slower than the preceding vehicle, and decelerates when it is faster than the preceding vehicle (labeled by "speed adaptation" in Fig. 11.3).

11.2.3 Simulation of Over-Acceleration Effect

A mathematical simulation of a discontinuous character of probability of over-acceleration like shown in Fig. 3.6 can be performed even without any vehicle lane changing through the choice of some model characteristics. Such an *implicit simulation* of the over-acceleration effect, which should not necessarily be associated with a realistic driver behavior, can nevertheless show all fundamental empirical macroscopic features of traffic breakdown at a bottleneck[3]. The value of such traffic flow simulations is the simplicity of the model.

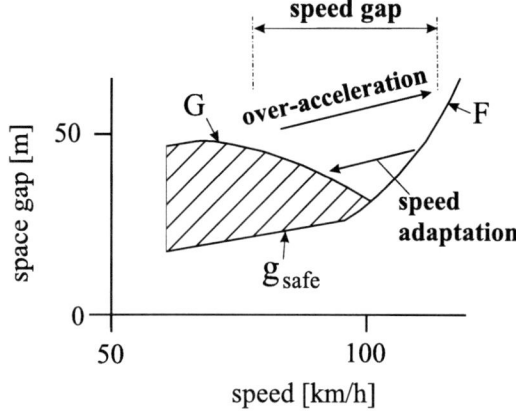

Fig. 11.4 Explanation of competition of over-acceleration and speed adaptation effects in ATD model. Parts of steady states of synchronized flow for ATD model at higher speeds (dashed region) and free flow speed (F) taken from Fig. 11.1 (a)

In the ADT model, an implicit simulation of the over-acceleration effect is made through the use of a model speed gap between steady states of synchronized flow (dashed region in Fig. 11.4) and free flow states (F). To simulate this speed gap, we use the synchronization space gap for steady states of synchronized flow $G(v) = v\tau_G(v)$ (see the synchronization time headway τ_G in Table 11.1) that begins to decrease at higher speeds as shown in Figs. 11.1 (a) and 11.4. This leads to

---
[3] An explicit simulation of the over-acceleration effect based on lane changing to a faster lane appears in Sect. 11.3.3.

a *speed gap* between states of free flow and steady states of synchronized flow of higher speeds (labeled by "speed gap" in Fig. 11.4).

This speed gap simulates a *discontinuous character* of over-acceleration needed for simulations of traffic breakdown (Sect. 3.2.3) as follows. In a dynamic model state of traffic flow, which lies within this speed gap, condition (11.16) is satisfied.

Therefore, in accordance with (11.3)–(11.5) in the dynamic flow state a driver accelerates to a gap-dependent free flow speed $V(g)$. This driver acceleration, which is determined by the first term in the right side of formula (11.5), i.e.,

$$A(V(g) - v) > 0, \qquad (11.26)$$

can be considered over-acceleration with probability that is a great enough value (labeled by arrow "over-acceleration" in Fig. 11.4). In contrast, when a dynamic state of traffic flow is associated with synchronized flow, i.e., the condition (11.16) is not satisfied, then in accordance with (11.3), (11.4), and (11.6) there is no driver acceleration to free flow at all. This can be considered over-acceleration with probability that is equal to zero. This model feature simulates a *discontinuous drop* in probability of over-acceleration from the great enough value under condition (11.16) to zero that occurs when condition (11.16) is not valid.

Vehicle deceleration to a lower speed of the preceding vehicle within a local disturbance in free flow, i.e., the speed adaptation effect (labeled by arrow "speed adaptation" in Fig. 11.4) is simulated through the second term in the right side of formula (11.5), i.e.,

$$K\Delta v < 0. \qquad (11.27)$$

A competition of the over-acceleration and speed adaptation effects within this disturbance in free flow simulates traffic breakdown.

11.3 Stochastic Three-Phase Traffic Flow Model

As in the ATD model (Fig. 11.1), in the Kerner-Klenov stochastic three-phase traffic flow model [1,28] steady states of synchronized flow cover a 2D-region in the space-gap–speed and flow–density planes (Fig. 11.5). A difference between the steady states in these models is that in the stochastic model there is no speed gap between free flow states and steady states of synchronized flow (Fig. 11.5) as that used in the ATD model for a simulation of the over-acceleration effect (Fig. 11.4). This is because in the stochastic model we use other mathematical approaches for simulations of the over-acceleration effect, in particular, an explicit simulation of the over-acceleration through lane changing to a faster lane.

Another difference between the ATD model and the stochastic model is as follows. In the ATD model different mean driver time delays are explicitly simulated in dynamic equations of vehicle motion. In contrast, in the stochastic model driver time delays associated with different driving situations are simulated as collective effects

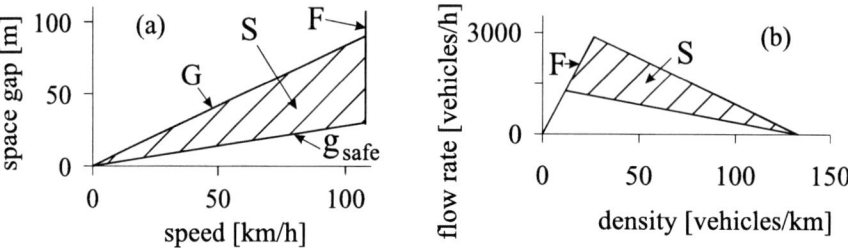

Fig. 11.5 Steady states for a stochastic three-phase traffic flow in the space-gap–speed (a) and flow–density planes (b). Taken from [29]

through the use of model fluctuations that change randomly vehicle acceleration and deceleration with some given probabilities.

11.3.1 Rules of Vehicle Motion

Update rules of vehicle motion in the stochastic model are as follows [1, 28]:

$$v_{n+1} = \max(0, \min(v_{\text{free}}, \tilde{v}_{n+1} + \xi_n, v_n + a_{\max}\tau_{\text{step}}, v_{\text{safe},n})), \tag{11.28}$$

$$x_{n+1} = x_n + v_{n+1}\tau_{\text{step}}, \tag{11.29}$$

where a discrete time $t = n\tau_{\text{step}}$ is used, the index $n = 0, 1, 2, ...$, τ_{step} is time step; v_n is the vehicle speed at time step n, v_{free} is the maximum speed in free flow that is constant, x_n is the vehicle co-ordinate, the lower index ℓ marks functions and values related to the preceding vehicle, $a_{\max}$ is the maximum acceleration; $v_{\text{safe},n}$ is a safe speed; ξ_n is a random speed change (model fluctuations), $\tilde{v}_{n+1}$ is the speed without model fluctuations ξ_n given by the equation

$$\tilde{v}_{n+1} = \max(0, \min(v_{\text{free}}, v_{c,n}, v_{\text{safe},n})), \tag{11.30}$$

$$v_{c,n} = \begin{cases} v_n + a_n\tau_{\text{step}} & \text{at } g_n > G_n \\ v_n + \Delta_n & \text{at } g_n \leq G_n, \end{cases} \tag{11.31}$$

$$\Delta_n = \max(-b_n\tau_{\text{step}}, \min(a_n\tau_{\text{step}}, \Delta v_n)), \tag{11.32}$$

$$\Delta v_n = v_{\ell,n} - v_n, \tag{11.33}$$

$g_n = x_{\ell,n} - x_n - d$ is the space gap, d is the vehicle length; b_n and a_n are random deceleration and acceleration, respectively; G_n is a synchronization space gap that depends on the vehicle speed v_n and on the speed of the preceding vehicle $v_{\ell,n}$

$$G_n = G(v_n, v_{\ell,n}), \tag{11.34}$$

11.3 Stochastic Three-Phase Traffic Flow Model

where the function $G(u, w)$ is chosen as

$$G(u, w) = \max(0, k\tau_{\text{step}} u + a_{\max}^{-1} u(u - w)), \tag{11.35}$$

$k > 1$ is constant (Table 11.2); random deceleration and acceleration ξ_n in (11.28) are applied depending on whether the vehicle decelerates or accelerates, or else maintains its speed:

$$\xi_n = \begin{cases} -\xi_b & \text{if } S_{n+1} = -1 \\ \xi_a & \text{if } S_{n+1} = 1 \\ 0 & \text{if } S_{n+1} = 0, \end{cases} \tag{11.36}$$

where S in (11.36) denotes the state of motion ($S_{n+1} = -1$ represents deceleration, $S_{n+1} = 1$ acceleration, and $S_{n+1} = 0$ motion at nearly constant speed)

$$S_{n+1} = \begin{cases} -1 & \text{if } \tilde{v}_{n+1} < v_n - \delta \\ 1 & \text{if } \tilde{v}_{n+1} > v_n + \delta \\ 0 & \text{otherwise}, \end{cases} \tag{11.37}$$

where δ is constant ($\delta \ll a_{\max} \tau_{\text{step}}$),

$$\xi_b = a_{\max} \tau_{\text{step}} \Theta(p_b - r), \tag{11.38}$$

p_b is probability of random deceleration, $\Theta(z) = 0$ at $z < 0$ and $\Theta(z) = 1$ at $z \geq 0$, $r = \text{rand}(0, 1)$, ξ_a is defined in Sect. 11.3.3.

Table 11.2 Functions and parameters of stochastic model as well as on-ramp bottleneck model used in simulations [28]

Functions
$p_0(v) = 0.575 + 0.125 \min(1, v/10)$ (v in m/s),
$p_2(v) = 0.48 + 0.32\Theta(v - 15)$ (v in m/s).
Model of on-ramp bottleneck (Fig. 11.2)
$\hat{v}_n^+ = \max(0, \min(v_{\text{free}}, v_n^+ + \Delta v_r^{(2)}))$;
the vehicle speed after merging: $\hat{v}_n = \min(v_n^+, v_n + \Delta v_r^{(1)})$;
vehicle merging rule (*): $g_n^+ > \min(v_n \tau_{\text{step}}, G_n^+)$, $g_n^- > \min(v_n^- \tau_{\text{step}}, G_n^-)$, where $G_n^+ = G(v_n, v_n^+)$, $G_n^- = G(v_n^-, v_n)$, $G(u, w)$ is given by (11.35), vehicle coordinates just before and after merging are equal to each other;
vehicle merging rules (**): $x_n^+ - x_n^- - d > g_{\text{target}}^{(\min)} = \lambda v_n^+ + d$ and the vehicle passes the midpoint between two neighboring vehicles in the target lane, after merging the vehicle coordinate is equal to the coordinate of this midpoint; superscripts + and − are related to the preceding and trailing vehicles in the target lane on the main road, respectively.
Parameters
$\tau_{\text{safe}} = \tau_{\text{step}} = 1$ s, $d = 7.5$ m, $v_{\text{free}} = 108$ km/h, $a_{\max} = 0.5$ m/s^2, $b = 1$ m/s^2, $k = 3$, $\delta = 0.01$ m/s, $p_1 = 0.3$, $p_a = 0.17$, $p_b = 0.1$, $\delta_1 = 1$ m/s, $p_c = 0.2$, $\lambda = 0.75$, $L_m = 300$ m, $L_r = 500$ m, $v_{\text{free on}} = 80$ km/h, $\Delta v_r^{(1)} = 10$ m/s, $\Delta v_r^{(2)} = 5$ m/s, $\Delta v^{(1)} = 2$ m/s.

Random deceleration and acceleration b_n and a_n in (11.31) and (11.32) are taken as

$$b_n = a_{\max}\Theta(P_1 - r_1), \tag{11.39}$$

$$a_n = a_{\max}\Theta(P_0 - r_1), \tag{11.40}$$

$r_1 = \text{rand}(0,1)$, P_1 and P_0 are probabilities of random deceleration and acceleration, respectively:

$$P_0 = \begin{cases} p_0 & \text{if } S_n \neq 1 \\ 1 & \text{if } S_n = 1, \end{cases} \tag{11.41}$$

$$P_1 = \begin{cases} p_1 & \text{if } S_n \neq -1 \\ p_2 & \text{if } S_n = -1, \end{cases} \tag{11.42}$$

where $p_0(v_n)$, p_1, and $p_2(v_n)$ are probabilities (Table 11.2).

The safe speed $v_{\text{safe},n}$ in (11.30) is chosen in the form

$$v_{\text{safe},n} = \min(v_n^{(\text{safe})}, g_n/\tau_{\text{step}} + v_\ell^{(a)}), \tag{11.43}$$

where $v_n^{(\text{safe})} = v^{(\text{safe})}(g_n, v_{\ell,n})$ is taken from Krauß's model (10.37), (10.38); $v_\ell^{(a)}$ is an "anticipation" speed of the preceding vehicle at the next time step that is chosen as

$$v_\ell^{(a)} = \max(0, \min(v_{\ell,n}^{(\text{safe})}, v_{\ell,n}, g_{\ell,n}/\tau_{\text{step}}) - a_{\max}\tau_{\text{step}}), \tag{11.44}$$

$v_{\ell,n}^{(\text{safe})}$ and $g_{\ell,n}$ are the safe speed and space gap for the preceding vehicle, respectively.

In the on-ramp lane at $x_{\text{on}}^{(b)} \leq x \leq x_{\text{on}}$ (Fig. 11.2) vehicles move in accordance with (11.28)–(11.44), however, at another maximum speed $v_{\text{free on}}$. In the on-ramp lane at $x_{\text{on}} < x \leq x_{\text{on}}^{(e)}$, vehicles move in accordance with Eqs. (11.28)–(11.44), however, in Eqs. (11.31)–(11.34) the speed and coordinate of the preceding vehicle are replaced by values $\hat{v}_n^+$ and x_n^+, respectively, which are given in Table 11.2; in addition, for vehicle merging onto the main road rules (∗) or (∗∗) are used (Table 11.2).

11.3.2 Speed Adaptation Effect

We consider lower vehicles speeds (Fig. 11.6) that satisfy the condition

$$0 < v_n < v_{\text{free}}. \tag{11.45}$$

If the condition

$$g_n > G_n \tag{11.46}$$

is satisfied and $v_n + a_n\tau_{\text{step}} < v_{\text{safe},n}, v_{\text{free}}$, then from (11.28), (11.30), (11.31), (11.36), (11.37), (11.40), (11.45) it follows that

$$v_{n+1} = v_n + a_n\tau_{\text{step}}, \tag{11.47}$$

11.3 Stochastic Three-Phase Traffic Flow Model

i.e., with some probability vehicle accelerates (labeled by "acceleration" in Fig. 11.6) with the maximum acceleration $a_{\max}$.

If the condition

$$g_n < g_{\text{safe},n} \tag{11.48}$$

is satisfied, where a safe space gap $g_{\text{safe},n}$ is determined from the equation

$$v_n = v_{\text{safe},n}(g_n, v_{\ell,n}, v_\ell^{(a)}) \tag{11.49}$$

in which g_n should be replaced by $g_{\text{safe},n}$, then the speed is higher than the safe speed:

$$v_n > v_{\text{safe},n}. \tag{11.50}$$

Then from (11.28), (11.30) we find that the vehicle decelerates (labeled by "deceleration" in Fig. 11.6).

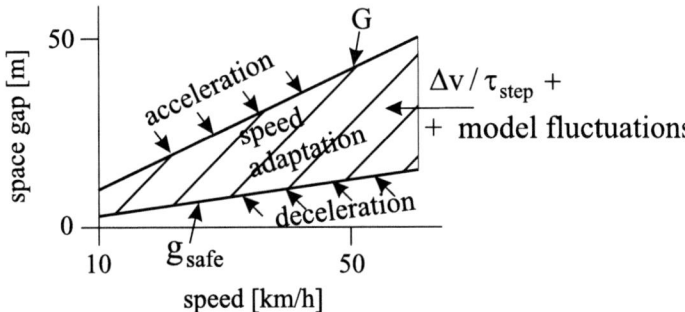

Fig. 11.6 Explanation of speed adaptation in stochastic model. A part of steady states of synchronized flow associated with (11.45) taken from Fig. 11.5 (a)

A different vehicle behavior we find within a 2D-region of steady states of synchronized flow (2D-dashed region in Fig. 11.6) that satisfy conditions

$$g_{\text{safe},n} < g_n \leq G_n. \tag{11.51}$$

Under conditions (11.51) with some probability, when

$$|\Delta v_n| \leq a_{\max} \tau_{\text{step}}, \ a_n = b_n = a_{\max}, \ v_n + \Delta_n + \xi_n \leq \min(v_{\text{safe},n}, v_n + a_{\max} \tau_{\text{step}}), \tag{11.52}$$

from (11.28), (11.30)–(11.32) we get

$$v_{n+1} = v_n + \Delta v_n + \xi_n. \tag{11.53}$$

This means that the vehicle acceleration (deceleration) is equal to

$$(\Delta v_n + \xi_n)/\tau_{\text{step}}. \tag{11.54}$$

This formula explains the speed adaptation effect in the stochastic model (labeled "speed adaptation" in Fig. 11.6): As in the ATD model (11.25), within the space gap range (11.51) the speed difference Δv_n in (11.54) describes speed adaptation to the speed of the preceding vehicle that occurs without caring what the precise space gap to the preceding vehicle is. In addition to this deterministic effect, in the stochastic model there is a random speed change ξ_n, which leads to a stochastic behavior of speed adaptation in the model. This model feature is used for simulations of driver time delays in acceleration and deceleration.

11.3.3 Simulation Approaches to Over-Acceleration Effect

Vehicle acceleration is often only a part of a variety of driver maneuver of the over-acceleration effect (Sect. 3.2.3), which can include various driver behavior, like lane change for passing or temporary closing to the preceding vehicle. For example, if a driver believes that she/he can overtake a slow moving preceding vehicle that does not accelerate, the driver can accelerate in an initial road lane before lane change. However, it can turn out that although the driver accelerates, nevertheless she/he cannot change lane and pass. In this case, the driver must decelerate to the speed of the preceding vehicle and wait for another possibility to pass. This vehicle acceleration with the subsequent deceleration to the speed of the preceding vehicle can be repeated several times before the driver can pass.

In the stochastic model, there are several different mathematical approaches for the simulation of the over-acceleration effect. In each of these approaches, a *dynamic* speed gap appears (gray region labeled by "dynamic speed gap" in Fig. 11.7) between free flow (line F) and steady states of synchronized flow of lower speeds. Within this dynamic speed gap, initially steady states of synchronized flow of higher speeds are destroyed and no long-time living synchronized flow states occur.

Thus rather than the use of a model speed gap between steady states of synchronized flow and free flow states as this made in the ATD model (Fig. 11.4), in the stochastic model a *discontinuous character* of over-acceleration discussed in Sect. 3.2.3 is modeled through the occurrence of the dynamic speed gap caused by stochastic model dynamics that destroys initially steady states of synchronized flow of higher speeds (gray region in Fig. 11.7). A competition between the speed adaptation and over-acceleration effects (labeled by "speed adaptation" and "over-acceleration" in Fig. 11.7), which occurs within a local disturbance in free flow, is associated with this dynamic speed gap. This simulates all fundamental empirical features of traffic breakdown.

11.3 Stochastic Three-Phase Traffic Flow Model

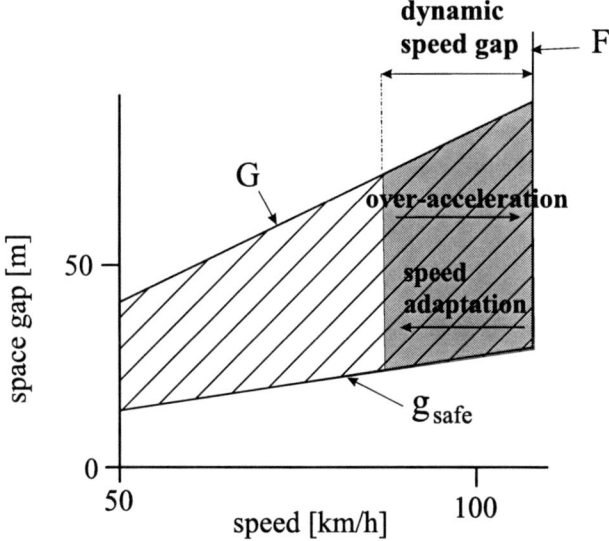

Fig. 11.7 Explanation of competition of over-acceleration and speed adaptation effects in stochastic model. Parts of steady states of synchronized flow at higher speeds (dashed region) and free flow speed taken from Fig. 11.5 (a). Gray region shows qualitatively a dynamic speed gap in which steady states of synchronized flow are destroyed and no long-time living synchronized flow states occur

11.3.3.1 Implicit Simulation of Over-Acceleration Effect through Driver Acceleration

The dynamic speed gap between states of free flow and synchronized flow resulting in a discontinuous character of over-acceleration can be simulated through driver acceleration even without vehicle lane changing. Under condition (11.51), the function ξ_a in (11.36) is taken as [28]:

$$\xi_a = a_{\max} \tau_{\text{step}} \Theta(p_a - r), \qquad (11.55)$$

where p_a is probability of random acceleration, $r = \text{rand}(0,1)$. In accordance with (11.36), these model fluctuations are applied only if the vehicle should accelerate without the fluctuations.

11.3.3.2 Simulation of Over-Acceleration Effect through Combination of Lane Changing to Faster Lane and Random Driver Acceleration

The discontinuous character of over-acceleration can also be simulated through the use of a combination of an implicit simulation through a random driver acceleration

(11.55) *and* an explicit simulation based on lane changing to a faster lane [28]. Rules for lane changing to a faster lane can be as follows.

If necessary lane changing rules together with some safety conditions are satisfied, a vehicle changes lane with probability p_c. For lane changing from the right lane to the left lane ($R \to L$) and a return change from the left lane to the right lane ($L \to R$) lane changing rules are as follows

$$R \to L: \ v_n^+ \geq v_{\ell,n} + \delta_1 \text{ and } v_n \geq v_{\ell,n}, \quad (11.56)$$

$$L \to R: \ v_n^+ > v_{\ell,n} + \delta_1 \text{ or } v_n^+ > v_n + \delta_1. \quad (11.57)$$

The safety conditions for lane changing are given by the inequalities:

$$g_n^+ > \min(v_n \tau_{\text{step}}, G_n^+), \quad (11.58)$$

$$g_n^- > \min(v_n^- \tau_{\text{step}}, G_n^-), \quad (11.59)$$

where functions G_n^+, G_n^- are given in Table 11.2; superscripts $+$ and $-$ in variables, parameters, and functions denote the preceding vehicle and the trailing vehicle in the "target" (neighboring) lane, respectively (the target lane is the lane into which the vehicle wants to change); δ_1 is constant (Table 11.2).

11.3.3.3 "Boundary" Over-Acceleration

In addition with the over-acceleration effect *within* a local disturbance in free flow discussed above that can be called as "bulk" over-acceleration, there can also be "boundary" over-acceleration. In contrast with bulk over-acceleration, boundary over-acceleration occurs at the downstream disturbance front within which a vehicle accelerates from a lower free flow speed to a higher one. Indeed, at the front the vehicle can reduce the time delay in acceleration due to passing of the preceding vehicle, i.e., due to over-acceleration. The lower the speed and the greater the density, the smaller the over-acceleration probability, i.e., the smaller the time delay reduction due to over-acceleration. Thus a boundary over-acceleration can be described by a decreasing speed function of the mean driver time delay in vehicle acceleration $\tau_{\text{del}}^{(a)}$ within a speed range between speeds in free and synchronized flows.

11.3.3.4 Explicit Simulation of Over-Acceleration Effect through Lane Changing to Faster Lane

Simulations show that the discontinuous character of over-acceleration can be simulated explicitly through lane changing to a faster lane *only*, i.e., when *no* other mathematical formulations for over-acceleration are used, specifically, when in (11.55) $p_a = 0$ and the time delay in acceleration $\tau_{\text{del}}^{(a)}$ does not change within a speed range between speeds in free and synchronized flows.

To reach this goal, we use much weaker safety conditions for lane changing in comparison with (11.58) and (11.59). In particular, when (11.58) and (11.59) are not satisfied, then under condition (11.56) or (11.57) a vehicle can nevertheless change to a faster lane with the abovementioned given probability p_c, if the following safety conditions are satisfied. The space gap between two neighboring vehicles in the target lane should satisfies the condition

$$x_n^+ - x_n^- - d > g_{\text{target}}^{(\text{min})}, \qquad (11.60)$$

where

$$g_{\text{target}}^{(\text{min})} = \lambda v_n^+ + d, \qquad (11.61)$$

λ is constant. In addition to (11.60), the condition that the vehicle passes the midpoint between two neighboring vehicles in the target lane should be satisfied. After lane changing, the coordinate of the vehicle is set to this midpoint and the vehicle speed v_n is set to $\hat{v}_n$:

$$\hat{v}_n = \min(v_n^+, v_n + \Delta v^{(1)}), \qquad (11.62)$$

where $\Delta v^{(1)}$ is constant that describes the increase in speed after lane changing (Table 11.2).

11.4 Cellular Automata Three-Phase Traffic Flow Model

Basic driver behavioral assumptions of the Kerner-Klenov-Wolf (KKW) CA model [2] are similar to those of the stochastic model [1,28] discussed in Sect. 11.3. In particular, steady states of synchronized flow cover a 2D-region in the space-gap–speed and flow–density planes (Fig. 11.5); however, these states do not form a continuum because the speed v and space gap g are integers in CA models. The rules of vehicle motion in the KKW CA model are as follows:

$$v_{n+1} = \max(0, \min(v_{\text{free}}, \tilde{v}_{n+1} + a_{\max} \tau_{\text{step}} \eta_n, v_n + a_{\max} \tau_{\text{step}}, v_{\text{safe},n})), \qquad (11.63)$$

$$x_{n+1} = x_n + v_{n+1} \tau_{\text{step}}, \qquad (11.64)$$

$$\tilde{v}_{n+1} = \max(0, \min(v_{\text{free}}, v_{c,n}, v_{\text{safe},n})), \qquad (11.65)$$

$$v_{c,n} = \begin{cases} v_n + a_{\max} \tau_{\text{step}} & \text{at } g_n > G_n, \\ v_n + a_{\max} \tau_{\text{step}} \text{sign}(\Delta v_n) & \text{at } g_n \leq G_n, \end{cases} \qquad (11.66)$$

where $\text{sign}(x) = 1$ for $x > 0$, 0 for $x = 0$, and -1 for $x < 0$; a synchronization space gap G_n is taken as

$$G(v_n) = kv_n \tau_{\text{step}}, \qquad (11.67)$$

$k > 1$ is constant; as in the Nagel-Schreckenberg CA model (Sect. 10.3.5) [36], a safe speed is equal to

$$v_{\text{safe},n} = g_n/\tau_{\text{step}}, \qquad (11.68)$$

accordingly, a safe space gap $g_{\text{safe},n}$ and a safe time headway τ_{safe} are

$$g_{\text{safe},n} = \tau_{\text{step}} v_n, \quad \tau_{\text{safe}} = \tau_{\text{step}}; \qquad (11.69)$$

random acceleration and random deceleration are simulated by adding of a random term $a_{\max} \tau_{\text{step}} \eta_n$ to vehicle speed,

$$\eta_n = \begin{cases} -1 & \text{if } r < p_b, \\ 1 & \text{if } p_b \leq r < p_b + p_a, \\ 0 & \text{otherwise}, \end{cases} \qquad (11.70)$$

p_a and p_b are probabilities of random acceleration and deceleration, respectively that are speed functions, $p_a + p_b \leq 1$, $r = \text{rand}(0,1)$; designations of other variables and functions are the same as those in Sect. 11.3.

In the KKW model, qualitatively the same approaches to simulations of speed adaptation and over-acceleration are used as those in the stochastic model discussed in Sects. 11.3.2 and 11.3.3.

11.5 Methodology of Empirical Test

For an adequate comparison of empirical and theoretical dynamic nonlinear features of spatiotemporal congested patterns in freeway traffic, the following methodology of the empirical test is used [29].

As in real traffic, in a model of a road with on-ramp and off-ramp bottlenecks vehicles appear at the upstream road boundary of the main road and of an on-ramp(s). In this open traffic process, all traffic variables downstream result from traffic demand at the upstream road boundary and on-ramp(s) as well as drivers' destinations (whether a vehicle leaves the main road to the off-ramp or it further follows the main road). At the downstream model boundary, free flow conditions are given. Spatiotemporal congested patterns emerge, develop, and dissolve in this open freeway model with the same types of bottlenecks as those in empirical observations.

The spatiotemporal dynamics of traffic breakdown and resulting congested patterns found in simulations should be compared with the associated spatiotemporal dynamics found in empirical observations. Only after the fundamental spatiotemporal features of traffic have been simulated and compared with the empirical data, *secondary characteristics* associated with these spatiotemporal patterns, which include for example fundamental diagrams, time headway distributions, speed and flow–density correlation functions and OV functions, can be compared with the associated empirical characteristics.

Such an empirical test made in [20–22] shows that three-phase traffic flow models can reproduce and predict all known microscopic and macroscopic empirical features of traffic breakdown and resulting congested traffic patterns.

11.6 What Three-Phase Traffic Flow Model is better to Use?

The following questions are often discussed when traffic flow models are compared:

- Does deterministic or stochastic microscopic traffic flow model describe better real traffic flow characteristics?
- Do explicit or implicit simulations of driver behavior describe better real measured traffic flow characteristics?
- Can nucleation features of traffic breakdown and moving jam emergence be simulated through the use of model instability with respect to infinitesimal fluctuations (so-called "linear" instability)?

11.6.1 Deterministic or Stochastic Models?

In deterministic microscopic traffic flow models, each driver can exhibit individual driver time delays in various traffic situations as well as individual lane changing and merging behavior on multi-lane roads with bottlenecks. In other words, driver behavioral characteristics can find a clear individual description in such a model.

In contrast, in stochastic microscopic traffic flow models that include also CA traffic flow models, many of driver behavioral characteristics are simulated on average, in particular, through the use of model fluctuations.

These model fluctuations have often no clear correspondence with driver behavioral characteristics. In particular, for simulations of driver time delays in various traffic situations with CA models, vehicles should decelerate randomly and sometimes without any obvious reason, for example when the space gap to the preceding vehicle is greater than the safe one and the preceding vehicle moves with a time-independent speed. Moreover, model fluctuations in vehicle deceleration and acceleration take often large non-realistic values for the current driving situation. For these reasons, the stochastic microscopic traffic flow models might seem to be much less applicable for the description of real traffic flow characteristics.

However, a high opinion about the deterministic traffic flow models and low or negative opinion about the stochastic models have no real reasons. This is because in reality, each individual driver exhibits *stochastic* driver characteristics, which can be simulated by *none* of the deterministic traffic flow models, which the author knows. For example, each of the driver time delays in acceleration and deceleration rather than a constant value is a stochastic variable, which takes different random values even in the *same driving situation*. In other words, in the deterministic models individual driver characteristics can be considered only as *mean* values, whereas all effects associated with real *stochastic* individual driver characteristics are neglected.

Thus the criterion like "clear mathematical description of an individual driver behavior" that is usually used to prefer a deterministic traffic flow model to a stochastic one does not necessarily ensure that the deterministic model can simulate spatiotemporal features of measured traffic patterns better than the stochastic traffic flow model.

In contrast, in many cases a stochastic traffic flow model, in which non-realistic driver acceleration and deceleration are used, can show empirical phase transitions and resulting congested patterns much more realistic than the deterministic one. One of the reasons for this is that in "pure" deterministic models none of the real stochastic driver behavior characteristics is taken into account.

11.6.2 Explicit or Implicit Simulations of Driver Behavior?

A similar question arises about a value of implicit simulations of driver behavior in comparison with explicit simulation approaches. An example is an explicit simulation of over-acceleration with lane changing to faster lane (Sect. 11.3.3) and an implicit simulation of over-acceleration through a speed gap between states of free flow and steady states of synchronized flow (Sect. 11.2.3).

In the first case, over-acceleration exhibits a discontinuous character because at a given density the resulting probability for lane changing to a faster lane in free flow is greater than that in synchronized flow.

In the second case, qualitatively the same discontinuous character of over-acceleration is simulated directly through the use of the speed gap, i.e., the discontinuity in speed at a given density. Although this direct simulation of the discontinuous character of over-acceleration is only a mathematical method, which is not related to the physical reason of the discontinuous character of over-acceleration, this mathematical method does lead to qualitatively and quantitatively (at an appropriate choice of model parameters) the same simulation results for spatiotemporal features of congested traffic patterns.

11.6.3 About Mathematical Approaches for Simulations of Nucleation of Traffic Breakdown and Moving Jam Emergence

In accordance with three-phase traffic theory, traffic breakdown in an initial metastable free flow can occur, if a nucleus required for traffic breakdown appears in the free flow (Chap. 3). The nucleus is associated with the occurrence of a local disturbance in free flow. Within this disturbance, i.e., within the nucleus the speed is lower than a critical speed (density is greater than a critical density) required for traffic breakdown in free flow. This critical speed is usually considerably lower than the speed (critical density is considerably greater than the density) is in the initial free flow.

In three-phase traffic theory is further assumed that wide moving jam emergence in an initial metastable synchronized flow can occur, if a nucleus required for wide moving jam emergence appears in synchronized flow (Chap. 5). The nucleus is associated with the occurrence of a local disturbance in the synchronized flow. Within this disturbance, i.e., within the nucleus the speed is lower than a critical speed (density is greater than a critical density) required for wide moving jam emergence in the synchronized flow. This critical speed is usually considerably lower than the speed (critical density is considerably greater than the density) is in the initial synchronized flow.

These assumptions made in three-phase traffic theory [25–27] about the nucleation nature of traffic breakdown and wide moving jam emergence in free and synchronized flows, respectively, are confirmed by all known up to now real measured traffic data:

- in the data traffic breakdown and wide moving jam emergence have always been nucleated by the occurrence of great enough disturbances in initial free and synchronized flows, respectively.

In three-phase traffic theory, there are three mathematical approaches for simulations of the nucleation nature of traffic breakdown and moving jam emergence:

1. The use of *discontinuities* in speed and/or density between steady states for free flow, synchronized flow, and wide moving jam incorporated in a traffic flow model. An example is a speed gap between steady states of free flow and synchronized flow incorporated in the ATD model (Fig. 11.4)[4].
2. An *explicit simulation* of driver behavior. An example is the explicit simulation of the over-acceleration effect through lane changing to a faster lane (Sect. 11.3.3.4).
3. The use of *model instabilities* with respect to infinitesimal disturbances (fluctuations).

At the first glance, an instability of traffic flow with respect to infinitesimal disturbances in traffic flow incorporated in a traffic flow model should be inconsistent with the abovementioned empirical results in which traffic breakdown and wide moving jam emergence have always been nucleated by the occurrence of great enough disturbances. Nevertheless, three-phase traffic flow models that incorporate model solution instabilities can simulate the nucleation effects as this is observed in real traffic [29]. This statement is independent of whether these model instabilities have a sense for real traffic or not.

This "paradox" can be explained as follows. Through an appropriated choice of a mathematical description of model instabilities, a three-phase traffic flow model can simulate both metastable states of free flow and metastable states of synchronized flow. If in addition, models of multi-lane roads with highway bottlenecks are associated with real roads, then traffic simulations show that there are always local disturbances in free and synchronized flows of great enough amplitudes. For this reason, *none* of instabilities with respect to infinitesimal disturbances incorporated into mathematical rules of the model plays a role for traffic breakdown and resulting congested traffic patterns [4, 29]. This is because due to these large local disturbances, nuclei for traffic breakdown and wide moving jam emergence occur in free

---
[4] Other examples appear in Sect. 12.1 (Fig. 12.1 (a, b)).

and synchronized flows *before* a threshold for a model instability with respect to infinitesimal fluctuations can be reached in simulations. Thus such model instabilities are associated only with a mathematical method for simulations of metastable free and synchronized flows observed in real traffic. In other words, the model instabilities have nothing to do with the question whether a traffic flow theory can explain traffic phenomena based on traffic flow instabilities or not.

Summarizing results of the discussion made in this section, we can conclude that rather than questions about (i) deterministic or stochastic models, (ii) explicit or implicit simulations of driver behavior, (iii) a correspondence to real traffic of an instability incorporated into mathematical rules of a model, the important question is what driver behavioral characteristics should be taken into account in a *traffic flow model* to call the model as the model for *traffic flow*. This means that the model must describe traffic breakdown at a bottleneck and resulting congested patterns as found in real measured traffic data. Only after driver behavior characteristics needed for the description of these real measured spatiotemporal traffic flow characteristics have been identified, the choice of a deterministic or stochastic model or else explicit or implicit simulations of these driver behavior characteristics is determined by the *objective* of traffic flow modeling. Examples of such objectives are macroscopic and microscopic traffic flow characteristics, motion of a specific vehicle in traffic flow, effectiveness and reliability of methods for traffic flow control or dynamic traffic assignment.

References

1. B.S. Kerner, S.L. Klenov, J. Phys. A: Math. Gen. **35**, L31–L43 (2002)
2. B.S. Kerner, S.L. Klenov, D.E. Wolf, J. Phys. A: Math. Gen. **35**, 9971–10013 (2002)
3. L.C. Davis, Phys. Rev. E **69**, 016108 (2004)
4. B.S. Kerner, S.L. Klenov, J. Phys. A: Math. Gen. **39**, 1775–1809 (2006); 7605
5. R. Jiang, Q.-S. Wu, J. Phys. A: Math. Gen. **37**, 8197–8213 (2004)
6. H.K. Lee, R. Barlović, M. Schreckenberg, D. Kim, Phys. Rev. Lett. **92**, 238702 (2004)
7. K. Gao, R. Jiang, S.-X. Hu, B.-H. Wang, Q.-S. Wu, Phys. Rev. E **76**, 026105 (2007)
8. J.A. Laval, in *Traffic and Granular Flow' 05. Proceedings of the International Workshop on Traffic and Granular Flow*, ed. by A. Schadschneider, T. Pöschel, R. Kühne, M. Schreckenberg, D.E. Wolf. (Springer, Berlin, 2007), pp. 521–526
9. S. Hoogendoorn, H. van Lint, V.L. Knoop, Trans. Res. Rec. **2088**, 102-108 (2008)
10. L.C. Davis, Physica A **368**, 541–550 (2006)
11. L.C. Davis, Physica A **361**, 606–618 (2006)
12. L.C. Davis, Physica A **379**, 274–290 (2006)
13. R. Jiang, M.-B. Hua, R. Wang, Q.-S. Wu, Phys. Lett. A **365**, 6–9 (2007)
14. R. Jiang, Q.-S. Wu, Phys. Rev. E **72**, 067103 (2005)
15. R. Jiang, Q.-S. Wu, Physica A **377**, 633–640 (2007)
16. X.G. Li, Z.Y. Gao, K.P. Li, X.M. Zhao, Phys. Rev. E **76**, 016110 (2007)
17. A. Pottmeier, C. Thiemann, A. Schadschneider, M. Schreckenberg, in *Traffic and Granular Flow' 05. Proceedings of the International Workshop on Traffic and Granular Flow*, ed. by A. Schadschneider, T. Pöschel, R. Kühne, M. Schreckenberg, D.E. Wolf. (Springer, Berlin, 2007), pp. 503–508
18. F. Siebel, W. Mauser, Phys. Rev. E **73**, 066108 (2006)

19. R. Wang, R. Jiang, Q.S. Wu, M. Liu, Physica A **378**, 475–484 (2007)
20. B.S. Kerner, S.L. Klenov, A. Hiller, H. Rehborn, Phys. Rev. E **73**, 046107 (2006)
21. B.S. Kerner, S.L. Klenov, A. Hiller, J. Phys. A: Math. Gen. **39**, 2001–2020 (2006)
22. B.S. Kerner, S.L. Klenov, A. Hiller, Non. Dyn. **49**, 525–553 (2007)
23. B.S. Kerner, J. Phys. A: Math. Theor. **41**, 215101 (2008); 369801 (2008)
24. R.M. Colombo, SIAM J. Appl. Math. **63**, 708–721 (2003)
25. B.S. Kerner, Phys. Rev. Lett. **81**, 3797–3400 (1998); in *Proceedings of the 3rd Symposium on Highway Capacity and Level of Service*, ed. by R. Rysgaard. Vol 2, (Road Directorate, Ministry of Transport – Denmark, 1998), pp. 621–642; J. Phys. A: Math. Gen. **33**, L221-L228 (2000).
26. B.S. Kerner, Trans. Res. Rec. **1678**, 160–167 (1999)
27. B.S. Kerner, in *Transportation and Traffic Theory*, ed. by A. Ceder. (Elsevier Science, Amsterdam 1999), pp. 147–171; Physics World **12**, 25–30 (August 1999).
28. B.S. Kerner, S.L. Klenov, Phys. Rev. E **68**, 036130 (2003)
29. B.S. Kerner, *The Physics of Traffic*, (Springer, Berlin, New York, 2004)
30. R. Herman, E.W. Montroll, R.B. Potts, R.W. Rothery, Oper. Res. **7**, 86–106 (1959)
31. D.C. Gazis, R. Herman, R.W. Rothery, Oper. Res. **9**, 545–567 (1961)
32. E. Kometani, T. Sasaki, J. Oper. Res. Soc. Jap. **2**, 11 (1958)
33. E. Kometani, T. Sasaki, Oper. Res. **7**, 704–720 (1959)
34. G.F. Newell, Oper. Res. **9**, 209–229 (1961)
35. D.C. Gazis, *Traffic Theory*, (Springer, Berlin, 2002)
36. K. Nagel, M. Schreckenberg, J. Phys. (France) I **2**, 2221–2229 (1992)
37. M. Takayasu, H. Takayasu, Fractals **1** 860–866 (1993)
38. B.S. Kerner, P. Konhäuser, Phys. Rev. E **48**, 2335–2338 (1993)
39. R. Barlović, L. Santen, A. Schadschneider, M. Schreckenberg, Eur. Phys. J. B **5**, 793–800 (1998)
40. P.G. Gipps, Trans. Res. B **15**, 105–111 (1981)
41. S. Krauß, P. Wagner, C. Gawron, Phys. Rev. E **55**, 5597–5602 (1997)

Chapter 12
Linking of Three-Phase Traffic Theory and Fundamental Diagram Approach to Traffic Flow Modeling

12.1 Three-Phase Traffic Models in the Framework of Fundamental Diagram Approach

A link between three-phase traffic theory and the fundamental diagram approach to traffic flow modeling can be created through the use of the *averaging* of an infinite number of steady states of synchronized flow shown in Figs. 11.1 and 11.5 to one synchronized flow speed for each density. In this case, we should find rules for vehicle motion in a traffic flow model whose steady states are associated with a fundamental diagram, however, the model should show the free flow, synchronized flow, and wide moving jam phases as well as the F→S→J transitions.

Recall that three-phase traffic theory is a qualitative theory. For this reason, the fundamental hypothesis of this theory is associated with features of hypothetical steady states of synchronized flow that should cover a 2D-region in the flow–density plane. Rather than a hypothesis about model steady states of synchronized flow, the fundamental hypothesis of a mathematical three-phase traffic flow model can be a hypothesis about some specific dynamic model features.

Although such three-phase traffic flow models are possible to develop, however, through the replacing of a 2D-region for steady states of synchronized flow by a 1D-region, i.e., by a curve in the flow–density plane, these three-phase traffic flow models lose a possibility of the description of very important features of synchronized flow found in empirical observations. This emphasizes the sense and importance of 2D steady states of synchronized flow for the development of a three-phase traffic flow model. We illustrate these general conclusions below through a brief discussion of speed adaptation (SA) three-phase traffic flow models [1].

In contrast with the deterministic ATD and stochastic three-phase traffic flow models discussed in Chap. 11, which incorporate the fundamental hypothesis of three-phase traffic theory (Figs. 11.1 and 11.5), in the SA models hypothetical steady states of synchronized flow are associated with a curve (curves *S* in Fig. 12.1), i.e., they cover a 1D-region in the flow–density plane.

The basis hypothesis of the SA models is associated with the F→S→J transitions of three-phase traffic theory. Based on an implementation of this hypothesis into the SA models, the models can show and explain both traffic breakdown (F→S transition) and moving jam emergence within synchronized flow (pinch effect and associated S→J transitions) [1] as found in empirical observations [2].

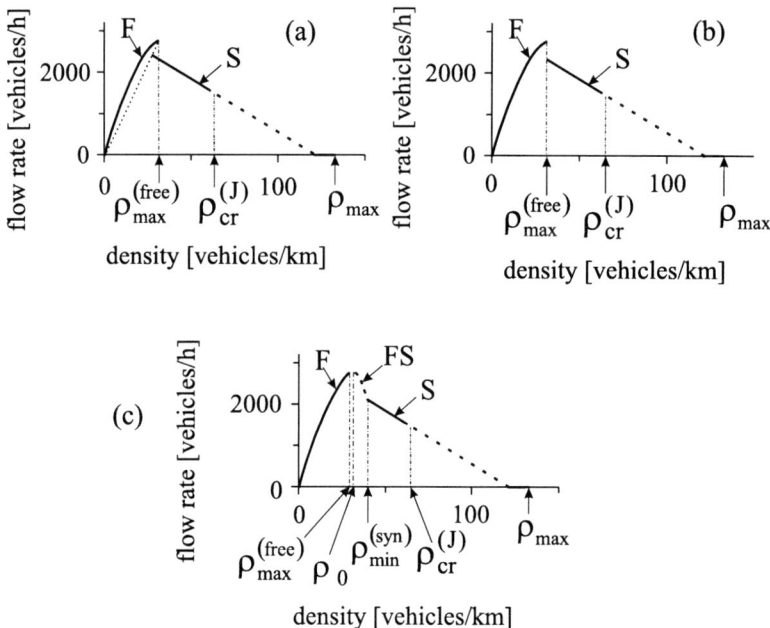

Fig. 12.1 Steady states for three different SA three-phase traffic flow models with discontinuity (a, b) and instability (c) of steady speed solutions between free flow (curves F) and synchronized flow states (curves S). Taken from [1]

Traffic breakdown is simulated in the SA models through the use of one of the following features of states in a neighborhood of the maximum point on the fundamental diagram:

(i) A *discontinuity* of states on the fundamental diagram between free flow and synchronized flow states (Fig. 12.1 (a, b)). This discontinuity determines the maximum (limit) point of free flow ($q_{\max}^{(\text{free})}$, $\rho_{\max}^{(\text{free})}$) on the fundamental diagram, i.e., at the densities $\rho \geq \rho_{\max}^{(\text{free})}$ an F→S transition should occur.

(ii) An *instability* of states on the fundamental diagram within a density range ($\rho_{\max}^{(\text{free})}$, $\rho_{\min}^{(\text{syn})}$) (curve FS in Fig. 12.1 (c)). This instability determines also the maximum (limit) point of free flow ($q_{\max}^{(\text{free})}$, $\rho_{\max}^{(\text{free})}$) on the fundamental diagram, i.e., at the densities $\rho \geq \rho_{\max}^{(\text{free})}$ an F→S transition should occur.

12.1 Three-Phase Traffic Models in the Framework of Fundamental Diagram Approach

In addition, a feature of vehicle motion is incorporated into the SA models that a vehicle tries to adjust the speed to the speed of the preceding vehicle while approaching synchronized flow states. This simulates the speed adaptation effect in synchronized flow. However, in contrast with three-phase traffic flow models discussed in Chap. 11, a driver adapts the speed to the speed of the preceding vehicle at a *particular* space gap associated with the fundamental diagram for synchronized flow (curves S in Fig. 12.1).

An S→J transition is simulated in the SA models through an instability of states of synchronized flow on the fundamental diagram whose density is greater than some critical density $\rho_{\text{cr}}^{(J)}$ (Fig. 12.1): at each density $\rho > \rho_{\text{cr}}^{(J)}$ synchronized flow is unstable (dashed parts of curves S in Fig. 12.1).

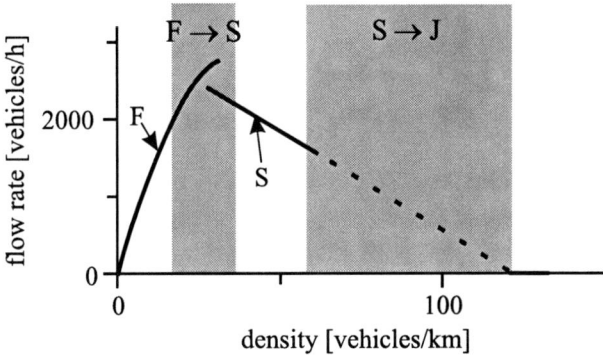

Fig. 12.2 Explanation of simulations of sequence of F→S→J transitions in the context of the fundamental diagram approach. Gray regions illustrate density ranges of traffic breakdown (labeled by F→S) and wide moving jam emergence (labeled by S→J), respectively. Fundamental diagram is taken from Fig. 12.1 (a). Taken from [1]

Thus the SA models exhibit the following features [1]:

- there are two separated density regions on the fundamental diagram in which phase transitions occur (Fig. 12.2);
- in accordance with these two regions, there are two different critical densities on the fundamental diagram:
  - the critical density $\rho_{\text{max}}^{(\text{free})}$ for traffic breakdown (F→S transition) and
  - the critical density $\rho_{\text{cr}}^{(J)}$ for an S→J transition, i.e., for wide moving jam emergence in synchronized flow (Fig. 12.1).

In other words, traffic breakdown is simulated in a density range in a neighborhood of the maximum point on the fundamental diagram, whereas the S→J transition is simulated at greater densities of synchronized flow (Fig. 12.2). This allows us to overcome drawbacks of earlier traffic flow models with the fundamental diagram in description of traffic breakdown and the pinch effect (Chap. 10).

On the one hand, the SA models are able to show and predict the F→S→J transitions. On the other hand, due to the averaging of the infinite number of steady states of synchronized flow to one synchronized flow speed for each density, the SA models are not able to show important features of synchronized flow as well as many features of coexistence of the free flow, synchronized flow, and wide moving jam phases found in real traffic [1]. These drawbacks of the SA models associated with the ignoring of the fundamental hypothesis of three-phase traffic theory are illustrated below.

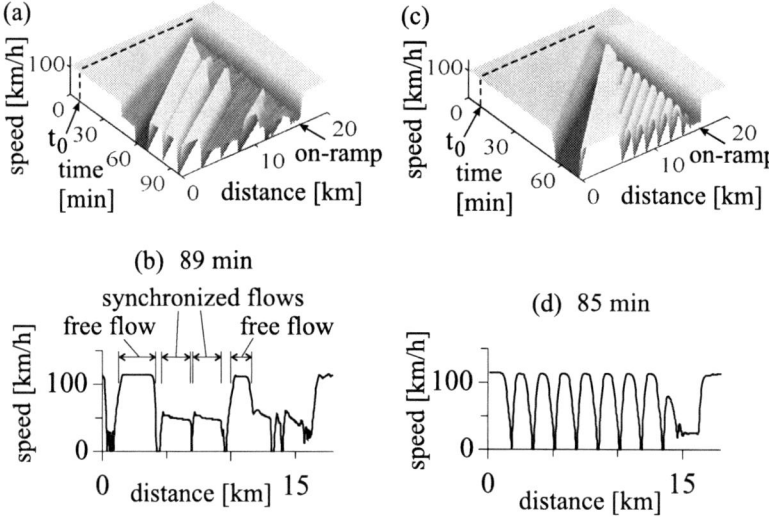

Fig. 12.3 Explanation of drawbacks of SA models: (a, b) A three-phase traffic flow model with 2D-region of steady states for synchronized flow. (c, d) SA models (Fig. 12.1). Taken from [1]

In accordance with empirical results [2], three-phase traffic flow models with a 2D-region of steady states for synchronized flow can show and predict *both* free flow *and* synchronized flow between wide moving jams (Fig. 12.3 (a, b)). In contrast, for the SA models, whose fundamental diagrams are shown in Fig. 12.1, *only* free flow can occur between wide moving jams, when the jams are moving at large enough distances from each other (Fig. 12.3 (c, d)).

To explain this, note that if a three-phase traffic flow model exhibits a 2D-region of steady states for synchronized flow (Fig. 12.4 (a)), in the wide moving jam outflow either a state of free flow with the density ρ_{min} or any state of synchronized flow from the infinite number of synchronized flow states that lie on the line J is possible. In contrast, there is only one unstable steady state of synchronized flow that lies on the intersection of the line J and curve S in the case of the SA models (Fig. 12.4 (b)). For this reason, when distances between wide moving jams are great enough, free flow can be formed in the jam outflows only (Fig. 12.3 (c, d)).

12.2 Features of Three-Phase Traffic Theory Missing in Earlier Theories

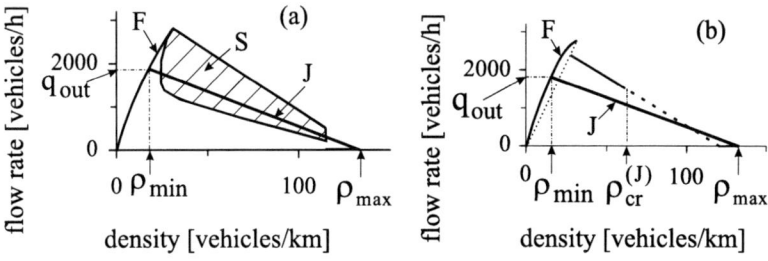

Fig. 12.4 Steady states for synchronized flow and line J for two classes of three-phase traffic flow models: (a) ATD model of Sect. 11.2. (b) SA model. Taken from [1]

We can conclude that

- the fundamental hypothesis of three-phase traffic theory about a 2D-region for steady states of synchronized flow permits the infinite number of synchronized flow states on, above, and below the line J. This allows us to simulate different synchronized flow states in a variety of diverse combinations with wide moving jams and free flows, as observed in empirical data. This is not possible to do based on the SA models with a fundamental diagram for steady states of synchronized flow.

12.2 What Features of Three-Phase Traffic Theory are Missing in Earlier Traffic Flow Theories and Models?

12.2.1 Common Missing Features

12.2.1.1 The Failure of Earlier Theories to Explain Traffic Breakdown: Missing Discontinuous Character of Over-Acceleration

Firstly, we should note that

- *none* of earlier traffic flow models [3–24] discussed in Chap. 10 exhibits a discontinuous character of over-acceleration.

Probably the discontinuous character of the over-acceleration, i.e., the discontinuous character of probability of passing is one of the most important features of three-phase traffic theory [25, 26], which is missing in all these traffic flow theories and models. The importance of this model feature is explained as follows: a competition between the over-acceleration effect, which exhibits the discontinuous character, and the speed adaptation effect incorporated in three-phase traffic models explains all fundamental empirical features of traffic breakdown at a highway bottleneck listed in Sect. 3.1.3 (see, e.g., numerical simulations of traffic breakdown with three-phase traffic flow models shown in Figs. 3.17, 3.18, and 4.2).

Thus we can make the following conclusion:

- The missing of a discontinuous character of passing probability, i.e., the discontinuous character of over-acceleration probability introduced in three-phase traffic theory [25, 26] is probably the main reason why none of the earlier traffic flow theories and models [3–24] can explain traffic breakdown at the highway bottleneck as observed in real measured traffic data.

12.2.1.2 The Failure of Earlier Theories to Explain The Coexistence of Free Flow, Synchronized Flow, and Wide Moving Jams: Missing of 2D-Region of Steady States of Synchronized flow

Another common feature of three-phase traffic flow models of Chap. 11 is that these models exhibit a 2D-region of steady states of synchronized flow in the flow–density plane (Figs. 11.1 and 11.5). In contrast, earlier traffic flow models assume the existence of a fundamental diagram. From the analysis of the SA models of Sect. 12.1 we can conclude that

- traffic flow models in the framework of the fundamental diagram fail to explain the coexistence of free flow, synchronized flow, and wide moving jams as observed in real measured traffic data.

12.2.2 From Three-Phase Traffic Flow Models back to GM Model Class

Traffic flow models in the framework of the LWR-theory take into account none of driver time delays, which are important for simulations of phase transitions in traffic flow. For this reason, we compare three-phase traffic flow models of Chap. 11 with some traffic flow models of the GM model class.

12.2.2.1 From ATD Model to Deterministic Models of GM Model Class

To come from the ATD three-phase traffic flow model of Sect. 11.2 to a deterministic traffic flow model of the GM model class, we neglect the following features of the ATD model:

- A speed gap between states of free and synchronized flows (labeled by "speed gap" in Fig. 11.4) that simulates a discontinuous character of over-acceleration in the ATD model.
- A 2D-region of steady states of synchronized flow and a synchronization space gap G of the ATD model (Fig. 11.1).

12.2 Features of Three-Phase Traffic Theory Missing in Earlier Theories

In this case, the ATD model transforms into a version of the GM model like given by Eq. (10.14).

12.2.2.2 From Stochastic Three-Phase Traffic Flow Model to Krauß's Model

The stochastic three-phase traffic flow model of Sect. 11.3 transforms into Krauß's stochastic traffic flow model (10.35)–(10.37) [27] of the GM model class, when in the stochastic model (11.28)–(11.44) the following changes are made:

- A discontinuous character of over-acceleration of the stochastic three-phase traffic flow (Sect. 11.3.3) is neglected.
- A synchronization space gap G is not taken into account. As a result, a 2D-region of steady states of synchronized flow (Fig. 11.5 (b)) transforms into the fundamental diagram of Krauß's model consisting of the line F for free flow and the line with a negative slope for congested traffic determined by the safe speed equation (10.37).

Firstly, we neglect the second term $v_n + \Delta_n$ in formula (11.31), consider the first term in (11.31) independent of the space gap, and replace a random acceleration a_n by the maximum one $a_{\max}$. This yields

$$v_{c,n} = v_n + a_{\max} \tau_{\text{step}}. \qquad (12.1)$$

Secondly, we replace model fluctuations (11.36) of the stochastic three-phase traffic flow model by a random vehicle deceleration used in Eq. (10.35) of Krauß's model. Then Eqs. (11.28)–(11.30) transform into Eq. (10.35) of Krauß's model.

In Krauß's model, when the speed is higher than the safe one, i.e., $v_n > v_{\text{safe},n}$, then a vehicle decelerates. In contrast, when the speed is lower than the safe one, then from Eq. (10.35) it follows that vehicle acceleration is equal to $a_{\max}(1-r)$. Thus there is no discontinuous character of over-acceleration in this model. For these reasons, Krauß's model cannot describe traffic breakdown and synchronized flow as found in empirical observations.

12.2.2.3 From KKW CA Three-Phase Traffic Flow Model to Nagel-Schreckenberg CA Model

To come from the KKW CA model (11.63)–(11.68) to the Nagel-Schreckenberg (NaSch) CA model (10.28)–(10.30), firstly, we neglect in Eq. (11.66) of the KKW CA model the second term $v_n + a_{\max} \tau_{\text{step}} \text{sign}(\Delta v_n)$ and assume that the first term in this equation is applied independent of the space gap. Secondly, we ignore random driver acceleration of the KKW CA model, i.e., neglect the second term in (11.70) for model fluctuations of the KKW CA model. Then rather than (11.70) we get

$$\eta_n = \begin{cases} -1 & \text{if } r < p_{\text{b}} \\ 0 & \text{otherwise}. \end{cases} \qquad (12.2)$$

If we use the maximum acceleration $a_{\max}$, time step τ_{step}, and probability $p_b = p$ the same as those of the NaSch CA model, then model fluctuations in deceleration $-\xi_n$ of the NaSch CA model (10.30) coincides with model fluctuations η_n given by (12.2). As the consequence, Eqs. (11.63) and (11.65) of the KKW CA model transform into Eq. (10.28) of the NaSch CA model [28].

In the NaSch CA model, when the speed is higher than the safe one, i.e., $v_n > g_n/\tau_{\text{step}}$, then a vehicle decelerates. In contrast, when the speed is lower than the safe one, then the vehicle accelerates or maintains the speed. Thus there is no discontinuous character of over-acceleration in this model. This is due to the ignoring of random driver acceleration and the synchronization distance of the KKW CA model (11.63)–(11.68). For these reasons, the NaSch CA model cannot describe traffic breakdown and synchronized flow as found in empirical observations.

References

1. B.S. Kerner, S.L. Klenov, J. Phys. A: Math. Gen. **39**, 1775–1809 (2006); 7605
2. B.S. Kerner, *The Physics of Traffic*, (Springer, Berlin, New York, 2004)
3. A.D. May, *Traffic Flow Fundamentals*, (Prentice-Hall, Inc., New Jersey, 1990)
4. F.A. Haight, *Mathematical Theories of Traffic Flow*, (Academic Press, New York, 1963)
5. W. Leutzbach *Introduction to the Theory of Traffic Flow*, (Springer, Berlin, 1988)
6. *Highway Capacity Manual 2000*, (National Research Council, Transportation Research Boad, Washington, D.C., 2000)
7. C.F. Daganzo, *Fundamentals of Transportation and Traffic Operations*, (Elsevier Science Inc., New York, 1997)
8. R. Wiedemann, *Simulation des Verkehrsflusses*, (University of Karlsruhe, Karlsruhe, 1974)
9. I. Prigogine, R. Herman, *Kinetic Theory of Vehicular Traffic*, (American Elsevier, New York, 1971)
10. G.B. Whitham, *Linear and Nonlinear Waves*, (Wiley, New York, 1974)
11. M. Cremer, *Der Verkehrsfluss auf Schnellstrassen*, (Springer, Berlin, 1979)
12. N.H. Gartner, C.J. Messer, A.K. Rathi (editors), *Traffic Flow Theory: A State-of-the-Art Report*, (Transportation Research Board, Washington DC, 2001)
13. D.C. Gazis, *Traffic Theory*, (Springer, Berlin, 2002)
14. G.F. Newell, *Applications of Queuing Theory*, (Chapman Hall, London, 1982)
15. M. Papageorgiou, *Application of Automatic Control Concepts in Traffic Flow Modeling and Control*, (Springer, Berlin, New York, 1983)
16. D.E. Wolf, Physica A **263**, 438–451 (1999)
17. D. Chowdhury, L. Santen, A. Schadschneider, Physics Reports **329**, 199 (2000)
18. D. Helbing, Rev. Mod. Phys. **73**, 1067–1141 (2001)
19. T. Nagatani, Rep. Prog. Phys. **65**, 1331–1386 (2002)
20. K. Nagel, P. Wagner, R. Woesler, Oper. Res. **51**, 681–716 (2003)
21. R. Mahnke, J. Kaupužs, I. Lubashevsky, Phys. Rep. **408**, 1–130 (2005)
22. N. Bellomo, V. Coscia, M. Delitala, Math. Mod. Meth. App. Sc. **12**, 1801–1843 (2002)
23. B. Piccoli, A. Tosin, in *Encyclopedia of Complexity and System Science*, ed. by R.A. Meyers. (Springer, Berlin, 2009), pp. 9727–9749
24. H. Rakha, P. Pasumarthy, S. Adjerid, Transportation Letters, **1**, 95–110 (2009)
25. B.S. Kerner, Trans. Res. Rec. **1678**, 160–167 (1999)
26. B.S. Kerner, Physics World **12**, 25–30 (August 1999)
27. S. Krauß, P. Wagner, C. Gawron, Phys. Rev. E **55**, 5597–5602 (1997)
28. K. Nagel, M. Schreckenberg, J. Phys. (France) I **2**, 2221–2229 (1992); R. Barlović, L. Santen, A. Schadschneider, M. Schreckenberg, Eur. Phys. J. B **5**, 793–800 (1998)

Chapter 13
Conclusions and Outlook

Freeway traffic flow can be free or congested. There are two phases in congested traffic, synchronized flow and wide moving jams. Thus there are three traffic phases: 1. Free flow. 2. Synchronized flow. 3. Wide moving jam. The synchronized flow and wide moving jam traffic phases are defined through spatiotemporal empirical criteria associated with propagation characteristics of the downstream fronts of these phases.

Empirical traffic breakdown at a highway bottleneck, i.e., the onset of congestion in an initial free flow at the bottleneck is associated with a phase transition from free flow to synchronized flow (F→S transition). At the same bottleneck, traffic breakdown can be either spontaneous or induced.

Traffic breakdown at the bottleneck is associated with a competition of the speed adaptation effect, which discribes a tendency towards synchronized flow, and the over-acceleration effect, which describes a tendency towards free flow. This competition occurs within a local disturbance in free flow at the bottleneck. Probability of over-acceleration exhibits a discontinuous character. This discontinuity of over-acceleration explains the fundamental empirical features of traffic breakdown.

Wide moving jams can emerge in synchronized flow (S→J transition). Thus wide moving jams emerge spontaneously due to a sequence of the F→S→J transitions associated with a double Z-characteristic for these phase transitions. Wide moving jams do not emerge spontaneously in free flow, i.e., spontaneous F→J transitions are not observed in real traffic flow.

The most of earlier traffic flow models used by traffic researchers are based on the fundamental diagram hypothesis. These models can be classified into two main classes referred to the classic LWR and GM models. None of the earlier traffic flow theories and models can explain and predict traffic breakdown at the bottleneck as observed in real measured traffic data. These traffic flow models are basic traffic flow models for simulations of freeway control and dynamic management strategies. However, we have to conclude that the related simulations of the control and dynamic management strategies cannot predict many of the freeway traffic phenomena that would occur through the use of a simulated dynamic management strategy.

B.S. Kerner, *Introduction to Modern Traffic Flow Theory and Control*,
DOI 10.1007/978-3-642-02605-8_13, © Springer-Verlag Berlin Heidelberg 2009

In contrast with the traffic flow models and theories, three-phase traffic theory can explain the empirical features of traffic breakdown and resulting congested patterns. For this reason, traffic flow models in the framework of this theory should be used for simulations of control and dynamic management strategies in freeway traffic before they are introduced to the market.

In the book, we explain why the application of a *particular* (either fixed or stochastic) highway capacity of free flow at the bottleneck as a control parameter for traffic control, dynamic traffic assignment, dynamic routing, and for other dynamic management methods and strategies is not consistent with features of real traffic. For this reason, qualitatively new methods for traffic control and dynamic traffic assignment in the context of three-phase traffic theory should be developed in the future.

Empirical microscopic features of synchronized flow and phase transitions in freeway traffic have not been sufficiently understood. Traffic flow models in the framework of three-phase traffic theory are only at the beginning of their development. There are almost no analytical results within the context of this theory. Although the FOTO and ASDA models for tracking of spatiotemporal congested patterns based on three-phase traffic theory are already successfully used in on-line installations and there are ideas about applications of the theory for feedback on-ramp metering, speed limit control, cooperative driving and adaptive cruise control, many other useful engineering applications can be expected. These and many other unsolved problems are an interesting and important field of the future empirical and theoretical investigations.

Glossary

Bottleneck The breakdown phenomenon leading to the onset of traffic congestion occurs mostly at a highway bottleneck. The bottleneck can be a result of road works, on- and off-ramps, a decrease in the number of road lanes, road curves and road gradients, etc. At great enough flow rates upstream of the bottleneck, it introduces a permanent disturbance in free flow, specifically, the disturbance is localized in a neighborhood of the bottleneck. On average the speed is lower and density is greater within this disturbance than these traffic variables are in free flow outside of the disturbance.

Catch Effect The catch effect is an induced traffic breakdown at a bottleneck caused by the propagation of a spatiotemporal synchronized flow pattern (SP) to the bottleneck. The term *catch effect* emphasizes that the initial SP is usually caught at the bottleneck as a result of the induced traffic breakdown. The catch effect can occur only, if free flow conditions have been at the bottleneck before the SP reaches it.

Characteristic Parameters of Wide Moving Jam Propagation Characteristic parameters of the propagation of a wide moving jam are traffic variables and jam parameters associated with the jam that do not depend on traffic variables in traffic flow upstream of the wide moving jam. The characteristic parameters are the flow rate, density, and speed in free flow formed in the jam outflow, the mean velocity of the downstream jam front as well as the jam density; the characteristic parameters are the same for different wide moving jams. However, the characteristic parameters can depend considerably on traffic parameters (weather, road conditions, etc.).

Congested Traffic Congested traffic can be defined as a state of traffic in which the average speed is *lower* than the minimum average speed that is still possible in free flow.

Congested Traffic Pattern A congested traffic pattern (congested pattern for short) is a spatiotemporal traffic pattern within which there is congested traffic.

Criteria [S] and [J] for Traffic Phases Criteria (definitions) [S] and [J] for traffic phases are spatiotemporal macroscopic criteria for traffic phases in congested traffic that define the synchronized flow and wide moving jam traffic phases, respectively.

Critical Speed and Density for Traffic Breakdown Critical speed and critical density required for traffic breakdown are, respectively, the vehicle speed and density within a local disturbance occurring in metastable free flow at which the tendency towards free flow due to the over-acceleration effect is on average as strong as the tendency towards synchronized flow due to the speed adaptation effect. It is equivalent to consider either the critical speed or critical density required for traffic breakdown in metastable free flow.

Critical Speed and Density for Wide Moving Jam Emergence A critical speed and critical density required for wide moving jam emergence (S→J transition) in metastable synchronized flow are, respectively, the vehicle speed and density within a local disturbance occurring in this flow at which the tendency towards a wide moving jam due to the over-deceleration effect is on average as strong as the tendency towards synchronized flow due to the speed adaptation effect. It is equivalent to consider either the critical speed or critical density required for wide moving jam emergence in metastable synchronized flow.

Effective Location of Effectual Bottleneck After traffic breakdown has occurred at an effectual bottleneck, synchronized flow is forming at the bottleneck. An effective location of the effectual bottleneck is a road location in a neighborhood of the bottleneck at which the downstream front of synchronized flow is fixed (effective location of bottleneck for short). The effective location of the bottleneck can be different from the location at which traffic breakdown has occurred leading to congested pattern emergence. Moreover, both location of traffic breakdown and effective location of the bottleneck are probabilistic values in real traffic. Even for the same type of congested pattern the effective location of the bottleneck can randomly change over time.

Effectual Bottleneck An effectual bottleneck (bottleneck for short) is a bottleneck at which traffic breakdown most frequently occurs on many different days.

Expanded Traffic Congested Pattern (EP) An expanded traffic congested pattern (EP) (expanded pattern for short) is a congested traffic pattern whose synchronized flow affects at least two effectual adjacent bottlenecks. An EP can be either an SP or an GP. The term *expanded pattern* emphasizes only that rather than the SP or GP occurs at an isolated effectual bottleneck, synchronized flow of the SP or GP affects two or more effectual adjacent highway bottlenecks.

First-Order Phase Transition in Vehicular Traffic A first-order phase transition is a phase transition occurring in a metastable state of an initial traffic phase. Examples of first-order phase transitions are F→S and S→J transitions. An F→S transition can occur in metastable free flow with respect to traffic breakdown. An S→J transition can occur in metastable synchronized flow with respect to wide moving jam emergence.

Free Traffic Flow Free traffic flow (free flow for short) is usually observed, when the vehicle density in traffic is small enough. The flow rate increases in free flow with increase in vehicle density, whereas the average vehicle speed is a decreasing density function. The increase in the flow rate with the density increase in free flow has a limit. At the associated limit (maximum) point of free flow, the flow rate and density reach their maximum values while the average speed has a minimum value that is still possible in free flow.

Front of Traffic Pattern A front of a traffic pattern is either a moving or motionless region within which one or several of the traffic variables change abruptly in space (and in time, when the front is a moving one). There are downstream front and upstream fronts of the traffic pattern. The downstream pattern front separates the pattern from other traffic patterns downstream. The upstream pattern front separates the pattern from other traffic patterns upstream.

F→S transition In all known observations, traffic breakdown is a phase transition from the free flow phase to synchronized flow phase called an F→S transition. The terms *traffic breakdown* and an *F→S transition* are synonyms related to the same phenomenon of the onset of congestion in free flow.

Fundamental Diagram Approach to Traffic Flow Theory and Modeling The fundamental hypothesis of the fundamental diagram approach to traffic flow theory and modeling assumes that steady states of both free flow and congested traffic lie on a one-dimension curve(s) (i.e., on a theoretical fundamental diagram of traffic flow) in the flow–density plane. At each given time-independent speed of the preceding vehicle, the theoretical fundamental diagram determines a single desired space gap at which a vehicle moves with this time-independent speed while following the preceding vehicle. This model vehicle behavior is related to a steady state of traffic flow associated with a hypothetical noiseless and acceleration less (deceleration less) model limit.

Fundamental Diagram of Traffic Flow The fundamental diagram of traffic flow is a relationship between the flow rate and density in vehicle traffic. Because the flow rate is the product of the average speed and density, the fundamental diagram is associated with a relationship between these traffic variables. In accordance with an obvious result of traffic measurements, on average the speed decreases when the density increases. Thus in the flow–density plane, the fundamental diagram should pass through the origin (when the density is zero so is the flow rate) and should have at least one maximum. The fundamental diagram gives also a connection between the average space gap (net distance) between vehicles and average speed.

Fundamental Hypothesis of Three-Phase Traffic Theory In contrast with the fundamental diagram approach, the fundamental hypothesis of three-phase traffic theory assumes that in *synchronized flow*, at each given time-independent speed of the preceding vehicle, there are the infinite number of space gaps to the preceding vehicle at which a vehicle can move with this time-independent speed, while following the preceding vehicle. Thus hypothetical steady states of synchronized flow cover a two-dimensional (2D) region in the flow–density plane, i.e., there is no desired space gap in hypothetical steady states of synchronized flow. Each of the steady states describes a hypothetical non-disturbed and noiseless vehicle motion in synchronized flow.

General Congested Traffic Pattern (GP) A general congested traffic pattern (GP) (general pattern for short) is a spatiotemporal congested pattern within which congested traffic consists of the synchronized flow and wide moving jam phases.

Highway Capacity of Free Flow at Bottleneck Highway (freeway) capacity of free flow at a bottleneck (called also bottleneck capacity) is limited by traffic breakdown, i.e., an F→S transition at the bottleneck. Highway capacity in free flow is equal to the flow rate downstream of the bottleneck at which free flow remains at the bottleneck with probability, which is less than one, during a given averaging time interval for traffic variables. There are the infinite number of highway capacities of free flow at bottleneck associated with different probabilities that free flow remains at the bottleneck during the averaging time interval for traffic variables.

Induced Traffic Breakdown (Induced F→S transition) at Bottleneck An induced traffic breakdown in metastable free flow at a bottleneck is traffic breakdown that is induced at the bottleneck by the propagation of a moving spatiotemporal congested traffic pattern. This congested pattern has occurred *earlier* than the time instant of traffic breakdown at the bottleneck and at a *different* road location (for example at another bottleneck) than the bottleneck location. When this congested pattern reaches the bottleneck, the pattern induces traffic breakdown at the bottleneck.

Induced Wide Moving Jam Emergence In contrast with spontaneous wide moving jam emergence in synchronized flow, induced wide moving jam emergence is caused by the upstream propagation of a moving jam, which has *initially* occurred within a *different* link of a road network connected with the road under consideration. Another peculiarity of induced wide moving jam emergence is that it can occur in either synchronized flow or free flow on the road, if the flow is in a metastable state with respect to wide moving jam emergence.

Line J The line J is a characteristic line in the flow–density plane representing a steadily propagation of the downstream front of a wide moving jam. The slope of the line J is determined by the mean velocity of the downstream jam front. All infinite number of states of traffic flow in the flow–density plane that lie on the line J are threshold states for wide moving jam existence and emergence. The line J separates all steady states of synchronized flow in the flow–density plane into two qualitatively different classes: (i) States below the line J are associated with homogeneous synchronized flow in which no wide moving jam can emerge spontaneously and persist. (ii) States on and above the line J are associated with metastable synchronized flow in which a wide moving jam can emerge and exist.

Metastable Free Flow with respect to Traffic Breakdown (F→S Transition) Metastable free flow with respect to traffic breakdown is free flow in which small enough speed (and density) disturbances decay over time, i.e., the small disturbances cause no traffic breakdown. In contrast with stable free flow with respect to traffic breakdown, metastable free flow is characterized by a critical speed (critical density) required for traffic breakdown: if in metastable free flow a local disturbance occurs within which the speed is lower than the critical speed (density is greater than the critical density), then the disturbance grows leading to traffic breakdown.

Metastable Synchronized Flow with respect to Wide Moving Jam Emergence (S→J Transition) Metastable synchronized flow with respect to wide moving jam emergence is synchronized flow in which small enough speed (and density) disturbances decay over time, i.e., the small disturbances cause no wide moving jam emergence. In contrast with stable synchronized flow with respect to wide moving jam emergence, metastable synchronized flow is characterized by a critical speed (critical density) required for wide moving jam emergence: if in metastable synchronized flow a local disturbance occurs within which the speed is lower than the critical speed (density is greater than the critical density), then the disturbance grows leading to wide moving jam emergence.

Microscopic Criterion for Wide Moving Jam Within wide moving jams, there are regions in which traffic flow is interrupted: the inflow into the jam has no influence on the jam outflow. The flow interruption effect determines the microscopic criterion for the wide moving jam phase. If in measured data congested traffic states associated with the wide moving jam phase have been identified, then with certainty all remaining congested states in the data set are related to the synchronized flow phase.

Moving Jam A moving traffic jam (moving jam for short) is a congested traffic pattern of great vehicle density spatially limited by two jam fronts. Within the downstream jam front vehicles accelerate escaping from the jam; within the upstream jam front, vehicles slow down approaching the jam. The jam as a whole localized structure propagates upstream in traffic flow. Within the jam (between the jam fronts) vehicle density is great and speed is very low (sometimes as low as zero).

Narrow Moving Jam A narrow moving jam is a moving jam, which consists of the jam fronts only. Narrow moving jams are associated with the synchronized flow phase. This is because there is no traffic flow interruption interval within a narrow moving jam; therefore, the narrow moving jam does not exhibit the characteristic jam feature [J]. The narrow moving jam can be considered a state of the synchronized flow traffic phase. During the upstream propagation of a narrow moving jam it can be surrounded upstream and downstream either by free flow or by other states of synchronized flow. A growing narrow moving jam can transform over time into a wide moving jam, i.e., the growing narrow moving jam is a nucleus for wide moving jam emergence.

Nucleus required for Moving Jam Emergence A nucleus required for wide moving jam emergence in metastable synchronized flow is a local speed (density) disturbance in the synchronized flow within which the speed is equal to or lower than the critical speed (density is equal to or greater than the critical density). A growing nucleus for wide moving jam emergence that propagates upstream in metastable synchronized flow is also called a growing narrow moving jam.

Nucleus required for Traffic Breakdown A nucleus required for traffic breakdown in metastable free flow is a local speed (density) disturbance in free flow within which the speed is equal to or lower than the critical speed (density is equal to or greater than the critical density).

Over-Acceleration Effect An over-acceleration effect is driver maneuver leading to a higher speed from an initial car-following at a lower speed, i.e., when the vehicle has initially been within the synchronization space gap. The over-acceleration effect is an opposite effect to the speed adaptation effect. In particular, following a slow moving preceding vehicle on a multi-lane road, a driver searches for the opportunity to accelerate and to pass. In this case, the over-acceleration effect is vehicle acceleration for passing in car-following, i.e., lane changing to a faster lane.

Over-Deceleration Effect An over-deceleration effect is as follows. If a vehicle begins to decelerate unexpectedly, then the following vehicle starts deceleration with a time delay. When this time delay is long enough, the driver decelerates stronger than it is needed to avoid collisions. As a result, the speed becomes lower than the speed of the preceding vehicle. If this over-deceleration effect is realized for the following drivers, a wave of vehicle speed reduction appears and increases in amplitude over time.

S→J Transition An S→J transition is a phase transition from the synchronized flow phase to wide moving jam phase.

Spatiotemporal Traffic Pattern A spatiotemporal traffic pattern (traffic pattern for short) is a specific distribution of traffic flow variables in space and time. To find real features of the spatiotemporal traffic pattern, a simultaneous measurements of traffic flow variables in space and time should be available. Examples of the traffic flow variables are the flow rate, speed, density, occupancy, space gaps and time headways between vehicles.

Speed Adaptation Effect An speed adaptation effect is vehicle speed adaptation to the speed of the preceding vehicle occurring usually when the vehicle cannot pass a slower moving preceding vehicle. The speed adaptation effect occurs within the synchronization space gap without caring, what the precise space gap is, as long as the gap is not smaller than a safe space gap.

Spontaneous Traffic Breakdown (Spontaneous F→S transition) at Bottleneck If before traffic breakdown occurs at a bottleneck there is free flow at the bottleneck as well as upstream and downstream in a neighborhood of the bottleneck, then traffic breakdown is called spontaneous traffic breakdown (spontaneous F→S transition) at the bottleneck.

In three-phase traffic theory, a spontaneous traffic breakdown occurs, if a nucleus required for traffic breakdown appears at the bottleneck. There can be various sources for the occurrence of such a nucleus leading to spontaneous traffic breakdown: unexpected braking of a vehicle in free flow, lane changing on the main road leading to time headway reduction (and as a result, to local speed reduction), fluctuations in flow rates upstream of the bottleneck, vehicle merging onto the road from other roads in the bottleneck neighborhood (e.g., at on-ramp bottlenecks), etc.

Spontaneous Wide Moving Jam Emergence in Synchronized Flow Spontaneous wide moving jam emergence in metastable synchronized flow is a phase transition from the synchronized flow phase to wide moving jam phase occurring due to the growth of a nucleus required for moving jam emergence that appears *within the synchronized flow*. There can be various sources for the nucleus required for moving jam emergence in metastable synchronized flow: unexpected braking of a vehicle in synchronized flow, lane changing on the main road leading to time headway reduction (and as a result, to local speed reduction), fluctuations in flow rates, vehicle merging from other roads, etc.

Stable Free Flow with respect to Traffic Breakdown Stable free flow with respect to traffic breakdown is free flow in which any time-limited speed (density) disturbances decay over time, i.e., no traffic breakdown (no F→S transition) can occur.

Stable Synchronized Flow with respect to Wide Moving Jam Emergence Stable synchronized flow with respect to wide moving jam emergence is synchronized flow in which any time-limited speed (density) disturbances decay over time, i.e., no wide moving jams can emerge or persist over time (no S→J transition can occur).

Steady States of Traffic Flow Steady states of traffic flow are hypothetical states of homogeneous (in time and space) traffic flow of identical vehicles (and identical drivers) in which all vehicles move with the same time-independent speed and have the same space gaps.

Synchronized Flow Traffic Pattern (SP) A synchronized flow traffic pattern (SP) is a spatiotemporal congested pattern in which congested traffic consists only of the synchronized flow traffic phase.

Synchronized Flow Traffic Phase In three-phase traffic theory, the following definition (criterion) [S] of the synchronized flow phase (synchronized flow for short) in congested traffic is made. In contrast to the wide moving jam phase, the downstream front of the synchronized flow phase does *not* maintain the mean velocity of the downstream front. In particular, the downstream front of synchronized flow is often *fixed* at a bottleneck. In other words, synchronized flow does not exhibit the characteristic jam feature [J].

Synchronization Space Gap and Safe Space Gap The lower boundary of the 2D-region of steady states of synchronized flow in the flow–density plane is associated with a *synchronization space gap*: if the space gap is greater than the synchronization gap, the vehicle accelerates. The upper boundary of the 2D-region of steady states of synchronized flow is associated with a *safe space gap* determined by a safe speed: if the gap is smaller than the safe gap, the vehicle decelerates.

Traffic Breakdown Traffic breakdown is the phenomenon of the onset of congestion in free flow. In all observations, traffic breakdown is an F→S transition. The terms *F→S transition*, *breakdown phenomenon*, *traffic breakdown*, and *speed breakdown* are synonyms related to the same effect: the onset of congestion in free flow.

Traffic Flow Model A traffic flow model is devoted to the explanation and simulation of traffic flow phenomena, which are observed in measured data of real traffic flow, and to the prediction of new traffic flow phenomena that could be found in real traffic flow. First of all, a traffic flow model should explain and predict empirical (measured) features of traffic breakdown.

Traffic Parameters Traffic parameters are those parameters of the dynamic spatiotemporal process "vehicular traffic," which can influence on traffic variables and traffic patterns. Examples of traffic parameters are a traffic network infrastructure (including, e.g., highway bottleneck types and their locations), weather (whether the day is sunny or rainy or else foggy, dry or wet road, or even ice and snow on road), percentage of long vehicles, day time, working day or week-end, other road conditions, and vehicle technology.

Three-Phase Traffic Flow Model A three-phase traffic flow model is a mathematical traffic flow model in the framework of three-phase traffic theory, i.e., the model that is based on the spatiotemporal criteria [S] and [J] for traffic phases as well as on hypotheses of three-phase traffic theory.

Three-Phase Traffic Theory In the author's three-phase traffic theory, besides the free flow phase there are two phases in congested traffic, the synchronized flow and wide moving jam phases. The synchronized flow and wide moving jam phases in congested traffic are defined through the use of empirical spatiotemporal criteria (definitions) [S] and [J], respectively. Three-phase traffic theory is a qualitative theory that explains traffic breakdown and resulting congested traffic patterns based on the criteria (definitions) [S] and [J] for traffic phases as well as on hypotheses of this theory.

Wide Moving Jam Traffic Phase In three-phase traffic theory, the following definition (criterion) [J] of the wide moving jam traffic phase (wide moving jam for short) in congested traffic is made. A wide moving jam is a moving jam that maintains the mean velocity of the downstream jam front, even when the jam propagates through any other traffic states or bottlenecks. This is the characteristic feature [J] of the wide moving jam phase.

Index

2D-region for synchronized flow states, 47

ALINEA on-ramp metering method of Papageorgiou *et al.*, 209, 214
ANCONA on-ramp metering method, 151, 214
average vehicle speed, 12
averaging time interval for traffic variables, 10
Aw-Rascle macroscopic model, 184

Bando *et al.* optimal velocity (OV) model, 183
bottleneck, 14, 255
bottleneck capacity, 73, 206, 258
bottleneck strength, 107
breakdown phenomenon, 21, 262

capacity drop, 208
car-following, 49
car-following model of Herman, Gazis *et al.*, 182
car-to-car communication, 162
catch effect, 43, 255
cellular automata (CA) model, 186, 237
characteristic jam feature [J], 20, 262
characteristic parameters of wide moving jam, 26, 255
congested pattern transformation, 129
congested traffic, 13, 255
congested traffic pattern, 14, 255
cooperative driving, 159
criteria [S] and [J] for traffic phases, 20
criterion for flow interruption, 31
critical density for S→J transition, 87, 256
critical density for traffic breakdown, 55, 256
critical density for wide moving jam emergence, 87, 256
critical speed for S→J transition, 87, 256
critical speed for traffic breakdown, 55, 256

critical speed for wide moving jam emergence, 87, 256

Daganzo's cell transmission model, 179
definitions [S] and [J] for traffic phases, 20
detector for measurements of traffic variables, 10
deterministic disturbance at bottleneck, 57
deterministic disturbance at off-ramp bottleneck, 60
deterministic disturbance at on-ramp bottleneck, 57
diagram of congested patterns, 108, 200
discharge flow rate, 16
discontinuous character of over-acceleration, 52
double Z-characteristic for phase transitions, 95
downstream front of congested pattern, 14
downstream front of synchronized flow, 20, 63
downstream front of traffic pattern, 9
downstream front of wide moving jam, 20
downstream jam front, 15
driver behavior, 159
driver time delay in acceleration, 28
duration of wide moving jam, 31
dynamic synchronization space gap, 49

effective location of bottleneck, 21, 63, 256
effective location of off-ramp bottleneck, 63
effective location of on-ramp bottleneck, 63
effectual bottleneck, 21, 256
expanded congested pattern (EP), 127
expanded general pattern (EGP), 128
expanded synchronized flow pattern (ESP), 128

F→J transition, 188

F→S transition, 21, 257, 262
F→S→J transitions, 24, 81
first-order phase transition, 102
flow interruption, 31
flow interruption effect, 31
flow interruption interval, 31
flow interruption region, 33
flow rate, 11
foreign wide moving jam, 126
FOTO and ASDA models, 143
free flow, 13, 257
front of traffic pattern, 9, 257
fundamental diagram approach, 175, 257
fundamental diagram of free flow, 13
fundamental diagram of traffic flow, 173, 257
fundamental hypothesis of three-phase traffic theory, 46, 257

General Motors (GM) model, 182
general pattern (GP), 26, 147
general pattern with non-regular pinch region, 122
Gipps's equation for safe speed, 187
growing narrow moving jam, 88

heavy bottleneck, 117
highway capacity, 73, 206, 258
homogeneous congested traffic (HCT), 193
homogeneous synchronized flow, 83

induced phase transition, 103
induced traffic breakdown, 41, 258
induced wide moving jam emergence, 90, 258
infinite number of highway capacities, 73
intelligent driving model (IDM) of Treiber et al., 184
intelligent transportation systems, 159
isolated bottleneck, 107

Kerner-Klenov ATD model, 222
Kerner-Klenov stochastic model, 230
Kerner-Konhäuser theory of jam propagation, 190
KKW cellular automata model, 237
Krauß's stochastic model, 187, 251

lane changing, 69, 90
Lighthill-Whitham-Richards (LWR) model, 178
Lighthill-Whitham-Richards (LWR) theory, 177
limit (maximum) point of free flow, 13
line J, 29
local disturbance in free flow, 55

local disturbance in synchronized flow, 86
localized synchronized flow pattern (LSP), 109

macroscopic characteristics of traffic pattern, 10
macroscopic traffic variables, 10
maximum flow rate for traffic breakdown, 66
maximum highway capacity, 76
mega-jam, 124
mesoscopic description of traffic phenomena, 10
metastable free flow, 66
metastable synchronized flow, 83
metastable traffic flow, 102
metastable traffic state, 102
methodology of three-phase traffic theory, 18
microscopic criterion for wide moving jam, 32, 259
microscopic description of traffic patterns, 10
microscopic driver navigation, 161
microscopic structure of wide moving jam, 37
microscopic traffic variables, 10
minimum highway capacity, 75
moving blank within wide moving jam, 37
moving jam, 15, 259
moving synchronized flow pattern (MSP), 109, 116

Nagel-Schreckenberg (NaSch) cellular automata (CA) model, 186, 251
narrow moving jam, 33, 81, 88, 259
nature of traffic breakdown, 45
Newell's optimal velocity (OV) model, 183
nucleation nature of S→J transition, 88
nucleation nature of traffic breakdown, 56
nucleation of phase transition, 103
nucleation-interruption effect, 133
nucleus for phase transition, 102
nucleus for S→J transition, 88
nucleus for traffic breakdown, 56, 260

occupancy, 12
off-ramp bottleneck, 60
on-ramp bottleneck, 57
onset of congestion in free flow, 21, 262
optimal velocity (OV) function, 183
oscillating congested traffic (OCT), 197
oscillations in congested traffic, 105
over-acceleration effect, 50, 228, 260
over-acceleration probability, 50
over-deceleration effect, 86, 260

passing probability, 50
Payne's macroscopic model, 184

Index

phase transition nucleation, 102
pinch effect in synchronized flow, 81
pinch region of synchronized flow, 81
pre-discharge flow rate, 16
prevention of moving jam emergence, 161
prevention of traffic breakdown, 159
probabilistic (stochastic) highway capacity, 206
probabilistic highway capacity of Elefteriadou et al., 206
probability of traffic breakdown at bottleneck, 75

road detectors, 10

S→J transition, 24, 260
safe space gap, 47, 261
safe time headway, 47
scattering of empirical data, 46
shock wave in traffic flow, 181
shock wave velocity, 178
single vehicle data, 10
single vehicle speed, 11
slow-to-start rule, 190, 192
space gap between vehicles, 10
spatiotemporal traffic pattern, 9, 260
speed adaptation effect, 49, 260
speed breakdown, 21, 262
spontaneous phase transition, 103
spontaneous traffic breakdown, 41, 260
spontaneous wide moving jam emergence, 90, 261
steady state of synchronized flow, 45
steady states of traffic flow, 261

Stokes's formula for shock wave velocity, 181
stop-and-go traffic, 105
synchronization space gap, 47, 261
synchronization time headway, 47
synchronized flow, 22, 261
synchronized flow pattern (SP), 25, 109

three-phase traffic theory, 262
threshold flow rate for traffic breakdown, 66
time headway between vehicles, 10
traffic breakdown, 21, 262
traffic breakdown probability of Persaud et al., 17, 206
traffic flow model, 178, 182, 222, 245, 262
traffic parameters, 9, 262
traffic pattern, 9, 260
traffic phase, 20
traffic phase nucleation, 103
traffic variables, 9
triggering of phase transition, 103

upstream front of congested pattern, 15
upstream front of traffic pattern, 9
upstream jam front, 15

vehicle density, 12
vehicle over-acceleration, 50, 260
vehicle over-deceleration, 260
vehicle trajectories, 10, 69, 90

wide moving jam, 23, 116, 262
widening synchronized flow pattern (WSP), 109
Wiedemann's psychophysical model, 185